WASTE BIOREFINERIES

Future Energy, Green Products, and Waste Treatment

WASTE BIOREFINERIES

Future Energy, Green Products, and Waste Treatment

Edited by

Abu Zahrim Yaser, PhD

Siew-Xian Chin, PhD

Leonardo Sepúlveda Torre, PhD

A. K. Haghi, PhD

AAP | APPLE ACADEMIC PRESS

First edition published 2025

Apple Academic Press Inc.
1265 Goldenrod Circle, NE,
Palm Bay, FL 32905 USA

760 Laurentian Drive, Unit 19,
Burlington, ON L7N 0A4, CANADA

CRC Press
2385 NW Executive Center Drive,
Suite 320, Boca Raton FL 33431

4 Park Square, Milton Park,
Abingdon, Oxon, OX14 4RN UK

Library and Archives Canada Cataloguing in Publication

Title: Waste biorefineries : future energy, green products, and waste treatment / edited by Abu Zahrim Yaser, PhD, Siew-Xian Chin, PhD, Leonardo Sepúlveda Torre, PhD, A.K. Haghi, PhD.
Names: Abu Zahrim Yaser, editor. | Siew-Xian Chin, editor. | Sepúlveda Torre, Leonardo, editor. | Haghi, A. K., editor.
Description: First edition. | Includes bibliographical references and index.
Identifiers: Canadiana (print) 20240311868 | Canadiana (ebook) 20240311884 | ISBN 9781774915462 (hardcover) | ISBN 9781774915479 (softcover) | ISBN 9781032685076 (ebook)
Subjects: LCSH: Biomass energy.
Classification: LCC TP339 .W37 2024 | DDC 662/.88—dc23

Library of Congress Cataloging-in-Publication Data

CIP data on file with US Library of Congress

ISBN: 978-1-77491-546-2 (hbk)
ISBN: 978-1-77491-547-9 (pbk)
ISBN: 978-1-03268-507-6 (ebk)

About the Editors

Abu Zahrim Yaser, PhD
Associate Professor in Waste Processing Technology at the Faculty of Engineering, University Malaysia Sabah (UMS), Malaysia

Abu Zahrim Yaser, PhD, is an Associate Professor in Waste Processing Technology at the Faculty of Engineering, Universiti Malaysia Sabah, Malaysia. He obtained his PhD from Swansea University. His inventions in composting systems have been adopted by the Tongod and Nabawan districts in Sabah. He is a visiting scientist at the University of Hull and a member of the Institutions of Chemical Engineers (United Kingdom), Board of Engineers (Malaysia), and MyBIOGAS.

Siew-Xian Chin, PhD
Senior Lecturer at University Kebangsaan Malaysia (UKM), Malaysia

Siew-Xian Chin, PhD, is currently a Senior Lecturer at Universiti Kebangsaan Malaysia (UKM). She received her PhD in Materials Science from UKM, Malaysia, in 2015. Her research interests include the development of new materials for wastewater treatment, such as membranes, beads, aerogel, etc. She has utilized agricultural by-products, including cellulose and chitosan, to produce adsorbent materials with high adsorption capacity by introducing various chemical modifications. She was awarded the title of Visiting Doctoral Researcher from the Chinese Academy of Science, China, in 2014.

Leonardo Sepúlveda Torre, PhD
Researcher, School of Chemistry at the Universidad Autónoma de Coahuila, Saltillo, Coahuila, México

Leonardo Sepúlveda Torre, PhD, is a Researcher at the School of Chemistry at the Universidad Autónoma de Coahuila, Saltillo, Coahuila, México, where he is also a member of the Bioprocesses and Microbial Biochemistry Group. His postgraduate studies were related to topics on food biotechnology in the Food Research Department of the UAdeC. In 2011, he made a research stay at the Institute of Biotechnology and Bioengineering at the University of Minho (Uminho), Braga, Portugal. He worked as a collaborator at the Center of Biological Engineering in Uminho, Braga, Portugal, with the link project "Biotechnologies for Regional Food Biodiversity in Latin America." He carried out his postdoctoral program at the same institution, writing on "Assisted Extraction by Fermentation of Polyphenols from Agro-industrial Waste."

A. K. Haghi, PhD
Professor Emeritus of Engineering Sciences and Senior Material Engineering Consultant, UK

A. K. Haghi, PhD, is a retired professor and has written, co-written, edited or co-edited more than 1000 publications, including books, book chapters, and papers in refereed journals with over 3800 citations and h-index of 32, according to the Google Scholar database. He is currently a research associate at the University of Coimbra, Portugal. Professor Haghi has received several grants, consulted for several major corporations, and is a frequent speaker to national and international audiences. He is Founder and former Editor-in-Chief of the *International Journal of Chemoinformatics and Chemical Engineering and Polymers Research Journal*. Professor Haghi has acted as an editorial board member of many international journals. He has served as a member of the Canadian Research and Development Center of Sciences & Cultures. He has supervised several PhD and MSc theses at the University of Guilan (UG) and co-supervised international doctoral projects. Professor Haghi holds a BSc in urban and environmental engineering from the University of North Carolina (USA) and holds two MSc degrees, one in mechanical engineering from North Carolina State University (USA) and another one in applied mechanics, acoustics, and materials from the Université de Technologie de Compiègne (France). He was awarded a PhD in engineering sciences at Université de *Franche-Comté* (France). He is a regular reviewer of leading international journals.

Contents

Contributors

Diana Nabila Chau Abdullah
Faculty of Business, Economics and Accountancy, University Malaysia Sabah, Kota Kinabalu, Sabah, Malaysia

Erika Acosta-Cruz
Department of Biotechnology, School of Chemistry, Autonomous University of Coahuila, Saltillo, Coahuila, México

Siti Aishah Mohd. Ali
Faculty of Science and Natural Resources, University Malaysia Sabah (UMS), Jalan UMS, Kota Kinabalu, Sabah, Malaysia

Muhd. Arshad Amin
Faculty of Chemical Engineering Technology, University Malaysia Perlis, Arau, Perlis, Malaysia

Siti Irbah Anuar
Natural Disaster Research Center (NDRC), University Malaysia Sabah (UMS), Jalan UMS, Kota Kinabalu, Sabah, Malaysia

Juan A. Ascacio-Valdés
Food Research Department, School of Chemistry, Autonomous University of Coahuila, Mexico

Mohd. Asyraf
Faculty of Science and Natural Resources, University Malaysia Sabah (UMS), Jalan UMS, Kota Kinabalu, Sabah, Malaysia

A. Danitza Casas-Rodríguez
Food Research Department, School of Chemistry, Autonomous University of Coahuila, Mexico

José de Jesús Vázquez Chabolla
Advisor, The Office of the Attorney General for Environmental Protection of the Subnational Government of Quintana Roo, México; Member, Latin-America and the Caribbean Community of Practice on Managing for Development Results (CoPLAC-MfDR)

Olga David
Faculty of Science and Natural Resources, University Malaysia Sabah (UMS), Jalan UMS, Kota Kinabalu, Sabah, Malaysia

Miriam Desirée Dávila-Medina
Food Research Department, School of Chemistry, Autonomous University of Coahuila, Mexico

Fumitaka Furuoka
Asia-Europe Institute, University of Malaya, Kuala Lumpur

M. A. Garcia-Lobato
Faculty of Chemical Sciences, Autonomous University of Coahuila, Saltillo Unit, Saltillo, Coahuila, Mexico

F. J. González
Faculty of Chemical Sciences, Autonomous University of Coahuila, Saltillo Unit, Saltillo, Coahuila, Mexico

Harry Chong Lye Hin
Faculty of Science and Natural Resources, University Malaysia Sabah (UMS), Jalan UMS, Kota Kinabalu, Sabah, Malaysia

Hazlinda Ibno
Natural Disaster Research Center (NDRC), University Malaysia Sabah (UMS), Jalan UMS, Kota Kinabalu, Sabah, Malaysia; Faculty of Science and Natural Resources, University Malaysia Sabah (UMS), Jalan UMS, Kota Kinabalu, Sabah, Malaysia

Rafidah Jalil
Forest Research Institute Malaysia (FRIM), Kepong, Selangor, Malaysia

S. E. Kayode
Faculty of Chemical Sciences, Autonomous University of Coahuila, Saltillo Unit, Saltillo, Coahuila, Mexico

Sonia Khanna
Advanced Polymeric Materials Research Laboratory, Department of Chemistry and Biochemistry, Sharda University, Greater Noida, Uttar Pradesh, India

Liliana Londoño-Hernández
Biotics Group, School of Basic Sciences, Technology, and Engineering, Universidad Nacional Abierta y a Distancia, Colombia

Muaz Mohd. Zaini Makhtar
Bioprocess Technology Division, School of Industrial Technology, University Sains Malaysia, Gelugor, Pulau Pinang, Malaysia

Carolyn Payus
Natural Disaster Research Center (NDRC), University Malaysia Sabah (UMS), Jalan UMS, Kota Kinabalu, Sabah, Malaysia; Faculty of Science and Natural Resources, University Malaysia Sabah (UMS), Jalan UMS, Kota Kinabalu, Sabah, Malaysia; Water Research Unit, Faculty of Science and Natural Resources, University Malaysia Sabah (UMS), Jalan UMS, Kota Kinabalu, Sabah, Malaysia

Wu Qichun
Asia-Europe Institute, University of Malaya, Kuala Lumpur

Wijaya Kamal Ramlan
Faculty of Business, Economics and Accountancy, University Malaysia Sabah, Kota Kinabalu, Sabah, Malaysia

Thoo Khye Sean
Faculty of Science and Natural Resources, University Malaysia Sabah (UMS), Jalan UMS, Kota Kinabalu, Sabah, Malaysia

Justin Sentian
Natural Disaster Research Center (NDRC), University Malaysia Sabah (UMS), Jalan UMS, Kota Kinabalu, Sabah, Malaysia; Faculty of Science and Natural Resources, University Malaysia Sabah (UMS), Jalan UMS, Kota Kinabalu, Sabah, Malaysia

Leonardo Sepúlveda-Torre
Food Research Department, School of Chemistry, Autonomous University of Coahuila, Mexico

Rossita Shapawi
Borneo Marine Research Institute, University Malaysia Sabah, Jalan UMS, Kota Kinabalu, Sabah, Malaysia

Noor Fazliani Shoparwe
Faculty of Bioengineering and Technology, University Malaysia Kelantan, Jeli Campus, Jeli, Kelantan, Malaysia

Hafiza Shukor
Center of Excellence for Biomass Utilization, Faculty of Chemical Engineering Technology, University Malaysia Perlis, Arau, Perlis, Malaysia

Narender Singh
Advanced Polymeric Materials Research Laboratory, Department of Chemistry and Biochemistry, Sharda University, Greater Noida, Uttar Pradesh, India

Ng Ching Wei
Faculty of Science and Natural Resources, University Malaysia Sabah (UMS), Jalan UMS, Kota Kinabalu, Sabah, Malaysia

Moh Pak Yan
Faculty of Science and Natural Resources, University Malaysia Sabah (UMS), Jalan UMS, Kota Kinabalu, Sabah, Malaysia; Water Research Unit, Faculty of Science and Natural Resources, University Malaysia Sabah (UMS), Jalan UMS, Kota Kinabalu, Sabah, Malaysia

Jain Yassin
Faculty of Business and Management, Universiti Teknologi Mara (UiTM), Sabah Branch, Kota Kinabalu, Sabah, Malaysia

Wong Sing Yun
Faculty of Business, Economics and Accountancy, University Malaysia Sabah (UMS), Kota Kinabalu, Sabah, Malaysia

Abbreviations

AADC	acetoacetate decarboxylase
ABE	acetone-butanol-ethanol
AC	activated carbon
$Al(OH)_3$	aluminum hydroxide
ALE	adaptive laboratory evolution
AMD	acid mine drainage
ARTP	atmospheric and room temperature plasmas
CBP	consolidated bioprocessing
CH_4	methane
CO_2	carbon dioxide
CoPLAC-MfDR	community of practice on managing for development results
COVID-19	coronavirus disease 2019
CV	cyclic voltammetry
CVD	chemical vapor deposition
DEA	data envelopment analysis
DMHF	dimethyl-4-hydroxy-3(2H)-furanone
DNA	deoxyribonucleic acids
DySAM	dynamic social accounting matrix
EC	electrical conductivity
EFFCH	efficiency change
EIS	electrochemical impedance spectroscopy
EMCP	energy management and climate change program
ESD	education for sustainable development
FRGS	fundamental research grant scheme
GDP	gross domestic product
GHG	greenhouse gas
GITA	green investment tax allowance
GITE	green income tax exemption
GMI	Global Moran's I
GRP	gross regional product
GTAC	Green Tech Agency
GTFS	green technology financing scheme
GTMP	green technology master plan
H_2S	hydrogen sulfide
ILMIA	Institute of Labor Market Information and Analysis
LCA	life cycle assessment
LCCF	low carbon cities framework

LF	liquid fermentation
LMI	Local Moran's I
MAE	microwave-assisted extraction
MEF	Malaysian Employers Federation
MFC	microbial fuel cell
MLPI	Malmquist-Luenberger productivity index
MOE	Ministry of Education Malaysia
MoHR	Ministry of Human Resources
MRV	monitoring report and verification
MTUC	Malaysian Trade Union Congress
MWCNT	multi-walled carbon nanotubes
N_2	nitrogen
NGOs	non-government organizations
NGTP	National Green Technology Policy
PBM	poultry by-product meals
PTS	phosphotransferase system
PVDF	polyvinylidene fluoride
RFBs	redox flow batteries
SDGs	sustainable development goals
SEDEA	super-efficiency data envelopment analysis
SEM	scanning electron microscope
SHF	separate hydrolysis and fermentation
SLAAS	Sustainable Schools Program Environmental Award
SmF	submerged fermentation
SSF	simultaneous saccharification and fermentation
SSF	solid-state fermentation
SSFR	simultaneous saccharification, fermentation, and recovery
SWCNT	single-walled carbon nanotubes
TDS	total dissolved solids
TEA	terminal electron acceptor
TECH	technical change
TH	total hardness
TWh	terawatt-hours
UAE	ultrasound-assisted extraction
UG	University of Guilan
VRFBs	vanadium redox flow batteries

Preface

Due to the growth in the world population and consequently waste production, we need to consider sustainable waste management to ensure better use of our resources. This could be achieved using waste biorefinery techniques and principles. This reference source presents the most recent advanced techniques available for different renewable waste streams. It covers proven case studies with information about the most advanced and innovative process and product development on waste biorefinery principles. This new book also demonstrates valuable data and technologies by considering environmental impacts.

This reference book is divided into three parts, providing a comprehensive outlook on technologies associated with waste biorefinery. In the first section, the reader is introduced to the recent steps toward renewable and clean energy along with waste-to-energy technologies. The second and third sections of this book are devoted to modern technologies for waste valorization that hold a valuable perspective for a sustainable green world.

This volume is a practical guide for postgraduate scholars and scientists interested in waste biorefinery. The readers will also gain a broader overview of the theories and applications of this important field.

PART I

STEPS TOWARD RENEWABLE AND CLEAN ENERGY

CHAPTER 1

Use of Biowaste for the Preparation of Carbon Materials and Its Application in Redox Flow Batteries and Other Energy Storage Devices

S. E. KAYODE, M. A. GARCIA-LOBATO, and F. J. GONZÁLEZ

Faculty of Chemical Sciences, Autonomous University of Coahuila, Saltillo Unit, Saltillo, Coahuila, Mexico

ABSTRACT

Due to the rapid and increasing growth in human population and global development, the continuous energy demand has led to the exhaustion of non-renewable energy resources (petroleum, natural gas, and coal). The need to provide potential solutions for the alternating renewable energy sources like wind and solar energy is on the increase. Therefore, it is important to develop environmentally friendly and low-cost alternative large-scale energy storage systems. Two of the most innovative systems that has emerged during last decades are electrochemical capacitors or supercapacitors and redox flow batteries (RFBs). Supercapacitors are electrochemical capacitors with much higher capacitance values than conventional capacitors. On the other hand, RFBs are rechargeable batteries, in which the electrolytes are contained in two different tanks separated by a selective ion-membrane. The main components of RFBs and supercapacitors are the active materials, electrolytes, conductive electrodes, and the separator-membrane. Regarding electrodes, the desired characteristics for most energy storage devices is a high porosity, electrical conductivity (EC), and electrochemical stability.

Waste Biorefineries: Future Energy, Green Products, and Waste Treatment.
Abu Zahrim Yaser, Siew-Xian Chin, Leonardo Sepúlveda Torre, & A. K. Haghi, (Eds.)

This chapter focusses on the utilization of biowaste residues to develop efficient electrodes for RFBs. It covers the fundaments of RFBs and the state of the art of the electrode materials prepared from different biowaste residues and their performance and perspectives of this technology.

1.1 INTRODUCTION

Due to the rapid and increasing growth in human population and global development, the continuous energy demand has led to the exhaustion of non-renewable energy resources (petroleum, natural gas, and coal) in addition to the increasing of waste generated by humans. The need to provide potential solutions to the alternating renewable energy sources like wind and solar energy is commendable. Therefore, it is important to develop environmentally friendly and low-cost alternative large scale energy storage systems (Zhang & Sun, 2021) and to incorporate our process and products to the circular economy model.

Battery energy storage systems have become a major research line in both academics and industries. Part of these storage systems are the redox flow batteries (RFBs) (*see* Iwakiri et al., 2021). Particularly, RFBs are basically rechargeable batteries, in which the electrolytes are contained in two different tanks separated by a selective ion-membrane. The size and number of electrochemical cells in an RFB define its power, while the volume of the redox liquid electrolytes in the cell(s) determines the energy capacity. RFBs have several advantages, which make them appealing for large scale stationary applications like in grid-scale energy storage systems, telecommunication companies, and as one of the main energy storage systems in the nearest future. Charging of RFBs is best late at night due to low electrical energy consumption during these hours, while discharging occurs during the day when electrical energy consumption is highest. This characteristic made it possible to uphold the leveling of electric power load, which is an important concern in the electric power industry. Moreover, it has helped to advance the effective operation of electric power facilities and improvements in the power generation efficiency. Other advantages include simple operating principle and long service life, simple installation layout, no standby loss, quick start-up, easy maintenance and environmentally friendly (Noack et al., 2021; Perry et al., 2016; Tokuda et al., 1985). A typical RFB set-up is made up of the following components as shown in Figure 1.1. Two electrolyte reservoirs (tanks) through which the redox-active materials (electrolytes) are distributed by pumps through an electrochemical cell stack, which is made up of a number of cells connected either in series or in parallel to enable reactions at the inert electrodes. The cells are made up of both anode and cathode separated by an ion-exchange membrane that permits the flow of ions across the membrane, preventing the cross-mixing of electrolyte solutions in the reservoirs (Parasuraman et al., 2012; Soc & R-r, 2011). Thus, the three main components of an RFB are redox-active materials, electrodes, and the membrane.

The ideal features of the components in typical RFB are summarized in Scheme 1.1. The ideal redox-active material should have high concentration, voltage stability, fast kinetics, soluble, readily available, low cost and environmentally friendly. Over 80% energy efficiency was recorded in the all-vanadium RFBs developed at the University of South Wales, based on the use of vanadium ions in both electrolyte reservoirs where half-cell reactions take place. This helps to prevent the challenge of cross-contamination, which occurs in other battery types (Skyllas-Kazacos & Robins, 1988). Based on the ion selection principle, membranes used in vanadium redox flow batteries (VRFBs) can be classified into ion-exchange (cation, anion, and amphoteric, depending on the type of ionic functional groups attached to the membrane) and non-ion porous membranes. The ideal membrane should have high ionic selectivity, high ionic conductivity, low water uptake, low swelling ratio, low area electrical resistance, chemical, and thermal stability, easy processing, low cost and environmentally friendly. On the other hand, the ideal electrode should have high electrical conductivity (EC), large surface area with active sites, readily available, low cost and environmentally friendly (Ding et al., 2019; Grey & Tarascon, 2017; Li & Liu, 2017; Shin et al., 2016; Skyllas-Kazacos & Robins, 1988). Most of the RFBs previously developed and commercialized across the world like the VRFBs are still limited due to their high cost, availability, solubility of redox-active materials, stability, and hydrogen evolution at the electrode (Mao et al., 2020).

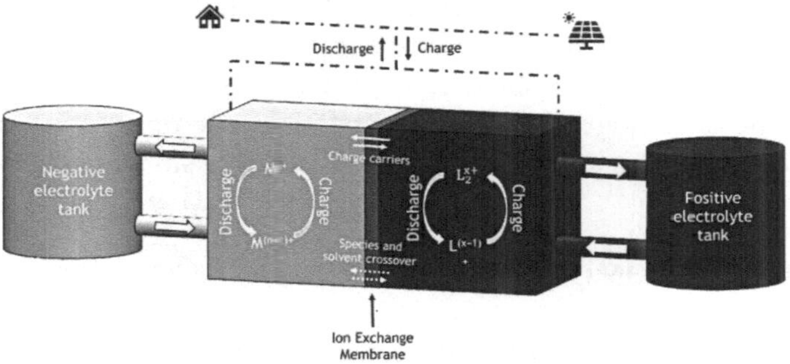

FIGURE 1.1 Schematic of a typical RFB.

Source: Reprinted with permission from Iwakiri et al. 2021. https://creativecommons.org/licenses/by/4.0/

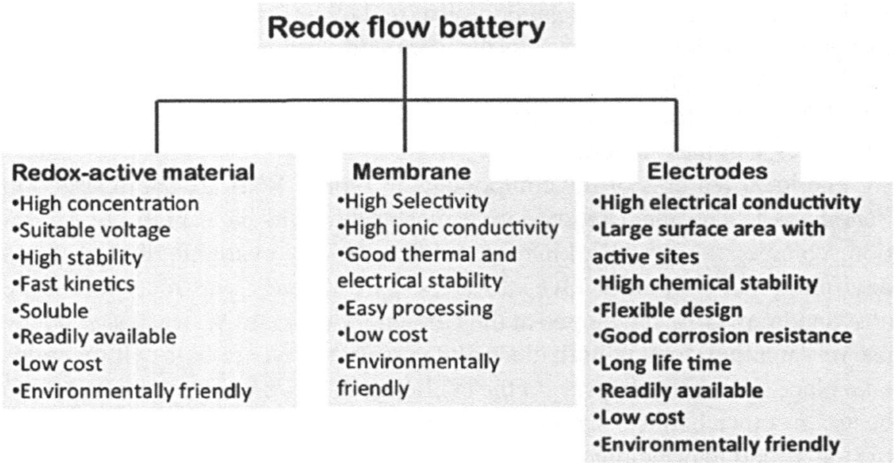

SCHEME 1.1 Summary of features of a typical RFB main components.

Supercapacitors are another type of energy storage device that stores and gives out energy at a very fast rate, i.e., they deliver a high current density within a short time duration (Mensah-Darkwa et al., 2019). Electrode materials for supercapacitors such as graphite and graphite-based materials have received much attention over the last years due to their high specific surface area, good EC, and excellent stability in strong chemical environments (Mensah-Darkwa et al., 2019). Activated carbon (AC) is the most used material used in supercapacitors due to its low cost and high surface area. Because of this, there has been an increasing interest in biowaste-derived carbon-based electrode materials such as AC and graphene (Mangisetti et al., 2019; Misnon et al., 2020).

This chapter covers fundamentals of carbon electrode materials for supercapacitors and RFBs. It comprises the most common carbon materials used for these applications, the processes and treatments carried out over the biowaste for the preparation of carbon materials, and the state of the art and developments applied as electrodes for these energy storage devices.

1.2 PREPARATION OF CARBON POROUS ELECTRODES

Porous electrodes for energy storage devices must possess the following properties: high surface area, high EC, high chemical stability, and preferably low cost (Bamgbopa et al., 2022; Chakrabarti et al., 2014). Due to this, carbon electrodes are the most convenient material, particularly for VRFBs (Ding et al., 2013). Different carbon-derived electrode materials have been used for RFBs such as carbon and graphite felt, carbon-polymer composites, PAN-based carbon felt, AC particles or felt, or carbon paper (Chakrabarti et al., 2014). Among them, the most applied materials used nowadays are carbon paper, carbon felt, and graphite felt. Unfortunately, carbon materials like carbon felt or carbon paper have the drawback of a poor catalytic activity (Zhou et al., 2016). In order to improve the performance of the carbon electrodes in RFBs, some chemical modifications have been achieved. This includes modification of carbon-based materials with metal oxides such as TiO_2, MnO_2, ZrO_2 as negative electrodes (Gencten & Sahin, 2020), modifications of carbon materials for positive electrodes include for example, the modification of a carbon felt coated with graphene dispersion to increase the electrocatalytic activity with the VO_2^+/VO^{2+} redox couple (Xia et al., 2019). Other methods include the surface modifications of the electrodes by oxidation, functionalization or heteroatom doping by thermal modification, the use of acids or other treatments like plasma or gamma ray irradiation (He et al., 2018; Kim et al., 2011).

Most widely used carbon electrode materials in commercial RFBs mostly come from petroleum sources. In the quest for sustainability and development of energy storage systems that are cost-friendly, the electrodes prepared from agricultural and industrial waste sources have turned-out to be actually promising. Due to the high carbon content in materials based on lignin and cellulose, biowaste derived from seeds, trees or plants, are considered promising materials for the preparation of electrodes for energy storage devices (Bamgbopa et al., 2022). Some examples as biowaste feedstock for carbon electrodes include orange peel (Maharjan et al., 2017), spent coffee beans (Krikstolaityte et al., 2018), coconut shell (Ulaganathan et al., 2015), Sal wood saw dust (Maharjan et al., 2019), and shaddock peel (Liu et al., 2015) among others used to prepare carbon electrodes for applications in RFBs and supercapacitors.

1.2.1 CARBON MATERIALS COMMONLY USED IN RFBS

1. **Carbon Nanotubes:** These are cylindrical molecules that consist of a rolled-up sheets of single layer carbon atoms (graphene). They can be single-walled (SWCNT) with a diameter of less than 1.0 nm or multi-walled (MWCNT), consisting of several concentrically

interlinked nanotubes, with diameters reaching more than 100 nm. Their length can reach several micrometers or millimeters. Similar to graphene, carbon nanotubes contain sp^2 bonds. They can rope together via Van der Waals forces, which make them to develop ultra-high strength, small size, low-weight material that possess highly conductive electrical and thermal properties. This property makes carbon nanotubes attractive for energy storage applications (Rathinavel et al., 2021; Venkataraman et al., 2019; Yang et al., 2019).

2. **Graphene:** It is defined as a single layer of atoms arranged in a two-dimensional honeycomb lattice nanostructure. Owing to its exceptional electrical, chemical, physical, thermal, structural, and optical properties, graphene has become a remarkable and emerging material in which researchers are exploiting its various forms (Jindal et al., 2022). Graphene-based materials such as graphene, graphene oxide and reduced graphene oxide have been widely applied in VRFBs. This is due to its outstanding conductivity, large specific surface area and excellent stability in strong acid solution (Long et al., 2021). At the moment, in order to reduce cost, and for human health concerns, researchers are shifting attention to employ different forms of natural resources like plants, biowastes, and other greener natural resources to synthesize graphene-based nanomaterial. Different methods such as chemical vapor deposition (CVD), hydrothermal method, thermal exfoliation, freeze-drying, Hummers' method, carbonization followed by activation, etc., are being employed for the synthesis (Jindal et al., 2022). These in-turn are used to fabricate electrodes for RFBs, supercapacitors, and lithium-ion batteries applications.

3. **Carbon Cloth:** It possesses good EC, high mechanical strength and high chemical resistance. The carbon cloth electrodes can operate over a wide potential range in aqueous electrolytes and molten salts. Polyacrylonitrile fibers are the most widely used precursors to produce commercial carbon cloth via heat treatment at 1,500–2,000°C (Liu et al., 2020; Tenny & Forner-Cuenca, 2020). Carbon cloth possesses unique flexibility, which can hardly be achieved by other carbonaceous matrices. Ordinary carbon cloth suffers from relatively small surface area, low electrochemical activity, and poor porosity. Various approaches have been developed to improving

the electrochemical performance of carbon cloth in recent years and substantial achievements have been recorded so far (Liu et al., 2020). Chemical and thermal methods can enhance the surface area and help to control the surface wettability of carbon cloth and make them a better candidate for applications in energy storage systems.

4. **Carbon Fiber:** These are materials that consist of thin, strong crystalline filaments of carbon. They are light-weighted, high temperature tolerant, and high chemical resistant which make them a potential candidate for applications in energy storage systems. In a recent publication by Sun et al. (2020), polyacrylonitrile was used as a precursor to synthesize electrospun carbon fiber by processing a high concentration polyacrylonitrile polymer solution. The electrospun carbon fiber obtained was used to fabricate electrodes for VRFB. The electrode was found to be a good potential electrode candidate for RFB system showing over 84% energy efficiency in VRFB which is comparable to those of the commercial carbon materials.
5. **Carbon Paper and Carbon Felt:** These are also electrode candidates for RFB systems. Though these materials differ in their morphologies, they possess good transport properties with high permeability. For instance, carbon felts exhibit a filamentary microscopic morphology with the diameter of fibers ranging from 10 to 20 μm and high porosity of over 90% (Abbas et al., 2018; Mehboob et al., 2019). Carbon paper, which is much thinner in thickness compared with carbon felts is also a carbon fiber-based material with overlapping of binder residue, forming a porous structure with a pore diameter of about 50 μm (Patel, 2019).

Abbas et al. (2019) obtained an improved energy density of 27 WhL^{-1} in VRFB using thin nanoporous functionalized carbon paper. This resulted in more effective electrode/electrolyte interaction and reduces mass transport resistance. This is comparable to the commercial pristine electrode. Tenny & Forner-Cuenca (2020) made use of a woven material to fabricate woven carbon cloth electrode and reported good electrochemical performance of the electrode. Cheng et al. (2020) reported advances in carbon fiber electrode for VRFB. They noted the need to improve on the surface activity by introducing metals, metallic oxides, carbon nanomaterials and heteroatoms into the electrospun carbon fiber. Also, to improve the structure by increasing graphitization and other structure optimization of the carbon fiber (Cheng, Li et al., 2020). Carbon felt electrode with an improved performance in VRFB has also been reported (Banerjee et al., 2019). Banerjee et al. reported an improved performance when the carbon-felt material was compressed to reduce its thickness. Decreased electrical resistivity and permeability due to reduced thickness was observed to have aided better performance of the electrode material.

1.2.2 PROCESSES AND TREATMENTS FOR PREPARATION OF BIOWASTE-BASED CARBON ELECTRODES

1.2.2.1 CARBONIZATION (PYROLYSIS)

The carbonization process is also referred to as pyrolysis process. In the process, dry raw organic materials (biowastes) like orange peel, corn cob, coffee shells, coconut shell, etc., are exposed to a temperature (less than 700°C) in a distillation apparatus, usually a furnace, to evaporate and remove the hydrocarbon contents in the absence of oxygen. The product obtained from this process is called char, biochar or carbonized material (Torres et al., 2021). The reactive carbon in this process escapes from the material on its oxidized form, and it is primarily accountable for the development of the porous structure of the carbonized material (Din et al., 2017). Carbonization of biowaste can either be a one-step process carried out at high temperature, or a two-step process which are pre-carbonization followed by activation (Sundriyal et al., 2021).

1.2.2.2 GRAPHITIZATION

The biochar obtained from carbonization is a form of amorphous carbon and can further be subjected to high temperature (above 800°C) for a long period of time in an inert (nitrogen) environment in order to rearrange the atomic structure to achieve an ordered crystalline structure that is typical of solids (graphite). During the process of graphitization, carbon atoms are rearranged to fill atom vacancies and improve atom layout. The key difference between carbonization and graphitization is that carbonization is a chemical change that involves the conversion of organic matter into carbon (biochar), whereas graphitization is a physical or micro-structural change which involves the conversion of biochar into graphite or graphite-like material (Delhaes, 2012; Mantell, 1968; Keit, 2020).

1.2.2.3 HYDROTHERMAL TREATMENT

A preventive measure to sudden thermal breakdown on biowaste is to treat the biowaste at a low temperature of between 300°C and 700°C before carbonization (pre-carbonizing). The high temperature carbonization approach increases the electrode density, which then decrease the surface area of the product (Agudosi et al., 2019). During hydrothermal treatment, the biowaste is heated in an airtight reactor in the presence of superheated water and self-generated pressure (Jung et al., 2018). Although the process helps to create pores, it is not practical to control the distribution of pores, the thermal conductivity and the specific surface area (Chen et al., 2021). Therefore, after the pre-carbonization, an activation process of the material is recommended. One of the advantages of hydrothermal treatment is that it produces lesser ash content. This is because of continuous washing condition into the liquid phase. This fact could be reflected on an improvement of the electrochemical performance of the activated carbon (AC), since ash can obstruct the pores (Reza et al., 2013). Lv et al. (2020) prepared porous carbon nanoparticles from *Scaphium scaphigerum*. Hydrothermal treatment was carried out at 180°C in an autoclave, followed by carbonization and activation at 800°C using K_2FeO_4 as the activating agent. The obtained char was further heated in a tube furnace at high temperature to obtain a product with microporous structure, large specific surface area and high graphitization degree. This demonstrated an excellent V^{2+}/V^{3+} redox reaction rate and good energy efficiency up to 72.4% at a current density of 50 mAcm^{-2}.

1.2.2.4 ACTIVATION

1. **Physical Activation:** In the physical activation method, the pre-carbonized biowaste passes through thermal conversion at about 400°C–850°C. This is then followed by activation in either CO_2, steam or air environment. As air and steam activation cannot be carried out at high temperature, CO_2 is the most useful agent because a controlled activation process at high temperatures of about 800°C can be operated. An advantage of physical activation process at high temperatures is that it helps to remove disordered and less porous inorganic materials (Bamgbopa et al., 2022).

2. **Chemical Activation:** In the chemical activation step, the pre carbonized biowaste is combined and mixed with chemical reagents such as $ZnCl_2$, KOH, NaOH, H_3PO_4, K_2CO_3, and $FeCl_3$ at desired

ratios and heated in a furnace at a temperature of 800°C and above to obtain ACs with desired properties (Tobi & Dennis, 2021). In this method, both carbonization and activation are being carried out simultaneously. This could account for a significant influence in the pyrolytic decomposition of the precursor and, therefore, resulting in the development of highly porous structures and functionalities on the carbon surface. $ZnCl_2$ and KOH are the commonly used chemical activating agents. Chemical activation has the advantage of producing a high degree of porosity (Balasubramanian, 2020).

1.2.2.5 HUMMERS' METHOD (GRAPHENE OXIDE SYNTHESIS)

Graphene oxide, a derivative of graphite consists of sp^2 and sp^3 hybridized carbon atoms with varying oxygen functional groups like carbonyl (-C=O), carboxylic acid (-COOH), hydroxyl (-OH) and alkoxy (R-O-) (Cittadini et al., 2013). The method of oxidation of graphite to obtain graphene oxide was developed by Hummers & Offeman in 1958 (Park, 2010). Hummers & Offeman in 1957 developed a method that has been widely employed for the synthesis of graphene oxide (Hummers & Offeman, 1957). They used H_2SO_4 to intercalate graphite with the assistance of $NaNO_3$ and $KMnO_4$ to oxidize the acid-intercalated graphite. Several modifications of the Hummers method have been reported in literature (Chen et al., 2013, 2014; Sohail et al., 2017). Modifications are done by optimizing the reaction conditions such as varying the reaction temperature, reactant ratios, oxidation time and the reducing agent (Jindal et al., 2022; Kurian, 2021; Torres et al., 2021).

The graphene oxide is a poor conductor; however, the reduction of graphene oxide yields reduced graphene oxide, a material with better solvent dispersibility, mechanical, thermal, and electrical performance (Tabish et al., 2019). High degree of reduction can increase the conducting performance of the reduced graphene oxide. The graphene oxide can be converted to reduced graphene oxide by chemical reduction using strong reducing agents like hydrazine or sodium borohydride ($NaBH_4$).

The reduced graphene oxide obtained using strong reducing agents often suffers from poor dispersion or irreversible agglomeration in water. In addition, the reducing agents are toxic and unstable (Xu et al., 2013). Moreover, the cost of the reducing agents is relatively high, increasing the cost of producing the electrode and thereby increasing the overall cost of the

RFB system. In order to overcome the problem of toxicity and high cost of the reducing agent, greener, and environmentally friendly reducing agents has been researched into (Kurian, 2021; Torres et al., 2021).

1.3 BIOWASTE-DERIVED CARBON-BASED ELECTRODES PERFORMANCE IN RFBS

In recent years, there has been a rising focus on the production of AC from biowaste for sustainable energy development (Tavasoli et al., 2018; Zhang et al., 2018). This helps in waste management, i.e., converting wastes into valuable resources (Mensah-Darkwa et al., 2019). These biowaste materials have the advantage to be readily available and at no cost (Singh et al., 2021). Agricultural and industrial biowastes have been reported in the literature as foundation materials for AC production for application as electrode materials for electrochemical energy systems (Chen et al., 2021; Surya & Michael, 2020). This is as a result of its relatively high EC, stability, and high specific surface area (Singh et al., 2021).

Preparation of biowaste derived carbon-based electrodes have been widely used in recent years for energy storage devices as it is shown in Table 1.1. Different methods detailed in Section 1.2 of this chapter were applied for the obtention of these devices. From the information contained in the table, several residues of different characteristics have been successfully tested as supercapacitors, such as residues of seeds, plants, fruits, oils, wastepaper, trees among others collected on the table. Some years ago, Jung et al. (2018) prepared a graphene-based carbon derived from biomass for supercapacitors application. They prepared a hierarchical porous (interconnected network of macro, micro, and mesopores) carbon-based using glucose syrup as biomass source. They carried out the polymerization of glucose in presence of ammonium chloride. This reaction allows the formation of 3D bubble networks due to the gases released from the decomposition reaction of NH_4Cl. After that, the product was carbonized and graphitized under an inert Argon gas or a mixture Ar/CO_2 flow at 1,100°C for graphitization and physical activation. The material was then chemically activated with KOH, and finally washed to obtain a hierarchical porous graphene-based carbon. A supercapacitor was prepared using the material prepared as electrodes and an ionic liquid EMIM-TFSI/AN electrolyte. The device prepared demonstrated a maximum energy density of 74 Wh kg^{-1} and a power density of 1.5 kW kg^{-1}.

TABLE 1.1 Biowaste-Derived Carbon-based Electrodes Performance in Supercapacitors

Carbon Source	Carbon Structure Formed	Synthesis Method	Performance Energy Density (Whkg–1)	References
Sword bean shell	Activated carbon	Carbonization	12.50	Chen et al. (2021)
Corn husk	Activated carbon	Carbonization	9.85	Surya & Michael (2020)
Oil palm fiber, empty fruit bunch and oil palm shell composite.	Activated carbon	Physical and chemical activation.	8.60	Tobi & Dennis (2021)
Oil tea shell	Activated carbon	Chemical activation	24.00	Zhou et al. (2017)
Sugarcane bagasse	Activated carbon	Hydrothermal method/ chemical activation.	32.00	Yadav & Hashmi (2022)
Waste office paper	Activated carbon	Hydrothermal method/ physical and chemical activation.	31.00	Sundriyal et al. (2021)
Stinky bean seedpod	Graphene	Carbonization/chemical activation.	10.58	Misnon et al. (2020)
Bombax malabaricum	Graphene	Chemical activation and carbonization.	19.40	Mangisetti et al. (2019)
Jute stick	Graphene	Carbonization	20.60	Nanaji et al. (2018)
Borassus flabellifer fruit flesh.	Graphene	Carbonization	50.00	Jung et al. (2018)

Very recently, Yadav & Hashmi (2022) developed solid-state supercapacitors using sugar cane bagasse as raw material to create a hierarchical porous AC. They used a hydrothermal treatment and soaking in ethanol followed by a physical and chemical activations. The material was tested in a capacitor using a solid-state gel polymer electrolyte composed of sodium triflate, the ionic liquid 1-ethyl-3-methylimidazolium trifluoro methane (CH_4) sulfonate in a poly(vinylidene fluoride-hexafluoropropylene) polymer matrix. The material showed a specific energy of around 32 Wh kg^{-1}, in addition to up to 10,000 charge-discharge cycles.

In contrast to supercapacitor technology, there still needs to be more literature about the use of biowaste as electrode materials for RFBs. Abbas et al. (2021) reported the synthesis of a graphene oxide obtained from coffee beans residues. The authors firstly prepared the carbonization of grinded residues under inert atmosphere at different conditions. The product was treated in a sulfuric acid solution in a typical modified Hummer's method to prepare graphene oxide and then reduced using hydrazine to obtain reduced graphene oxide. The electrode was prepared by preparing a slurry of polyvinylidene fluoride (PVDF) with N-methyl-2-pyrrolidone. The slurry was coated on a TF6 bipolar graphite plate and then heated to evaporate the solvent. The performance of the material was evaluated using electrochemical impedance spectroscopy (EIS), cyclic voltammetry (CV), galvanostatic charge-discharge tests to evaluate the potential of this technology in RFBs. CV curves indicate a higher redox reversibility compared to a graphite electrode and shows the feasibility of the use of these material for electrode production.

The electrochemical performance of biowaste-derived carbon-based electrodes in RFBs applications is given in Table 1.2. Maharjan et al. (2017), reported the synthesis of high surface area AC from orange peel as potential electrode material in VRFB application. The result of the electrochemical properties was compared with that of the commercial pristine electrode. Orange peel AC was reported to exhibit improved electrochemical properties toward the vanadium redox couple reactions as compared with the pristine electrode. The improved electro-catalytic activity of orange peel AC was attributed to joint effect of high surface area and the presence of rich oxygen-containing functional groups. The VRFB performance in a static cell was reported to show higher voltage and energy efficiencies for orange peel-AC than the pristine electrode. Chang et al. (2020) synthesized graphene-nitrogen-doped electrode material from kiwifruit. The graphene-N-doped electrode was reported to show excellent electro-catalytic activity towards vanadium redox couple reaction when compared with the commercial pristine electrode. This was noted to be as a result of the joint effect of large specific surface area, high graphitization, more active sites and excellent wettability. Wu et al. (2019) used pomelo peel to synthesize high specific surface area AC. The pomelo peel AC shows an excellent electrochemical in zinc-bromine flow battery. The result was also attributed to the high specific surface area, abundant mesopores (fast mass transport) and appropriate degree of graphitization. Maharjan et al. (2019) compared the electrochemical performance of electrode material from wood residue to that of the commercial pristine electrode. The result showed that the wood residue electrode gives better electrochemical performance compared to the pristine electrode. The electrochemical performance of the wood residue

TABLE 1.2 Biowaste-Derived Carbon-based Electrodes Performance in RFBs

Carbon Source	Carbon Structure Formed	Synthesis Method	Energy Efficiency (%)	Coulombic Efficiency (%)	Voltage Efficiency (%)	References
Orange peel	Activated carbon	Chemical activation	97.80	99.90	97.80	Maharjan et al. (2017)
Wood residue	Activated carbon	Chemical activation	91.20	95.60	95.40	Maharjan et al. (2019)
Coconut shell	Activated carbon	Hydrothermal method/ carbonization	85.00	90.00	–	Ulaganathan et al. (2015)
Bermuda grass	Activated carbon	Chemical activation/carbonization.	52.28	98.03	53.33	Anarghya et al. (2020)
Kiwi fruits	Graphene	Hydrothermal method/ freeze-drying/carbonization	80.00	–	–	Cheng, Tian et al. (2020)
Sugarcane bagasse	Graphene	Carbonization	72.00	94.00	77.00	Mahanta et al. (2019)
Pomelo peel	Graphene	Carbonization	82.20	–	–	Wu et al. (2019)
Spent coffee beans	Reduced graphene oxide	Modified Hummers' method	93.00	98.00	94.00	Abbas et al. (2021)

electrode was attributed to high surface area of the synthesized AC and the presence of surface oxygen functionalities which improved electro-catalytic activity of the electrode. In 2015, Ulganathan's group reported improved cell performance of VRFB using coconut shell derived mesoporous carbon as electrode material. The characteristic microporosity and high surface area of the coconut shell derived AC enhances the better performance of the flow battery. The performance of shaddock peel AC electrode in VRFB was reported by Liu et al. (2015). The AC was doped with nitrogen and phosphorus for enhanced electro-catalytic performance. High surface area, abundant oxygen containing functional groups, doped with N and P was reported to have greatly influenced the electro-catalytic performance of the shaddock peel AC. From the reports, it is clear that factors such as high surface area, porosity, presence of heteroatom functional groups (O, N, P, and S, etc.) greatly influence the electrochemical/electro-catalytic performance of biowaste-derived carbon electrodes. These factors need to be considered in the choice of synthesis method for the conversion of biowaste materials to carbon materials for the fabrications of electrode materials used in RFBs.

1.4 CONCLUSIONS

Developing high performance, efficient, and preferable zero-waste energy storage technologies like supercapacitors and RFBs are important issues to solve from the scientific community, due to the energy crisis we are heading nowadays. Carbon materials have shown promising characteristics as ideal electrode materials for supercapacitors and RFBs over the last decades. Furthermore, the use of biowaste or residues as carbon source for these applications is a very novel and relevant topic, heading to zero waste technologies following circular economy principles in concordance with the scope of the book.

From the reports, it is clear that factors such as high surface area, porosity, presence of heteroatoms greatly influences the electrochemical and electro-catalytic characteristics of biowaste-derived carbon electrodes. These factors are highly influenced by the choice of preparation method for the conversion of the biowaste feedstock materials to carbon materials and the following electrode fabrication, as these properties critically influence the electrochemical performance of the device. This review only focusses on the electrode materials for RFBs, and supercapacitors derived from biowaste. However, there are other elements and parameters which highly influence the performance of these devices like the catholyte, the anolyte and the choice of and adequate separator.

Despite the preparation of AC from biowaste and its application in supercapacitors is a very relevant and novel topic, very few works have been done and applied for RFBs. Furthermore, most of the papers discussed only present preliminary electrochemical measurements like EIS, CV, and galvanostatic charge-discharge profiles where the authors state the feasibility of these materials in VRFs. Technological advances should be headed towards RFBs prototypes and scale-up as stationary energy storage devices. This issue requires multidisciplinary actions from experts from biowaste and bioprocesses, materials science, organic synthesis, electrochemistry, and electrical and chemical engineering to improve the performance of these devices and try to solve some of the most critical humanity problems.

KEYWORDS

- **biowaste**
- **carbonization**
- **electrodes**
- **graphene oxide synthesis**
- **graphitization**
- **Hummers' method**
- **hydrothermal treatment**
- **pyrolysis**

REFERENCES

Abbas, A., Eng, X. E., Ee, N., Saleem, F., Wu, D., Chen, W., Handayani, M., et al., (2021). Development of reduced graphene oxide from biowaste as an electrode material for vanadium redox flow battery. *Journal of Energy Storage*, *41*, 102848. https://doi.org/10.1016/j.est.2021.102848.

Abbas, S., Lee, H., Hwang, J., Mehmood, A., Shin, H. J., Mehboob, S., Lee, J. Y., & Ha, H. Y., (2018). A novel approach for forming carbon nanorods on the surface of carbon felt electrode by catalytic etching for high-performance vanadium redox flow battery. *Carbon*, *128*, 31–37. https://doi.org/10.1016/j.carbon.2017.11.066.

Agudosi, E. S., Abdullah, E. C., Numan, A., Mubarak, M., Khalid, M., & Omar, N., (2019). A review of the graphene synthesis routes and its applications in electrochemical energy storage. *Critical Reviews in Solid State and Materials Sciences*, *45*(5), 1–39. https://doi.org/10.1080/10408436.2019.1632793.

Anarghya, D., Anantha, M. S., Venkatesh, K., Santosh, M. S., Priya, M. G., & Muralidhara, H. B., (2020). Bermuda grass derived nitrogen-doped carbon as electrocatalyst in graphite felt electrode to increase the efficiency of all-iron redox flow batteries. *Journal of Electroanalytical Chemistry*, *878*, 114577. https://doi.org/10.1016/j.jelechem.2020.114577.

Balasubramanian, N., (2020). *Graphene and Graphene-Like Structure from Biomass for Electrochemical Energy Storage Application: A Review*, 1–16. https://doi.org/10.1002/elsa.202000028.

Bamgbopa, M. O., Fetyan, A., Vagin, M., & Adelodun, A. A., (2022). Towards eco-friendly redox flow batteries with all bio-sourced cell components. *Journal of Energy Storage*, *50*, 104352. https://doi.org/10.1016/j.est.2022.104352.

Banerjee, R., Bevilacqua, N., Mohseninia, A., Wiedemann, B., Wilhelm, F., Scholta, J., & Zeis, R., (2019). Carbon felt electrodes for redox flow battery: Impact of compression on transport properties. *Journal of Energy Storage*, *26*, 100997. https://doi.org/10.1016/j.est.2019.100997.

Chakrabarti, M. H., Brandon, N. P., Hajimolana, S. A., Tariq, F., Yufit, V., Hashim, M. A., Hussain, M. A., et al., (2014). Application of carbon materials in redox flow batteries. *Journal of Power Sources*, *253*, 150–166. https://doi.org/10.1016/j.jpowsour.2013.12.038.

Chen, J., Li, Y., Huang, L., Li, C., & Shi, G., (2014). High-yield preparation of graphene oxide from small graphite flakes via an improved hummers method with a simple purification process. *Carbon*, 1–9. https://doi.org/10.1016/j.carbon.2014.10.033.

Chen, J., Yao, B., Li, C., & Shi, G., (2013). An improved hummers method for eco-friendly synthesis of graphene oxide. *Carbon*, *64*(1), 225–229. https://doi.org/10.1016/j.carbon.2013.07.055.

Chen, T., Luo, L., Luo, L., Deng, J., Wu, X., Fan, M., Du, G., & Weigang, Z., (2021). High energy density supercapacitors with hierarchical nitrogen-doped porous carbon as active material obtained from bio-waste. *Renewable Energy*, *175*, 760–769. https://doi.org/10.1016/j.renene.2021.05.006.

Cheng, D., Li, Y., Zhang, J., Tian, M., Wang, B., He, Z., Dai, L., & Wang, L., (2020). Recent advances in electrospun carbon fiber electrode for vanadium redox flow battery: Properties, structures, and perspectives. *Carbon*, *170*, 527–542. https://doi.org/10.1016/j.carbon.2020.08.058.

Cheng, D., Tian, M., Wang, B., Zhang, J., Chen, J., Feng, X., He, Z., Dai, L., & Wang, L., (2020). One-step activation of high-graphitization N-doped porous biomass carbon as advanced catalyst for vanadium redox flow battery. *Journal of Colloid and Interface Science*, *572*, 216–226. https://doi.org/10.1016/j.jcis.2020.03.069.

Cittadini, M., Bersani, M., Perrozzi, F., Ottaviano, L., Wlodarski, W., & Martucci, A., (2013). Graphene oxide coupled with gold nanoparticles for localized surface plasmon resonance-based gas sensor. *Carbon*, *69*, 452–459. https://doi.org/10.1016/j.carbon.2013.12.048.

Delhaes, P., (2012). *Carbon Science and Technology-from Energy to Materials* (Vol. 2012, pp. 35–38). Hoboken, NJ: John Wiley & Sons, Inc.

Din, M. I., Ashraf, S., & Intisar, A., (2017). *Comparative Study of Different Activation Treatments for the Preparation of Activated Carbon: A Mini-Review*, *100*, 299–312.

Ding, C., Zhang, H., Li, X., Liu, T., & Xing, F., (2013). Vanadium flow battery for energy storage: Prospects and challenges. *Journal of Physical Chemistry Letters*, *4*(8), 1281–1294. https://doi.org/10.1021/jz4001032.

Ding, Y., Zhang, C., Zhang, L., Zhou, Y., & Yu, G., (2019). Pathways to widespread applications: Development of redox flow batteries based on new chemistries. *Chempr.*, 1–24. https://doi.org/10.1016/j.chempr.2019.05.010.

Dreyer, D. R., Park, S., Bielawski, C. W., & Ruoff, R. S. (2010). The chemistry of graphene oxide. *Chem. Soc. Rev., 39*(1), 228–240. https://doi.org/10.1039/B917103G.

Gencten, M., & Sahin, Y., (2020). A critical review on progress of the electrode materials of vanadium redox flow battery. *International Journal of Energy Research*, *44*(10), 7903–7923. https://doi.org/10.1002/er.5487.

Grey, C. P., & Tarascon, J. M., (2017). *Sustainability and in Situ Monitoring in Battery Development* (Vol. 16, No. 1, pp. 45–56). Nature Publishing Group. https://doi.org/10.1038/nmat4777.

He, Z., Jiang, Y., Li, Y., Zhu, J., Zhou, H., Meng, W., Wang, L., & Dai, L., (2018). Carbon layer-exfoliated, wettability-enhanced, SO3H-functionalized carbon paper: A superior positive electrode for vanadium redox flow battery. *Carbon*, *127*, 297–304. https://doi.org/10.1016/j.carbon.2017.11.006.

Hummers, W. S., & Offeman, R. E. (1958). Preparation of Graphitic Oxide. *Journal of the American Chemical Society, 80*(6), 1339–1339. https://doi.org/10.1021/ja01539a017.

Iwakiri, I., Antunes, T., Almeida, H., Sousa, J. P., Figueira, R. B., & Mendes, A., (2021). Redox flow batteries: Materials, design and prospects. *Energies, 14*(18), 5643. https://doi.org/10.3390/en14185643.

Jindal, S., Anand, R., Sharma, N., Yadav, N., Mudgal, D., Mishra, R., & Mishra, V., (2022). Sustainable approach for developing graphene-based materials from natural resources and biowastes for electronic applications. *ACS Applied Electronic Materials*. https://doi.org/10.1021/acsaelm.2c00097.

Jung, S., Myung, Y., Kim, B. N., Kim, I. G., You, I., & Kim, T., (2018). Activated biomass-derived graphene-based carbons for supercapacitors with high energy and power density. *Scientific Reports*, *2017*, 1–8. https://doi.org/10.1038/s41598-018-20096-8.

Kim, K. J., Kim, Y. J., Kim, J. H., & Park, M. S., (2011). The effects of surface modification on carbon felt electrodes for use in vanadium redox flow batteries. *Materials Chemistry and Physics*, *131*(1), 547–553. https://doi.org/https://doi.org/10.1016/j.matchemphys.2011.10.022.

Krikstolaityte, V., Joshua, O. E. Y., Veksha, A., Wai, N., Lisak, G., & Lim, T. M., (2018). Conversion of spent coffee beans to electrode material for vanadium redox flow batteries. *Batteries*, *4*(4), 1–11. https://doi.org/10.3390/batteries4040056.

Kurian, M., (2021). Recent progress in the chemical reduction of graphene oxide by green reductants – A mini-review. *Carbon Trends*, *5*, 100120. https://doi.org/10.1016/j.cartre.2021.100120.

Li, B., & Liu, J., (2017). *Progress and Directions in Low-Cost Redox Flow Batteries for Large-Scale Energy Storage*, 91–105. https://doi.org/10.1093/nsr/nww098.

Liu, J., Wang, Z. A., Wu, X. W., Yuan, X. H., Hu, J. P., Zhou, Q. M., Liu, Z. H., & Wu, Y. P., (2015). Porous carbon derived from disposable shaddock peel as an excellent catalyst toward VO^{2+}/VO_2^+ couple for vanadium redox battery. *Journal of Power Sources*, *299*, 301–308. https://doi.org/10.1016/j.jpowsour.2015.09.004.

Liu, X., Xu, W., Zheng, D., Li, Z., Zeng, Y., & Lu, X., (2020). Carbon cloth as an advanced electrode material for supercapacitors: Progress and challenges. *Journal of Materials Chemistry A., 8*(35), 17938–17950. https://doi.org/10.1039/d0ta03463k.

Long, Y., Ding, M., & Jia, C., (2021). *Application of Nanomaterials in Aqueous Redox Flow Batteries*, 1–15. https://doi.org/10.1002/cnma.202100124.

Mahanta, V., Raja, M., & Kothandaraman, R., (2019). Activated carbon from sugarcane bagasse as a potential positive electrode catalyst for vanadium redox flow battery. *Materials Letters*, *247*, 63–66. https://doi.org/10.1016/j.matlet.2019.03.045.

Maharjan, M., Bhattarai, A., Ulaganathan, M., Wai, N., Ohnmar, M., Wang, J., & Mariana, T., (2017). High surface area bio-waste based carbon as a superior electrode for vanadium redox flow battery. *Journal of Power Sources*, *362*, 50–56. https://doi.org/10.1016/j.jpowsour.2017.07.020.

Maharjan, M., Wai, N., Veksha, A., Giannis, A., Lim, T. M., & Lisak, G., (2019). Sal wood sawdust derived highly mesoporous carbon as prospective electrode material for vanadium redox flow batteries. *Journal of Electroanalytical Chemistry*, *834*(2018), 94–100. https://doi.org/10.1016/j.jelechem.2018.12.049.

Mangisetti, S. R., Kamaraj, M., & Sundara, R., (2019). Green approach for synthesizing three different carbon microstructures from a single biowaste *Bombax malabaricum* for fully biocompatible flexible supercapacitors and their performance in various electrolytes [Research-article]. *ACS Omega*, *4*(4), 6399–6410. https://doi.org/10.1021/acsomega.9b00263.

Mantell, C., (1968). *Carbon and Graphite Handbook* (pp. 6–14). John Wiley & Sons, Inc.

Mao, J., Ruan, W., & Chen, Q., (2020). *Understanding the Aqueous Solubility of Anthraquinone Sulfonate Salts: The Quest for High-Capacity Electrolytes of Redox Flow Batteries*. https://doi.org/10.1149/1945–7111/ab7550.

Mehboob, S., Ali, G., Abbas, S., Chung, K. Y., & Ha, H. Y., (2019). Elucidating the performance-limiting electrode for all-vanadium redox flow batteries through in-depth physical and electrochemical analyses. *Journal of Industrial and Engineering Chemistry*, *80*, 450–460. https://doi.org/10.1016/j.jiec.2019.05.045.

Mensah-darkwa, K., Zequine, C., Kahol, P. K., & Gupta, R. K., (2019). *Supercapacitor Energy Storage Device Using Biowastes: A Sustainable Approach to Green Energy*. https://doi.org/10.3390/su11020414.

Misnon, I. I., Zain, N. K. M., Lei, T. S., Vijayan, B. L., & Jose, R., (2020). Activated carbon with graphitic content from stinky bean seedpod biowaste as supercapacitive electrode material. *Ionics*, *26*(8), 4081–4093. https://doi.org/10.1007/s11581–020–03565-x.

Nanaji, K., Varadaraju, U. V., Rao, T. N., & Anandan, S., (2018). *Robust, Environmentally Benign Synthesis of Nanoporous Graphene Sheets from Bio-waste for Ultrafast Supercapacitor Application*. https://doi.org/10.1021/acssuschemeng.8b05419.

Noack, J., Berkers, M., Ortner, J., & Pinkwart, K., (2021). *The Influence of Some Electrolyte Additives on the Electrochemical Performance of Fe / Fe^{2+} Redox Reactions for Iron / Iron Redox Flow Batteries the Influence of Some Electrolyte Additives on the Electrochemical Performance of Fe / Fe^{2+} Redox Reacti*. https://doi.org/10.1149/1945–7111/abf5a3.

Parasuraman, A., Lim, T. M., Menictas, C., & Skyllas-kazacos, M., (2012). Review of material research and development for vanadium redox flow battery applications. *Electrochimica Acta*. https://doi.org/10.1016/j.electacta.2012.09.067.

Patel, (2019). *Direct Visualization of Electrochemical Reactions and Comparison of Commercial Carbon Papers*, *77*(11), 9–25.

Perry, M. L., Weber, A. Z., Soc, J. E., Perry, M. L., & Weber, A. Z., (2016). *Advanced Redox-Flow Batteries: A Perspective Jes Focus Issue on Redox Flow Batteries – Reversible Fuel Cells Advanced Redox-Flow Batteries: A Perspective*. https://doi.org/10.1149/2.0101601jes.

Rathinavel, S., Priyadharshini, K., & Panda, D., (2021). A review on carbon nanotube: An overview of synthesis, properties, functionalization, characterization, and the application. *Materials Science and Engineering B: Solid-State Materials for Advanced Technology*, *268*, 115095. https://doi.org/10.1016/j.mseb.2021.115095.

Reza, M. T., Lynam, J. G., Uddin, M. H., & Coronella, C. J., (2013). Hydrothermal carbonization: Fate of inorganics. *Biomass and Bioenergy*, *49*, 86–94. https://doi.org/10.1016/j.biombioe.2012.12.004.

Shin, D. W., Guiver, M. D., & Lee, Y. M., (2016). *Hydrocarbon-Based Polymer Electrolyte Membranes: Importance of Morphology on Ion Transport and Membrane Stability*. https://doi.org/10.1021/acs.chemrev.6b00586.

Singh, P., Bhargava, C., Kumar, M., Singh, A., Singh, P. K., Kumar, P., & Sachdeva, A., (2021). Materials today: Proceedings extraction and comparative study of green energy using different types of biowaste material. *Materials Today: Proceedings*, *XXXX*. https://doi.org/10.1016/j.matpr.2021.05.468.

Skyllas-Kazacos, M., & Robins, R., (1988). *All Vanadium Redox Battery Patent*, *54*.

Soc, J. E., & R-r, P., (2011). *Progress in Flow Battery Research and Development*, *158*(8). https://doi.org/10.1149/1.3599565.

Sohail, M., Saleem, M., Ullah, S., Saeed, N., Afridi, A., Khan, M., & Arif, M., (2017). Modified and improved hummer's synthesis of graphene oxide for capacitors applications. *Modern Electronic Materials*, *3*(3), 110–116. https://doi.org/10.1016/j.moem.2017.07.002.

Sun, J., Jiang, H. R., Wu, M. C., Fan, X. Z., Chao, C. Y. H., & Zhao, T. S., (2020). A novel electrode formed with electrospun nano- and micro-scale carbon fibers for aqueous redox flow batteries. *Journal of Power Sources*, *470*. https://doi.org/10.1016/j.jpowsour.2020.228441.

Sundriyal, S., Shrivastav, V., Kaur, A., Dubey, P., Mishra, S., Deep, A., & Dhakate, S., (2021). Waste office papers as a cellulosic material reservoir to derive highly porous activated carbon for solid-state electrochemical capacitor. *IEEE Transactions on Nanotechnology*, *20*(c), 481–488. https://doi.org/10.1109/TNANO.2021.3080589.

Surya, K., & Michael, M. S., (2020). Novel interconnected hierarchical porous carbon electrodes derived from bio-waste of corn husk for supercapacitor applications. *Journal of Electroanalytical Chemistry*, *878*, 114674. https://doi.org/10.1016/j.jelechem.2020.114674.

Tabish, T. A., Pranjol, Z. I., Horsell, D. W., Rahat, A. A. M., Whatmore, J. L., Winyard, P. G., & Zhang, S., (n.d.). *Graphene Oxide-Based Targeting of Extracellular Cathepsin D and Cathepsin L As a Novel Anti-Metastatic Enzyme Cancer Therapy*. https://doi.org/10.3390/cancers11030319.

Tavasoli, A., Aslan, M., Salimi, M., Balou, S., Pirbazari, S. M., Hashemi, H., & Kohansal, K., (2018). Influence of the blend nickel / porous hydrothermal carbon and cattle manure hydrochar catalyst on the hydrothermal gasification of cattle manure for H_2 production. *Energy Conversion and Management*, *173*, 15–28. https://doi.org/10.1016/j.enconman.2018.07.061.

Tenny, K. M., & Forner-cuenca, A., (2020). *Comparing Physical and Electrochemical Properties of Different Weave Patterns for Carbon Cloth Electrodes in Redox Flow Batteries*, *17*. https://doi.org/10.1115/1.4046661.

Tobi, A. R., & Dennis, J. O., (2021). Activated carbon from composite of palm bio-waste as electrode material for solid-state electric double layer capacitor. *Journal of Energy Storage*, *42*, 103087. https://doi.org/10.1016/j.est.2021.103087.

Tokuda, N., Kanno, T., Hara, T., Shigematsu, T., Tsutsui, Y., Ikeuchi, A., Itou, T., & Kumamoto, T., (1985). *Development of a Redox Flow Battery System*, 88–94.

Torres, F. G., Troncoso, O. P., Rodriguez, L., & De-la-torre, G. E., (2021). Sustainable synthesis, reduction, and applications of graphene obtained from renewable resources. *Sustainable Materials and Technologies*, *29*, e00310. https://doi.org/10.1016/j.susmat.2021.e00310.

Ulaganathan, M., Jain, A., Aravindan, V., Jayaraman, S., Ling, W. C., Lim, T. M., Srinivasan, M. P., et al., (2015). Bio-mass derived mesoporous carbon as superior electrode in all vanadium redox flow battery with multicouple reactions. *Journal of Power Sources*, *274*, 846–850. https://doi.org/10.1016/j.jpowsour.2014.10.176.

Venkataraman, A., Amadi, E. V., Chen, Y., & Papadopoulos, C., (2019). Carbon nanotube assembly and integration for applications. *Nanoscale Research Letters*, *14*(1). https://doi.org/10.1186/s11671-019-3046-3.

Wu, M. C., Zhang, R. H., Liu, K., Sun, J., Chan, K. Y., & Zhao, T. S., (2019). Mesoporous carbon derived from pomelo peel as a high-performance electrode material for zinc-bromine flow batteries. *Journal of Power Sources*, *442*. https://doi.org/10.1016/j.jpowsour.2019.227255.

Xia, L., Zhang, Q., Wu, C., Liu, Y., Ding, M., Ye, J., Cheng, Y., & Jia, C., (2019). Graphene coated carbon felt as a high-performance electrode for all vanadium redox flow batteries. *Surface and Coatings Technology*, *358*, 153–158. https://doi.org/10.1016/j.surfcoat.2018.11.024.

Xu, C., Xu, B., Gu, Y., Xiong, Z., Sun, J., & Zhao, X. S., (2013). Graphene-based electrodes for electrochemical energy storage. *Energy and Environmental Science*, *6*(6), 1388–1414. https://doi.org/10.1039/c3ee23870a.

Yadav, N., & Hashmi, S. A., (2022). High energy density solid-state supercapacitors based on porous carbon electrodes derived from pre-treated bio-waste precursor sugarcane bagasse. *Journal of Energy Storage*, *55*(PA), 105421. https://doi.org/10.1016/j.est.2022.105421.

Yang, Z., Tian, J., Yin, Z., Cui, C., Qian, W., & Wei, F., (2019). Carbon nanotube- and graphene-based nanomaterials and applications in high-voltage supercapacitor: A review. *Carbon*, *141*, 467–480. https://doi.org/10.1016/j.carbon.2018.10.010.

Zhang, H., & Sun, C., (2021). *Cost-Effective Iron-Based Aqueous Redox Flow Batteries for Large-Scale Energy Storage Application: A Review*, *493*.

Zhang, Y., Song, X., Xu, Y., Shen, H., Kong, X., & Xu, H., (2018). Utilization of wheat bran for producing activated carbon with high specific surface area via NaOH activation using industrial furnace. *Journal of Cleaner Production*. https://doi.org/10.1016/j.jclepro.2018.11.041.

Zhou, M., Gomez, J., Li, B., Jiang, Y. B., & Deng, S., (2017). Oil tea shell derived porous carbon with an extremely large specific surface area and modification with MnO_2 for high-performance supercapacitor electrodes. *Applied Materials Today*, *7*, 47–54. https://doi.org/10.1016/j.apmt.2017.01.008.

Zhou, X. L., Zhao, T. S., Zeng, Y. K., An, L., & Wei, L., (2016). A highly permeable and enhanced surface area carbon-cloth electrode for vanadium redox flow batteries. *Journal of Power Sources*, *329*, 247–254. https://doi.org/10.1016/j.jpowsour.2016.08.085.

CHAPTER 2

Coupling Efficiency Measurement and Spatial Correlation of Energy Investment, Economic, and Pollution Nexus in China

WU QICHUN,[1] FUMITAKA FURUOKA,[1] DIANA NABILA CHAU ABDULLAH,[2] and WIJAYA KAMAL RAMLAN[2]

[1]*Asia-Europe Institute, University of Malaya, Kuala Lumpur*

[2]*Faculty of Business, Economics and Accountancy, University Malaysia Sabah, Kota Kinabalu, Sabah, Malaysia*

ABSTRACT

This chapter investigates the relationship between energy investment, economics, and water pollution in different regions of China. This study will provide suggestions for sustainable and stable regional development. In this study, we construct a network based on the data envelopment analysis (DEA) model for the energy investment-economic-water pollution input-output index system to evaluate the efficiency of decision-making units. This study was also applied to 31 provinces in China from 2011 to 2017 to measure energy investment-economic-pollution coupling efficiency in spatial correlation analysis. The result shows that most of the eastern and central provinces are highly efficient economically, and most low-economically efficient provinces are located in western China. Thus, policies and funding support for the economic dimension are recommended. The eastern area had lower efficiency values than the average in environmental aspects, and government should pay more attention to pollution treatment. Our result will disclose the correlation relationship between energy investment-economic-water pollution, reveal that there is spatial autocorrelation between the closed area,

Waste Biorefineries: Future Energy, Green Products, and Waste Treatment.
Abu Zahrim Yaser, Siew-Xian Chin, Leonardo Sepúlveda Torre, & A. K. Haghi, (Eds.)

and give suggestions for policymakers about the regional synergies in investment, economic development, and water pollution.

2.1 INTRODUCTION

The relationship between economic growth and environmental conservation is one of the thorniest issues in environmental economics. There is a political division between rich and poor countries to cope with this delicate relationship. The rich countries use cheap and dirty fossil fuels for their economic development. However, they have discouraged poor countries to do the same when climate change becomes a global agenda. Furthermore, climate change would have more serious negative effects on economic growth in poor countries. Nevertheless, the rich countries are reluctant to increase foreign aid to support the poor countries' concerted effects for climate change mitigation (Willett, 2015; Nixon, 2011).

There is an ongoing debate on the economic consequence of environmental degradation. Some scholars claim that economic growth is the main cause of environmental deterioration. By contrast, other scholars asserted that economic growth is not a cause, but a solution to improve environmental conservation (Beckerman, 1992; Furuoka, 2015). For example, Beckerman (1992) claimed that there is a need to enhance the favorable association between economic growth and the environmental condition by alleviating poverty and decreasing income inequality, which would contribute to reducing the environmentally damaging usage of natural resources. At the same time, there is also a need to break the unfavorable growth-environment association by investing in more environmentally friendly technologies and creating strong institutions to protect the environment.

From a theoretical perspective, the relationship between economic growth and the environment is known as the environmental Kuznet curve. According to this interesting hypothesis on environmental conservation, there is an inverted U-shape relationship between economic growth and environmental conservation. In other words, there is a threshold income level in the environment-development relation. If a country's income level is less than the threshold, there is a positive association between economic growth and environmental deterioration. In other words, as the country's economic growth accelerated, the environmental condition would deteriorate. By contrast, if a country's income level is greater than the threshold, there is a positive environment-growth relationship. It means that the economic growth in the country would contribute to improving environmental conditions (Rothman & de Bruyn, 1998; Furuoka, 2015).

In the case of China, as Figure 2.1 indicated, there is an increasing trend in CO_2 emissions for the period of 1990–2019. There is a gradual increase in CO_2 emissions in China in the 1990s. For example, China's CO_2 emission amounted to 2.173 million kilotons in 1990 and increased to 2.645 million kilotons in 1993, further to 3.088 million kilotons in 1995. Furthermore, the country's CO_2 emission amounted to 3.070 million kilotons in 1996 and increased to 3.153 million kilotons in 1999. By contrast, there is a relatively rapid expansion in China's CO_2 emission in the 2000s. For instance, the country's CO_2 emission amounted to 3.383 million kilotons in 2000 and increased to 4.415 million kilotons in 2003, further to 5.824 kilotons in 2005. Furthermore, the CO_2 emission in China increased from 6.437 million kilotons in 2006 to 7.419 million kilotons in 1995. The country's CO emissions continued increasing in the 2010s. China's CO_2 emission increased from 8.974 million kilotons in 2010 to 9.541 million kilotons in 2012. Interestingly, the country's CO_2 emission decreased slightly from 10.006 million kilotons in 2014 to 9.861 million kilotons in the following year. The country's CO_2 emission increased marginally from 10.096 million kilotons in 2017 to 10.502 million kilotons in 2018, further to 10.707 million kilotons in the following year.

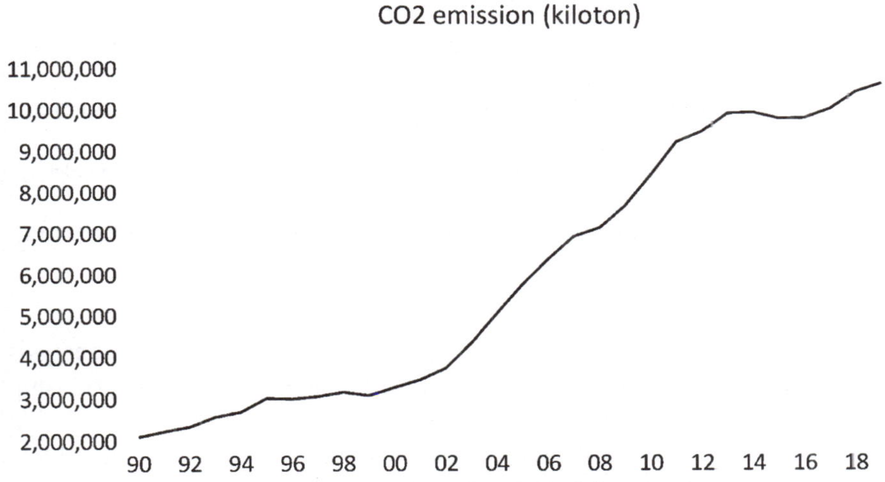

FIGURE 2.1 CO_2 emission in China from 1990 to 2019 (kiloton).

Source: World Bank (2022).

Because of its rapid increase in CO_2 emission, China has emerged as the top CO_2 emission country in the world. Table 2.1 indicates the top five CO_2 emission countries in the world. In 1990, the top CO_2 emission country was the United States whose emission amounted to 4.804 million kilotons in 1990, followed by China, whose emission amounted to 2.173 million kilotons in the same year. Similarly, the top CO_2 emission country in 2000 was still the United States, whose emissions amounted to 5.775 million kilotons in 2000, followed by China, whose emissions amounted to 3.345 million kilotons in the same year. However, China's CO_2 emission jumped to 8.474 million kilotons in 2010, and the country overtook the United States to become a top emitter. By contrast, CO_2 emissions from the United States decreased to 5.392 million kilotons in 2010. Remarkably, China is the dominant country in CO_2 emission, whose CO_2 emission amounted to 10.707 million kilotons in 2019. In other words, China emitted more than one-third of global CO_2 emissions in the same year, 5.392 million kilotons in 2010 (World Bank, 2022).

TABLE 2.1 Top Five CO_2 Emission Countries (Million Kilotons)

Sl. No.	1990	2000	2010	2019
1.	United States (4.804 million kiloton)	United States (5.775 million kiloton)	China (8.474 million kiloton)	China (10.707 million kiloton)
2.	China (2.173 million kiloton)	China (3.345 million kiloton)	United States (5.392 million kiloton)	United States (4.817 million kiloton)
3.	Russia (2.163 million kiloton)	Russia (1.563 million kiloton)	India (1.659 million kiloton)	India (2.458 million kiloton)
4.	Japan (1.090 million kiloton)	Japan (1.182 million kiloton)	Russia (1.615 million kiloton)	Russia (1.703 million kiloton)
5.	Germany (0.935 million kiloton)	India (0.937 million kiloton)	Japan (1.156 million kiloton)	Japan (1.081 million kiloton)
6.	World (20.625 million kiloton)	World (23.445 million kiloton)	World (31.043 million kiloton)	World (34.384 million kiloton)

Source: World Bank (2022).

Against such background, this chapter applied the data envelopment analysis (DEA) method and the spatial correlation method to examine the relationship between energy investments, economic growth, and environmental pollution in 31 provinces in China for the period 2011–2017. The DEA method is a highly popular method to examine the efficiency of resource allocation by estimating systematically total factor productivity (Li & Wang, 2013; Zhao et al., 2022). Furthermore, the spatial correlation is also a popular method to examine the dependence of resource allocation, which is known as the cross-efficiency of resource allocation (Li & Wang, 2013; Li & Liu, 2021; Zhao et al., 2022).

This chapter consists of five sections. Following this introductory section, Section 1.2 reviews previous empirical research which applied the DEA method and the spatial correlation method for their analysis. Section 1.3 explains data and research methods. Section 1.4 reports the empirical findings. The final section is the conclusion.

2.2 LITERATURE REVIEW

Furthermore, Tian et al. (2016) employed the DEA model and the spatial correlation analysis to examine the resource allocation of the grape production industry in 119 Chinese districts in 2014. They claimed that the grape industry resource allocation is the most efficient in the northern part of China, while the allocation is the least efficient in the southwestern part of China. The spatial correlation analysis indicated that there is indeed dependency on grape resource allocation in China. Wang & Du (2017) also employed the output-based DEA to examine the efficiency of power industries in 30 provinces in China for period 2003–2012, their finding was that the power industry in eastern China is the most efficient in the country. By contrast, the power industry in the western region is less efficient in China. The spatial correlation analysis revealed that there is an increasing trend of dependency on power generation resource allocation in China. Meanwhile, Xia et al. (2018) used the DEA to examine the efficiency of resource allocation in tourism industries in 31 provinces in China from the period of 2005–2014. They claimed that there is a decreasing trend in the efficiency of the tourism sector from 2005 to 2014. The researchers also pointed out that there is a decrease in the dependency on resource allocation in the tourism industry during the period in China. In addition, Bai et al. (2018) also applied the super-efficiency data envelopment analysis (SEDEA) to examine the efficiency of ecological resource allocation in 281 cities in China for the period of 2006–2014. They claimed the eastern cities are more ecologically efficient than central cities in the country. They revealed that there is a decreasing trend of dependencies on ecological resource allocation from 2006 to 2014.

More recently, Zhou et al. (2019) applied the DEA to examine the efficiency of ecological resource allocation in 30 provinces in China for the period of 2005–2014. They claimed three provinces are the most efficient in terms of ecological resource allocations, namely Beijing, Tianjin, and Shanghai. In contrast, three provinces are less efficient in terms of ecological resource allocation, namely Gansu, Qinghai, and Ningxia. The researchers also highlighted that there is a decreasing trend in ecological resource allocation between 2005 and 2014. Interestingly, they showed that there is a decrease in the dependence on ecological resource allocation for the period in China. Li & Liu (2021) also used the DEA to analyze the efficiency of resource allocation in the health sector in 29 provinces in China for the period of 2004–2015. They claimed that there is an increasing trend of efficiency of health resource allocation from 2004 to 2015. Among these provinces, six provinces are most efficient in terms of health resource allocations, namely Tianjin, Jiangsu, Guangdong, Guizhou, Tibet, and Ningxia. They also pointed out that there is an improvement in the efficiency of the rural health resource allocation during the period. They found that there is a significant dependency on health resource allocation in China.

In addition, Guo et al. (2020) used the spatial correlation analysis to examine the dependence of resource allocation in the agricultural sector in 316 cities in China for the period of 2000–2015. They claimed that there is an increase in the resource allocation dependency in these Chinese cities from 2000 to 2015. Gao et al. (2021) applied the three-stage DEA to examine the efficiency of the resource allocation in the forestry sector in 31 provinces in China for the period of 2001–2018. The result of the study shows that there is an increase in the efficiency of forest resource allocation in these provinces from 2001 to 2018. It is also highlighted that there is a gradual increase in the dependence on forest resource allocation among these provinces in the same period. Liao et al. (2021) also used the data envelopment method to examine the efficiency of resource allocation in the agriculture sector in the 30 provinces in China for the period of 2006–2018. They presented that there is a decreasing trend in the efficiency of agricultural resource allocation in these provinces from 2006 to 2018. Among three regions, namely west, east, and central, the eastern region is the most efficient agricultural resource allocation. By contrast, the western region is the least efficient in the country's agricultural resource allocation. Interestingly, they pointed out that there is no significant dependence on agriculture resource allocation in China. Wang et al. (2021) applied DEA to analyze the efficiency of resource allocations in the five districts in China. Among these five districts, the Ping district is the most efficient in resource allocation. By contrast, the Bai district is the least efficient in resource allocation. Ultimately, Zhao et al. (2022) employed the DEA to examine the efficiency of resource allocation in the transportation sector in 30 provinces in China for the period 2010–2016. They claimed that there is an improvement in the transportation resource allocation from 2010 to 2016. It was also emphasized that there is a slight decrease in the dependence on transportation resource allocation during the period in China.

2.3 DATA AND METHODOLOGY

2.3.1 DATA ENVELOPMENT ANALYSIS (DEA) MODEL

In this study, we assume that there are *n* decision units, each decision unit is using *m* input and *s* output.

$$x_j^t = \left(x_{1j}^t, x_{2j}^t, \ldots, x_{mj}^t\right)^T$$

where x_j^t means the input index value of *j* decision unit at period *t*.

$$y_j^t = \left(y_{1j}^t, y_{2j}^t, \ldots, y_{nj}^t\right)^T$$

where; y_j^t means the good output index value of *j* decision unit at period *t*.

$$b_j^t = \left(b_{1j}^t, b_{2j}^t, \ldots, b_{nj}^t\right)^T$$

where; b_j^t means the bad output index value of *j* decision unit at period *t*.

All of these are positive, $t = 1, 2, ..., T$.

Under the same scale and compensation, we assume the distance function of (x^t, y^t, y^t) is $D_0^t(x^t, y^t)$ at period *t*, the distance function of (x^t, y^t) is $D_0^{t+1}(x^t, y^t)$ at period $t + 1$; the distance function of (x^{t+1}, y^{t+1}) is $D_0^t(x^{t+1}, y^{t+1})$ at period *t*, the distance function of (x^{t+1}, y^{t+1}) is $D_0^{t+1}(x^{t+1}, y^{t+1})$ at period $t + 1$.

Malmquist productivity index (TFP) is defined by:

$$M_t^{t+1} = \left[\frac{D_0^t\left(x^{t+1}, y^{t+1}, b^{t+1}\right)}{D_0^t\left(x^t, y^t, b^t\right)} \frac{D_0^{t+1}\left(x^{t+1}, y^{t+1}, b^{t+1}\right)}{D_0^{t+1}\left(x^t, y^t, b^t\right)}\right]^{1/2}$$

The Malmquist index can be decomposed into two component measures, one accounting for efficiency change (EFFCH), and one measuring technical change (TECH).

$$EFFCH_t^{t+1} = \frac{D_0^t\left(x^{t+1}, y^{t+1}, b^{t+1}\right)}{D_0^t\left(x^t, y^t, b^t\right)}$$

$$TECH_t^{t+1} = \left[\frac{D_0^t\left(x^{t+1}, y^{t+1}, b^{t+1}\right)}{D_0^{t+1}\left(x^{t+1}, y^{t+1}, b^{t+1}\right)} \frac{D_0^t\left(x^t, y^t, b^t\right)}{D_0^{t+1}\left(x^t, y^t, b^t\right)}\right]^{1/2}$$

$$M_t^{t+1} = EFFCH_t^{t+1} * TECH_t^{t+1}$$

2.3.2 SPATIAL CORRELATION ANALYSIS

The spatial econometrics literature is interested in the dependence among observations across space and uses the so-called spatial weight matrix W to describe the spatial arrangement of the geographical units in a given sample. The choice of appropriate spatial weights is a central component of spatial models. In particular, we set $W_{ij} = 1$ if province i and city j share a border line and 0 otherwise.

$$W_{ij} = \begin{cases} 1, bnd(i) \cap bnd(j) \neq \varnothing \\ 0, bnd(i) \cap bnd(j) = \varnothing \end{cases}$$

According to Tobler's first law of geography (Tobler, 1970), everything is related to everything else, but near things are more related to each other. Spatial weight matrix (W), containing the spatial relations among observations, is commonly used to express Tobler's law. We then estimate two Moran's I index, Global Moran's I (GMI) and Local Moran's I (LMI), respectively test for global and local spatial correlation. The GMI and LMI are as follows:

$$GMI = \frac{n\sum_{i=1}^{n}\sum_{j=1}^{n} W_{ij}\left(Y_i - \overline{Y}\right)}{\left(\sum_{i=1}^{n}\sum_{j=1}^{n} W_{ij}\left(Y_i - \overline{Y}\right)^2 \sum_{i=1}^{n}\left(Y_i - \overline{Y}\right)^2\right)}, i \neq j$$

$$LMI = \frac{n\left(Y_i - \overline{Y}\right)}{\sum_{i=1}^{n}\left(Y_i - \overline{Y}\right)^2} \sum_{i \neq j}^{n} W_{ij}\left(Y_i - \overline{Y}\right)$$

where; n is the total number of spatial units; Yi and Yj stand for two output indicators of i and j province, respectively; and Wij is the binary spatial weight matrix.

2.4 EMPIRICAL FINDINGS

2.4.1 DESCRIPTIVE ANALYSIS

We have conducted a descriptive statistic related to the variables used in our model. The data is taken from 2011 to 2017 in 31 provinces of China. The result shows that the mean of energy investment is 6.57, the mean of gross regional product (GRP) is 9.636 and mean of water pollution is 11.951. The result of Figures 2.2 and 2.3 shows that as energy investment and province GRP significant growth, water pollution remains at the same stage (Table 2.2 and Figure 2.4).

TABLE 2.2 The Descriptive Analysis

Variable	Obs	Mean	Std. Dev.	Min.	Max.
$\ln_{invest}$	217	6.5700	0.7837	4.1589	8.1265
$\ln_{grp}$	217	9.6360	0.9746	6.4066	11.4043
$\ln_{water}$	217	11.9510	1.0135	8.4413	13.7518

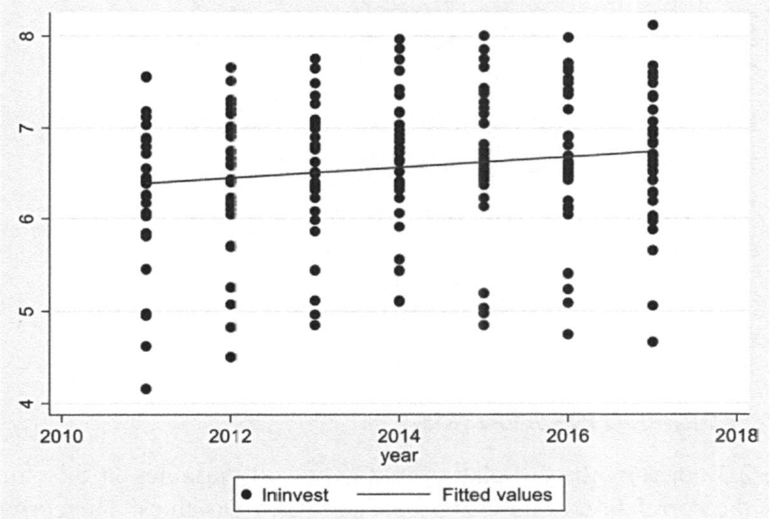

FIGURE 2.2 The scatter plot of energy investment.

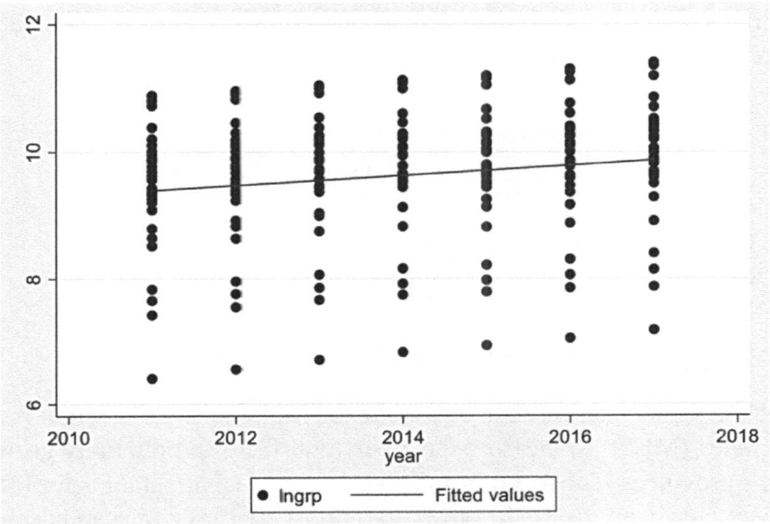

FIGURE 2.3 The scatter plot of gross regional product.

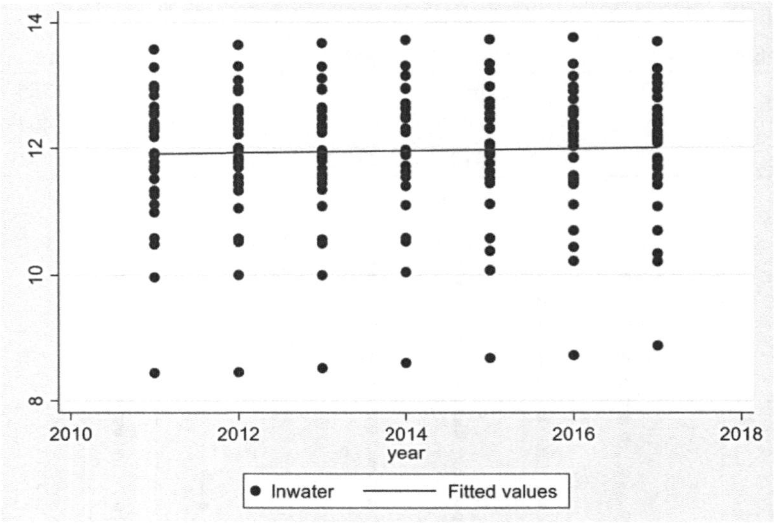

FIGURE 2.4 The scatter plot of water pollution.

2.4.2 CORRELATION ANALYSIS

Table 2.3 contains the correlation matrix for all variables in this study to check the correlation relationship between these variables. The correlation between energy investment and ln GRP is 0.506 (significant at 1% level), and the correlation between energy investment and water pollution is 0.487 (significant at 1% level). We can see that a positive correlation exists between energy investment, province GDP, and water pollution.

TABLE 2.3 The Correlation Matrix

	$\ln_{invest}$	$\ln_{grp}$	$\ln_{water}$
$\ln_{invest}$	1	–	–
$\ln_{grp}$	0.506***	1	–
$\ln_{water}$	0.487***	0.963***	1

2.4.3 DATA ENVELOPMENT ANALYSIS (DEA) MODEL

As the DEA analysis, we firstly estimate the Malmquist-Luenberger productivity index (MLPI) to measure the green total-factor productivity growth of China's provinces. The result from Table 2.4 shows that Total Factor Productivity are increasing from 2011 to 2014 in most areas of China and decreased from 2015 to 2017 in most areas of China. Meanwhile, the decreasing trend is evident in Middlee East provinces and spread to the adjacent provinces. From Figures 2.5–2.7, the result shows that the total factor productivity change is relying on the technical efficiency change (EFFCH).

TABLE 2.4 The DEA Result of Investment-Economic-Pollution Productivity

Province	id	Pdwise	TFPCH	EFFCH	TECH
Anhui	1	2011~2012	1.0013	1.0012	1.0001
Anhui	1	2012~2013	1.0046	1.0037	1.0009
Anhui	1	2013~2014	1.0027	1.0079	0.9949
Anhui	1	2014~2015	0.9995	1.0012	0.9983
Anhui	1	2015~2016	0.9884	0.9888	0.9996
Anhui	1	2016~2017	0.9965	0.9979	0.9986
Beijing	2	2011~2012	0.9933	1.0000	0.9933
Beijing	2	2012~2013	0.9959	1.0000	0.9959
Beijing	2	2013~2014	0.9930	0.9988	0.9942
Beijing	2	2014~2015	1.0096	1.0012	1.0085
Beijing	2	2015~2016	1.0082	1.0000	1.0082
Beijing	2	2016~2017	1.0036	0.9994	1.0043
Chongqing	3	2011~2012	1.0006	1.0005	1.0001
Chongqing	3	2012~2013	1.0072	1.0054	1.0018
Chongqing	3	2013~2014	1.0042	1.0032	1.0010
Chongqing	3	2014~2015	1.0030	1.0031	0.9999
Chongqing	3	2015~2016	1.0261	1.0257	1.0004
Chongqing	3	2016~2017	0.9978	0.9991	0.9987
Fujian	4	2011~2012	0.9901	0.9896	1.0005
Fujian	4	2012~2013	1.0034	1.0016	1.0018
Fujian	4	2013~2014	1.0025	1.0012	1.0012
Fujian	4	2014~2015	0.9998	0.9988	1.0010
Fujian	4	2015~2016	0.9966	0.9942	1.0024
Fujian	4	2016~2017	1.0024	1.0024	1.0000
Gansu	5	2011~2012	1.0076	1.0061	1.0014
Gansu	5	2012~2013	1.0063	1.0045	1.0018
Gansu	5	2013~2014	1.0034	1.0027	1.0007
Gansu	5	2014~2015	1.0008	1.0042	0.9966
Gansu	5	2015~2016	1.0010	0.9995	1.0016
Gansu	5	2016~2017	1.0016	1.0021	0.9994
Guangdong	6	2011~2012	1.0045	1.0032	1.0012
Guangdong	6	2012~2013	1.0037	1.0019	1.0018
Guangdong	6	2013~2014	1.0040	1.0026	1.0014

TABLE 2.4 *(Continued)*

Province	id	Pdwise	TFPCH	EFFCH	TECH
Guangdong	6	2014~2015	1.0006	0.9993	1.0012
Guangdong	6	2015~2016	1.0018	1.0006	1.0012
Guangdong	6	2016~2017	0.9963	0.9978	0.9986
Guangxi	7	2011~2012	1.0072	1.0085	0.9987
Guangxi	7	2012~2013	0.9939	0.9953	0.9985
Guangxi	7	2013~2014	0.9951	1.0004	0.9947
Guangxi	7	2014~2015	0.9984	0.9988	0.9996
Guangxi	7	2015~2016	0.9895	0.9904	0.9991
Guangxi	7	2016~2017	1.0126	1.0243	0.9885
Guizhou	8	2011~2012	1.0176	1.0180	0.9996
Guizhou	8	2012~2013	0.9999	1.0002	0.9997
Guizhou	8	2013~2014	1.0134	1.0184	0.9950
Guizhou	8	2014~2015	1.0011	1.0050	0.9961
Guizhou	8	2015~2016	1.0043	0.9948	1.0096
Guizhou	8	2016~2017	1.0194	1.0592	0.9624
Hainan	9	2011~2012	0.9960	1.0000	0.9960
Hainan	9	2012~2013	1.0328	1.0000	1.0328
Hainan	9	2013~2014	0.8960	1.0000	0.8960
Hainan	9	2014~2015	0.9854	0.9960	0.9894
Hainan	9	2015~2016	1.1078	1.0040	1.1034
Hainan	9	2016~2017	0.8986	0.9936	0.9044
Hebei	10	2011~2012	1.0070	1.0049	1.0020
Hebei	10	2012~2013	1.0025	1.0007	1.0018
Hebei	10	2013~2014	1.0002	0.9986	1.0016
Hebei	10	2014~2015	0.9998	0.9984	1.0014
Hebei	10	2015~2016	0.9962	0.9945	1.0017
Hebei	10	2016~2017	0.9931	0.9931	1.0000
Heilongjiang	11	2011~2012	1.0073	1.0059	1.0013
Heilongjiang	11	2012~2013	0.9987	0.9969	1.0018
Heilongjiang	11	2013~2014	1.0002	0.9990	1.0012
Heilongjiang	11	2014~2015	1.0003	1.0004	0.9998
Heilongjiang	11	2015~2016	1.0519	1.0163	1.0350
Heilongjiang	11	2016~2017	0.9441	0.9843	0.9591
Henan	12	2011~2012	1.0045	1.0061	0.9984
Henan	12	2012~2013	1.0019	1.0026	0.9994
Henan	12	2013~2014	1.0308	1.0736	0.9601
Henan	12	2014~2015	0.9679	0.9519	1.0168
Henan	12	2015~2016	0.9969	0.9809	1.0163
Henan	12	2016~2017	0.9832	0.9997	0.9835

TABLE 2.4 *(Continued)*

Province	id	Pdwise	TFPCH	EFFCH	TECH
Hubei	13	2011~2012	1.0027	1.0020	1.0007
Hubei	13	2012~2013	1.0037	1.0023	1.0014
Hubei	13	2013~2014	1.0035	1.0046	0.9989
Hubei	13	2014~2015	1.0021	1.0012	1.0009
Hubei	13	2015~2016	0.9933	0.9913	1.0020
Hubei	13	2016~2017	1.0010	1.0016	0.9994
Hunan	14	2011~2012	1.0071	1.0091	0.9980
Hunan	14	2012~2013	0.9990	0.9996	0.9994
Hunan	14	2013~2014	0.9921	1.0035	0.9886
Hunan	14	2014~2015	0.9994	0.9966	1.0028
Hunan	14	2015~2016	0.9961	0.9980	0.9980
Hunan	14	2016~2017	0.9960	1.0022	0.9938
Inner Mongolia	15	2011~2012	1.0065	0.9977	1.0089
Inner Mongolia	15	2012~2013	1.0053	1.0024	1.0029
Inner Mongolia	15	2013~2014	1.0040	1.0015	1.0026
Inner Mongolia	15	2014~2015	0.9994	0.9973	1.0021
Inner Mongolia	15	2015~2016	0.9981	0.9950	1.0031
Inner Mongolia	15	2016~2017	0.9933	0.9911	1.0022
Jiangsu	16	2011~2012	1.0034	0.9989	1.0045
Jiangsu	16	2012~2013	1.0045	1.0019	1.0026
Jiangsu	16	2013~2014	1.0045	1.0010	1.0035
Jiangsu	16	2014~2015	1.0041	1.0014	1.0027
Jiangsu	16	2015~2016	1.0073	1.0009	1.0064
Jiangsu	16	2016~2017	1.0082	1.0020	1.0062
Jiangxi	17	2011~2012	1.0014	1.0035	0.9979
Jiangxi	17	2012~2013	0.9995	0.9997	0.9998
Jiangxi	17	2013~2014	0.9889	1.0187	0.9707
Jiangxi	17	2014~2015	0.9863	0.9831	1.0032
Jiangxi	17	2015~2016	0.9983	1.0002	0.9981
Jiangxi	17	2016~2017	0.9945	1.0043	0.9903
Jilin	18	2011~2012	1.0061	1.0048	1.0013
Jilin	18	2012~2013	1.0025	1.0007	1.0018
Jilin	18	2013~2014	1.0041	1.0025	1.0016
Jilin	18	2014~2015	1.0029	1.0017	1.0012
Jilin	18	2015~2016	0.9861	0.9833	1.0028
Jilin	18	2016~2017	1.0163	1.0169	0.9993
Liaoning	19	2011~2012	1.0063	1.0022	1.0041
Liaoning	19	2012~2013	1.0026	1.0014	1.0011
Liaoning	19	2013~2014	1.0075	1.0051	1.0023

TABLE 2.4 *(Continued)*

Province	id	Pdwise	TFPCH	EFFCH	TECH
Liaoning	19	2014~2015	0.9998	0.9961	1.0037
Liaoning	19	2015~2016	1.0624	1.0151	1.0466
Liaoning	19	2016~2017	0.9315	0.9825	0.9481
Ningxia	20	2011~2012	0.9997	1.0000	0.9997
Ningxia	20	2012~2013	0.9983	1.0000	0.9983
Ningxia	20	2013~2014	0.9947	1.0000	0.9947
Ningxia	20	2014~2015	0.9867	1.0000	0.9867
Ningxia	20	2015~2016	1.0048	1.0000	1.0048
Ningxia	20	2016~2017	0.9922	0.9952	0.9970
Qinghai	21	2011~2012	1.0009	0.9996	1.0013
Qinghai	21	2012~2013	1.0037	1.0019	1.0018
Qinghai	21	2013~2014	1.0051	1.0054	0.9997
Qinghai	21	2014~2015	1.0015	1.0112	0.9904
Qinghai	21	2015~2016	1.0137	1.0089	1.0047
Qinghai	21	2016~2017	1.0007	1.0000	1.0007
Shaanxi	22	2011~2012	1.0093	1.0063	1.0030
Shaanxi	22	2012~2013	1.0056	1.0042	1.0014
Shaanxi	22	2013~2014	1.0083	1.0059	1.0023
Shaanxi	22	2014~2015	1.0089	1.0062	1.0026
Shaanxi	22	2015~2016	0.9999	0.9987	1.0012
Shaanxi	22	2016~2017	1.0052	1.0068	0.9984
Shandong	23	2011~2012	1.0124	1.0052	1.0072
Shandong	23	2012~2013	1.0021	0.9987	1.0034
Shandong	23	2013~2014	1.0034	1.0004	1.0030
Shandong	23	2014~2015	1.0053	1.0026	1.0026
Shandong	23	2015~2016	1.0004	0.9954	1.0049
Shandong	23	2016~2017	1.0024	0.9995	1.0028
Shanghai	24	2011~2012	0.9905	1.0000	0.9905
Shanghai	24	2012~2013	1.0150	1.0000	1.0150
Shanghai	24	2013~2014	0.9825	1.0000	0.9825
Shanghai	24	2014~2015	1.0171	1.0000	1.0171
Shanghai	24	2015~2016	0.9923	1.0000	0.9923
Shanghai	24	2016~2017	1.0076	1.0000	1.0076
Shanxi	25	2011~2012	1.0103	1.0090	1.0013
Shanxi	25	2012~2013	1.0025	1.0007	1.0018
Shanxi	25	2013~2014	1.0024	1.0016	1.0009
Shanxi	25	2014~2015	0.9996	1.0026	0.9970
Shanxi	25	2015~2016	0.9968	0.9941	1.0027

TABLE 2.4 *(Continued)*

Province	id	Pdwise	TFPCH	EFFCH	TECH
Shanxi	25	2016~2017	1.0004	1.0025	0.9979
Sichuan	26	2011~2012	1.0047	1.0035	1.0013
Sichuan	26	2012~2013	1.0073	1.0055	1.0018
Sichuan	26	2013~2014	1.0060	1.0047	1.0013
Sichuan	26	2014~2015	1.0021	1.0021	1.0000
Sichuan	26	2015~2016	1.0028	1.0011	1.0017
Sichuan	26	2016~2017	1.0031	1.0050	0.9982
Tianjin	27	2011~2012	1.0098	1.0000	1.0098
Tianjin	27	2012~2013	1.0044	1.0000	1.0044
Tianjin	27	2013~2014	1.0048	1.0000	1.0048
Tianjin	27	2014~2015	1.0002	1.0000	1.0002
Tianjin	27	2015~2016	1.0069	1.0000	1.0069
Tianjin	27	2016~2017	1.0049	1.0000	1.0049
Tibet	28	2011~2012	0.9955	0.9957	0.9998
Tibet	28	2012~2013	1.0101	1.0088	1.0013
Tibet	28	2013~2014	1.0096	1.0090	1.0006
Tibet	28	2014~2015	1.0069	1.0098	0.9971
Tibet	28	2015~2016	1.0031	1.0008	1.0023
Tibet	28	2016~2017	1.0148	1.0169	0.9979
Xinjiang	29	2011~2012	1.0104	1.0089	1.0015
Xinjiang	29	2012~2013	1.0068	1.0050	1.0018
Xinjiang	29	2013~2014	1.0026	1.0021	1.0005
Xinjiang	29	2014~2015	0.9949	0.9985	0.9963
Xinjiang	29	2015~2016	0.9939	0.9912	1.0027
Xinjiang	29	2016~2017	1.0052	1.0075	0.9977
Yunnan	30	2011~2012	1.0034	1.0027	1.0007
Yunnan	30	2012~2013	1.0049	1.0035	1.0014
Yunnan	30	2013~2014	1.0017	1.0020	0.9997
Yunnan	30	2014~2015	1.0066	1.0122	0.9944
Yunnan	30	2015~2016	1.0059	1.0046	1.0013
Yunnan	30	2016~2017	1.0054	1.0060	0.9995
Zhejiang	31	2011~2012	1.0027	1.0009	1.0018
Zhejiang	31	2012~2013	1.0023	1.0006	1.0017
Zhejiang	31	2013~2014	1.0019	0.9999	1.0020
Zhejiang	31	2014~2015	1.0031	1.0000	1.0031
Zhejiang	31	2015~2016	1.0015	0.9993	1.0022
Zhejiang	31	2016~2017	1.0036	1.0048	0.9988

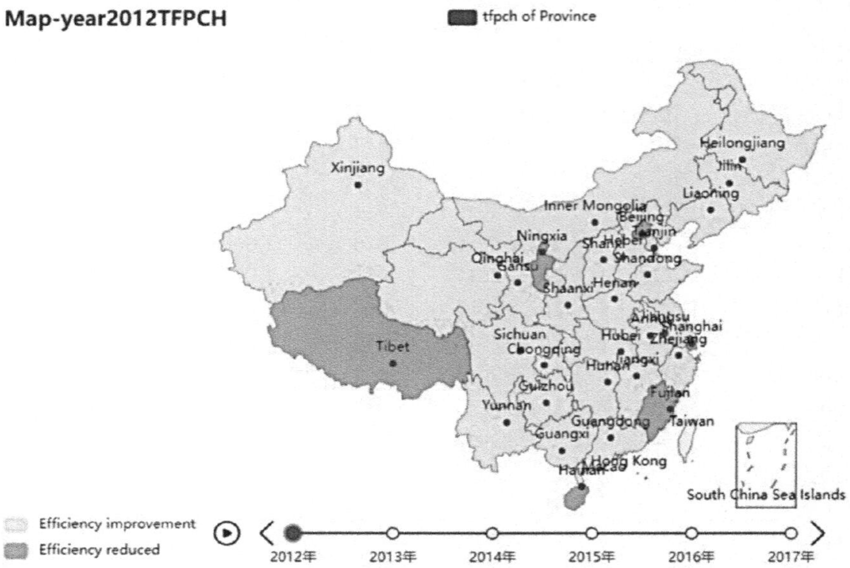
Map-year2012TFPCH
tfpch of Province
Xinjiang
Heilongjiang
Jilin
Liaoning
Inner Mongolia
Beijing
Tianjin
Ningxia
Hebei
Qinghai
Gansu
Shandong
Shaanxi
Henan
Shanghai
Sichuan
Tibet
Hubei
Chongqing
Hunan
Jiangxi
Guizhou
Fujian
Yunnan
Guangdong
Taiwan
Guangxi
Hong Kong
South China Sea Islands
Efficiency improvement
Efficiency reduced
2012年
2013年
2014年
2015年
2016年
2017年

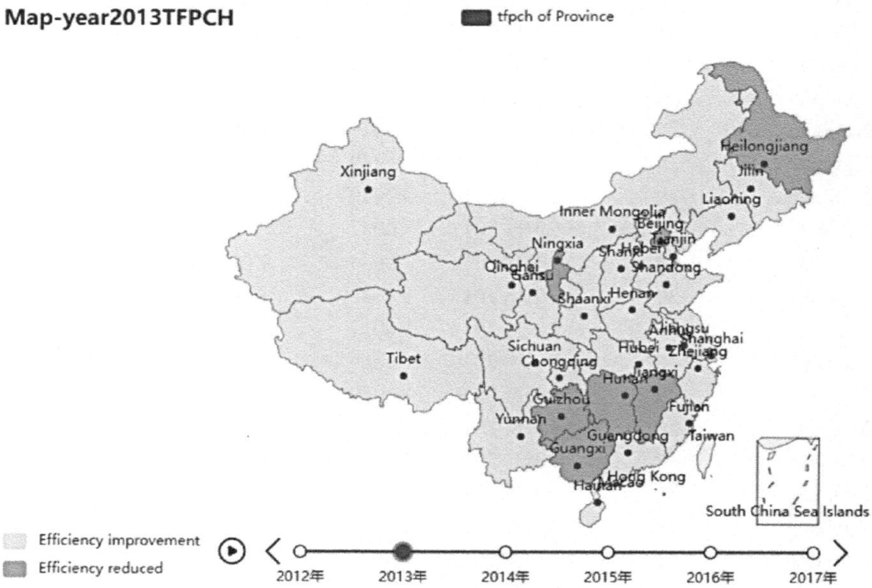
Map-year2013TFPCH
tfpch of Province
Heilongjiang
Xinjiang
Jilin
Liaoning
Inner Mongolia
Beijing
Tianjin
Ningxia
Qinghai
Gansu
Shandong
Shaanxi
Henan
Shanghai
Sichuan
Hubei
Tibet
Chongqing
Hunan
Jiangxi
Guizhou
Fujian
Yunnan
Guangdong
Taiwan
Guangxi
Hong Kong
South China Sea Islands
Efficiency improvement
Efficiency reduced
2012年
2013年
2014年
2015年
2016年
2017年

(Continued))

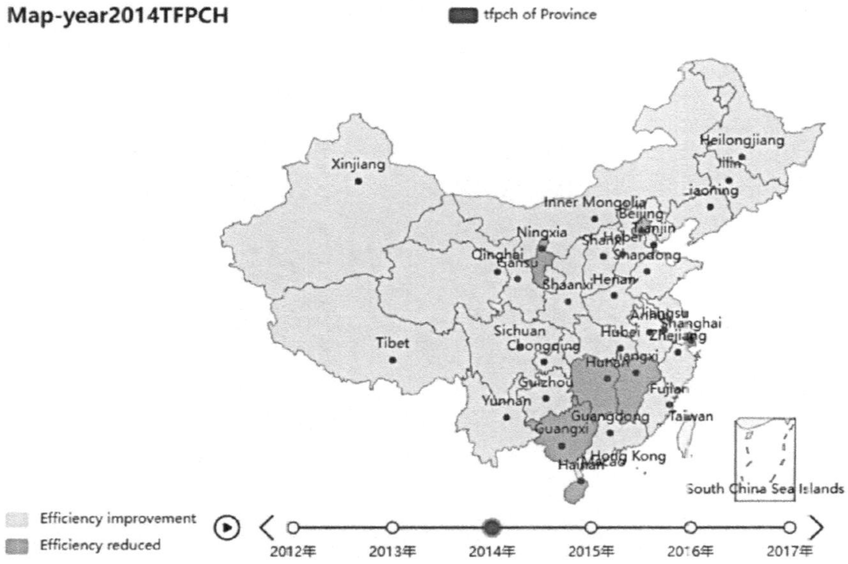
Map-year2014TFPCH
tfpch of Province
Heilongjiang
Xinjiang
Inner Mongolia
Ningxia
Qinghai
Gansu
Shandong
Henan
Tibet
Sichuan
Hubei
Zhejiang
Shanghai
Jiangxi
Fujian
Yunnan
Guangdong
Taiwan
Guangxi
Hong Kong
South China Sea Islands
Efficiency improvement
Efficiency reduced
2012年
2013年
2014年
2015年
2016年
2017年

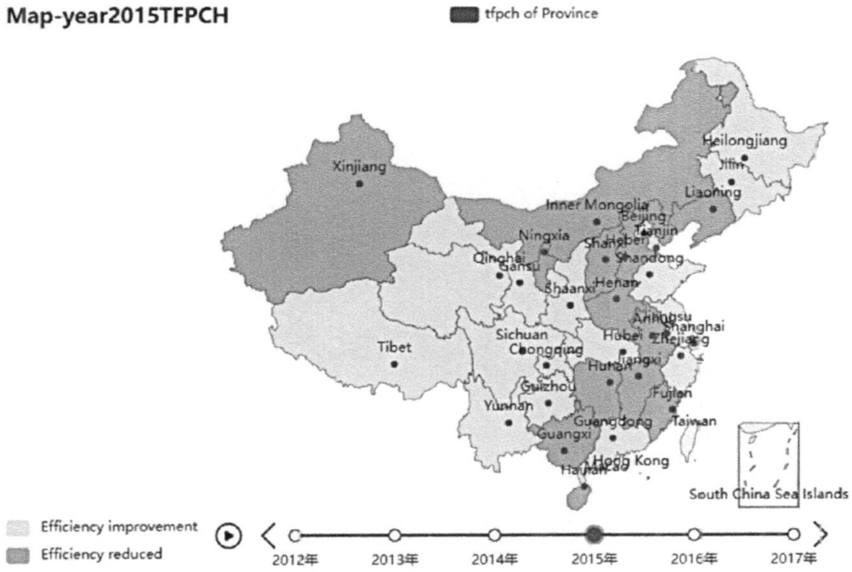
Map-year2015TFPCH
tfpch of Province
Heilongjiang
Liaoning
Xinjiang
Inner Mongolia
Ningxia
Qinghai
Gansu
Shandong
Henan
Tibet
Sichuan
Hubei
Zhejiang
Shanghai
Jiangxi
Fujian
Yunnan
Guangdong
Taiwan
Guangxi
Hong Kong
South China Sea Islands
Efficiency improvement
Efficiency reduced
2012年
2013年
2014年
2015年
2016年
2017年

(Continued))

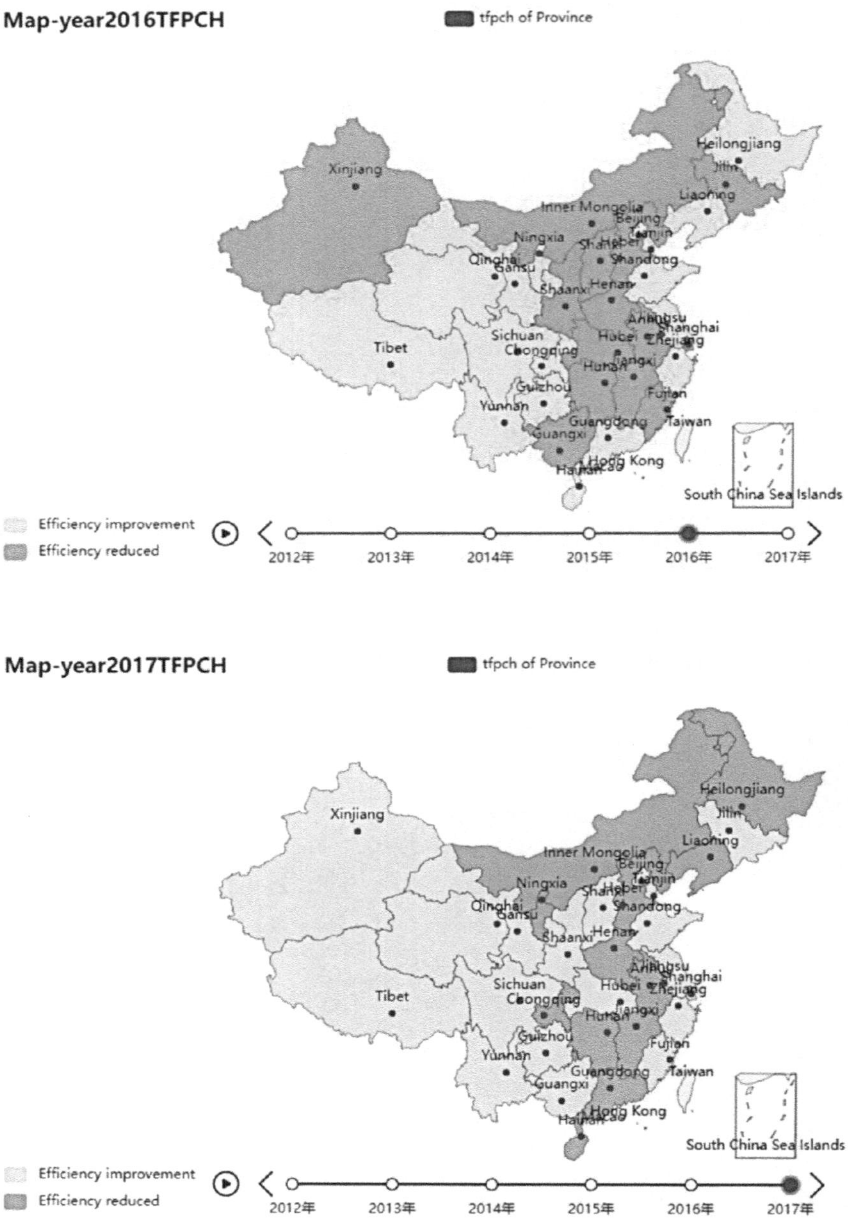

FIGURE 2.5 The total factor productivity change.

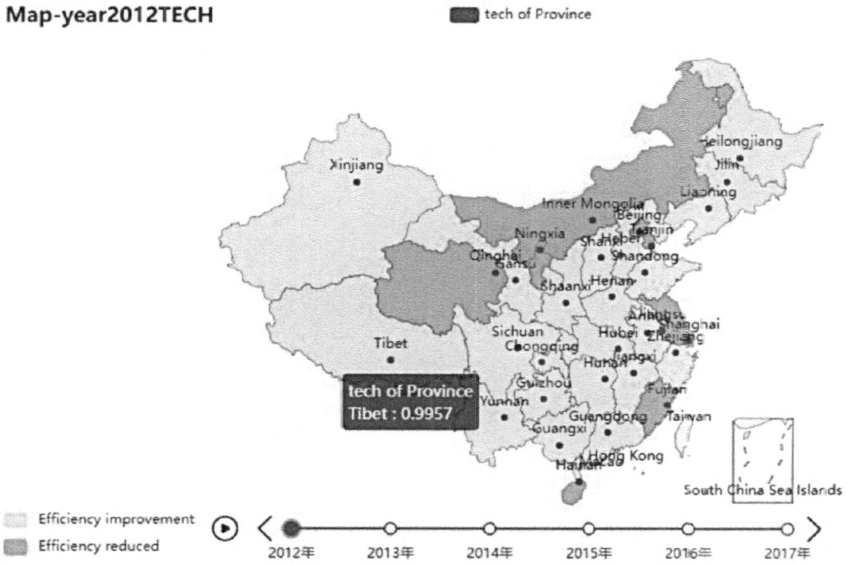
Map-year2012TECH
tech of Province
Xinjiang
Inner Mongolia
Ningxia
Qinghai
Shandong
Sichuan
Tibet
tech of Province
Tibet : 0.9957
Yunnan
Guangxi
Hong Kong
Taiwan
South China Sea Islands
Efficiency improvement
Efficiency reduced
2012年
2013年
2014年
2015年
2016年
2017年

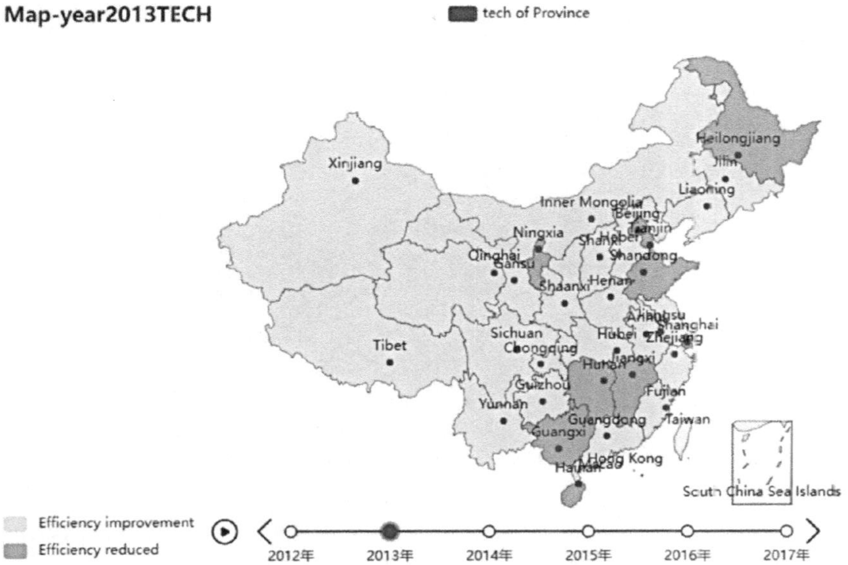
Map-year2013TECH
tech of Province
Xinjiang
Heilongjiang
Liaoning
Inner Mongolia
Ningxia
Qinghai
Shandong
Sichuan
Tibet
Yunnan
Guangxi
Hong Kong
Taiwan
South China Sea Islands
Efficiency improvement
Efficiency reduced
2012年
2013年
2014年
2015年
2016年
2017年

(Continued))

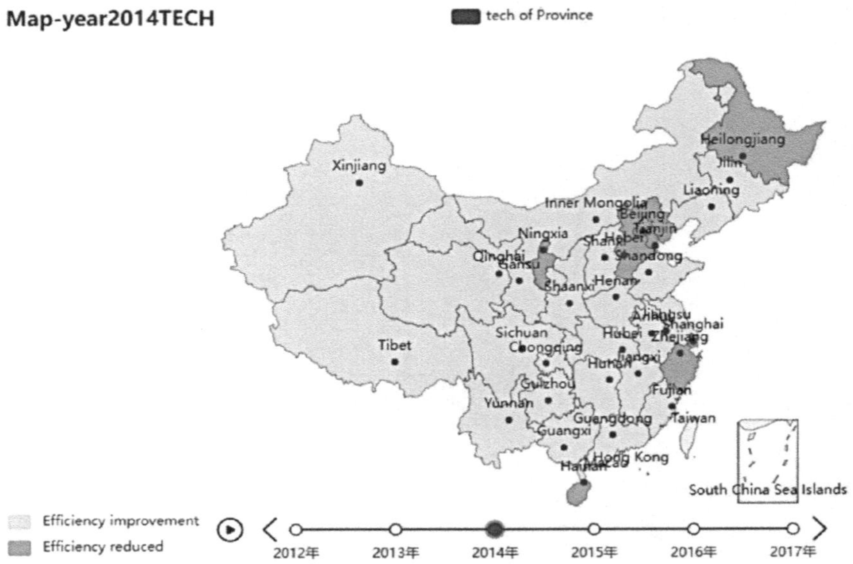
Map-year2014TECH
tech of Province
Xinjiang
Heilongjiang
Liaoning
Inner Mongolia
Ningxia
Qinghai
Shandong
Henan
Tibet
Sichuan
Shanghai
Guizhou
Yunnan
Fujian
Taiwan
Guangxi
South China Sea Islands
Efficiency improvement
Efficiency reduced
2012年
2013年
2014年
2015年
2016年
2017年

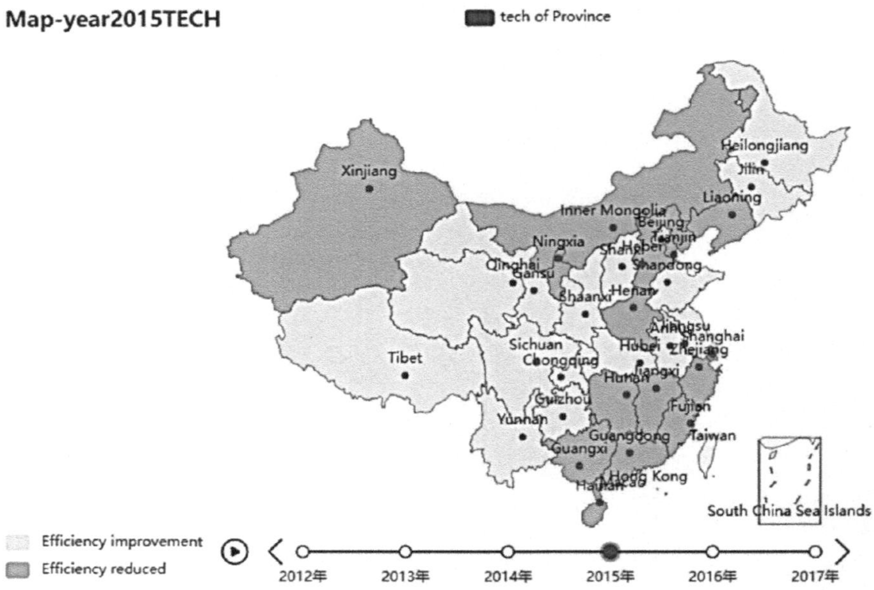
Map-year2015TECH
tech of Province
Xinjiang
Heilongjiang
Liaoning
Inner Mongolia
Ningxia
Qinghai
Shandong
Tibet
Sichuan
Shanghai
Yunnan
Fujian
Taiwan
Guangxi
South China Sea Islands
Efficiency improvement
Efficiency reduced
2012年
2013年
2014年
2015年
2016年
2017年

(Continued))

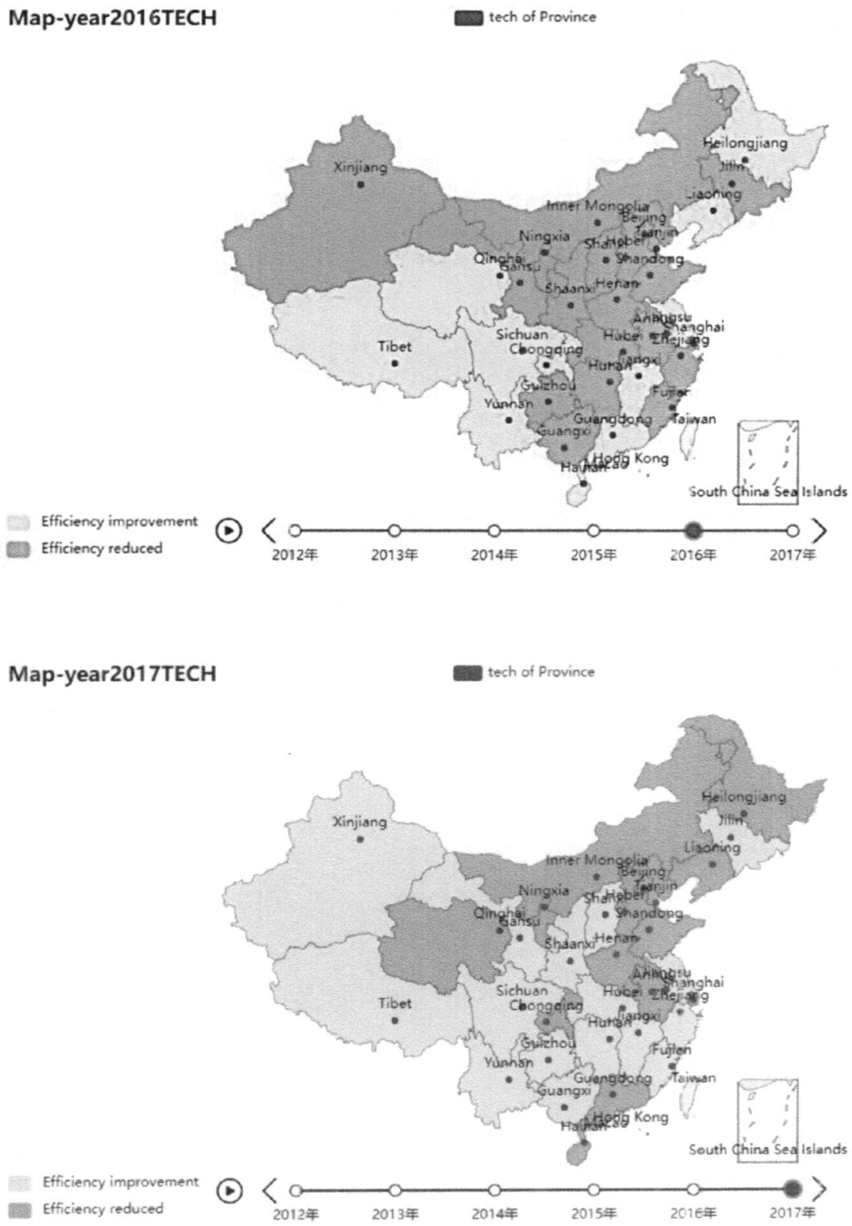

FIGURE 2.6 The technical efficiency change.

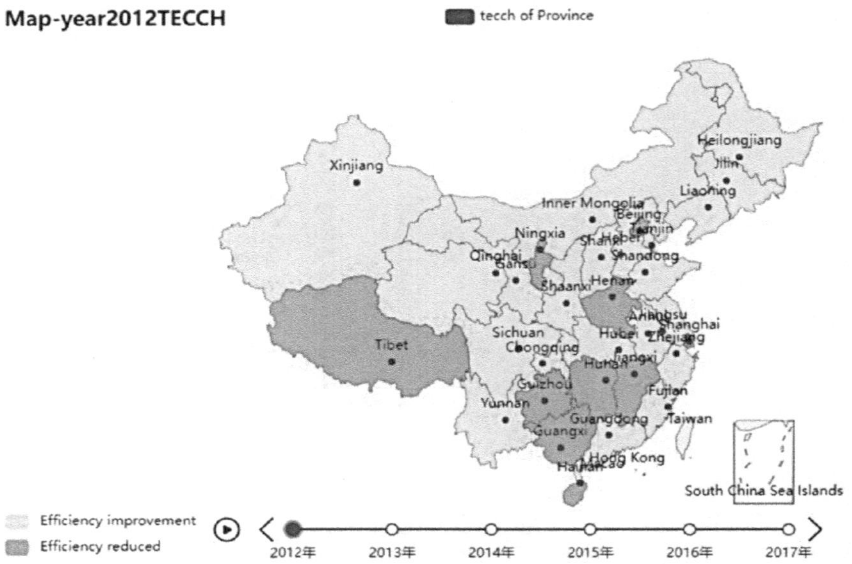
Map-year2012TECCH
tecch of Province
Xinjiang
Heilongjiang
Liaoning
Inner Mongolia
Ningxia
Qinghai
Shandong
Tibet
Sichuan
Chongqing
Hubei
Guizhou
Yunnan
Fujian
Taiwan
Guangdong
Guangxi
Hong Kong
South China Sea Islands
Efficiency improvement
Efficiency reduced
2012年
2013年
2014年
2015年
2016年
2017年

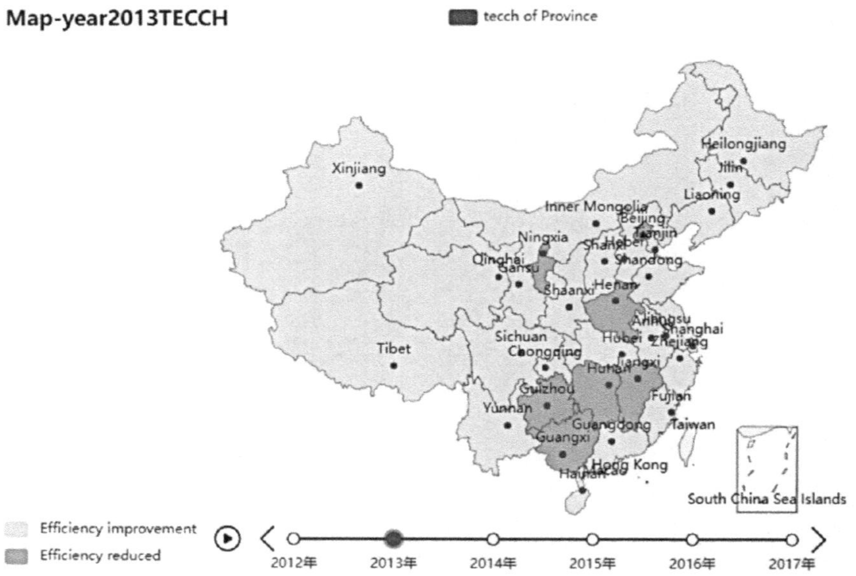
Map-year2013TECCH
tecch of Province
Xinjiang
Heilongjiang
Liaoning
Inner Mongolia
Ningxia
Qinghai
Shandong
Tibet
Sichuan
Chongqing
Hubei
Guizhou
Yunnan
Fujian
Taiwan
Guangdong
Guangxi
Hong Kong
South China Sea Islands
Efficiency improvement
Efficiency reduced
2012年
2013年
2014年
2015年
2016年
2017年

(Continued))

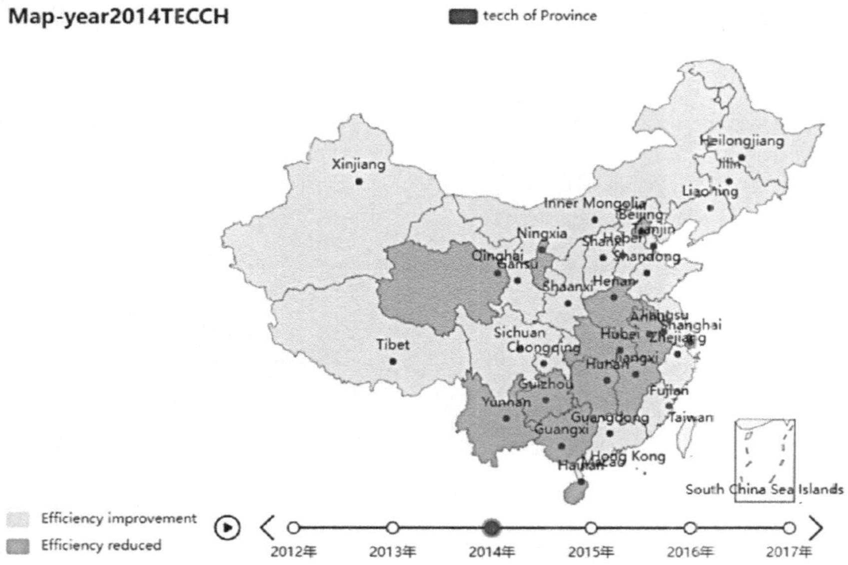
Map-year2014TECCH
tecch of Province
Xinjiang
Heilongjiang
Inner Mongolia
Ningxia
Qinghai
Shandong
Sichuan
Tibet
Yunnan
Guangxi
Fujian
Taiwan
Hong Kong
South China Sea Islands
Efficiency improvement
Efficiency reduced
2012年
2013年
2014年
2015年
2016年
2017年

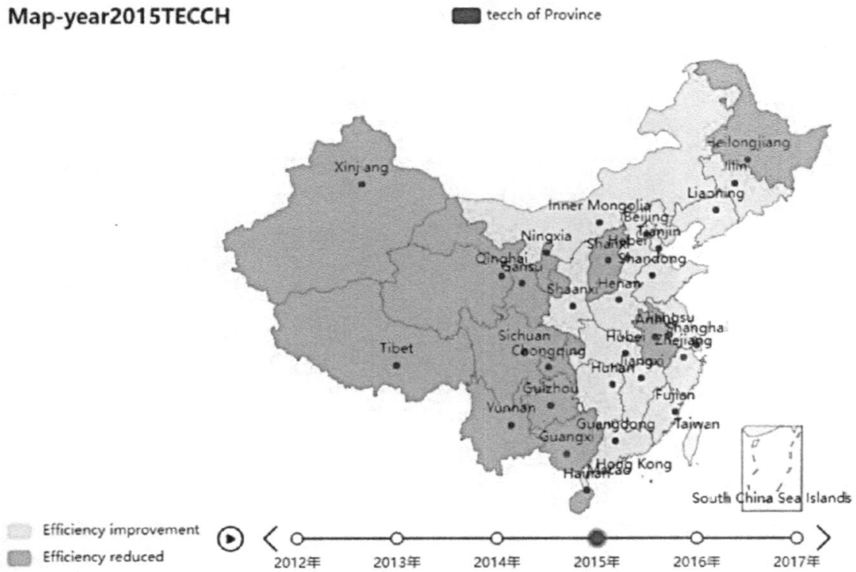
Map-year2015TECCH
tecch of Province
Heilongjiang
Inner Mongolia
Ningxia
Qinghai
Shandong
Sichuan
Tibet
Yunnan
Guangxi
Fujian
Taiwan
Hong Kong
South China Sea Islands
Efficiency improvement
Efficiency reduced
2012年
2013年
2014年
2015年
2016年
2017年

(Continued))

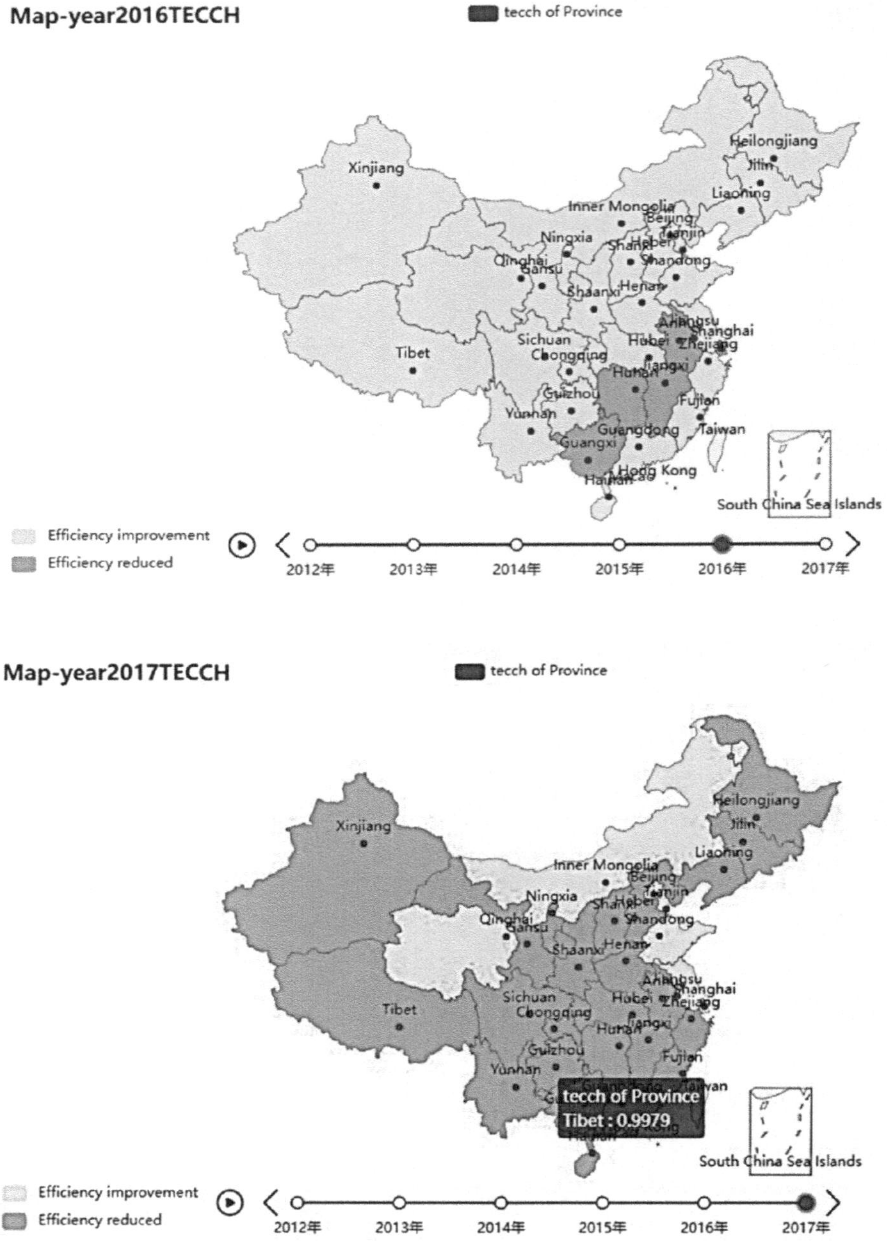

FIGURE 2.7 The technological change.

2.4.4 SPATIAL CORRELATION ANALYSIS

Table 2.5 shows the value of global correlation Moran's I index of China's GDP from 2011 to 2017. The result shows 31 provinces of China's global spatial correlation analysis. The function in the GMI index in Table 2.5 is caused by the function of the GDP value of China's province in these years, which is caused by the long-time imbalanced development among these provinces. Table 2.5 demonstrate that the GDP in different provinces in China has a marked agglomeration phenomenon at a significant level of 1%, meaning the GDP of different provinces in China has a significant positive correlation with the spatial distribution. The high GDP provinces are adjacent, and the low provinces are adjacent. This study also uses the LMI index to study whether the local GDP of provinces in China was partially concentrated. The Local Indicators of Spatial Association aggregation map of GDP in China's provinces from 2011 to 2017 is shown in Figure 2.8. The Local Indicators of Spatial Association aggregation map of water pollution in China's provinces from 2011 to 2017 is shown in Figure 2.9.

TABLE 2.5 Global Autocorrelation Moran's I Index of GDP in China's Province

Years	Moran's I	E(I)	sd(I)	z	p-Value*
2011	0.242	–0.033	0.115	2.396	0.008
2012	0.284	–0.033	0.105	3.016	0.001
2013	0.283	–0.033	0.105	3.002	0.001
2014	0.284	–0.033	0.105	3.013	0.001
2015	0.295	–0.033	0.106	3.106	0.001
2016	0.311	–0.033	0.106	3.245	0.001
2017	0.321	–0.033	0.106	3.338	0.000

From Figure 2.5, we can see that there was not much change in provinces. The result shows that most of the GDP HH area is from southeast provinces of China, which means that GDP in the southeast of China is high and clustering. And LL area is from the west of China, which means that the GDP in west north China spatial correlation is low and clustering. The HL area in the northeast such as Heilongjiang and Liaoning, and southwest provinces of China such as Sichuan and Yunnan, are high GDP spatial correlation areas and are surrounded by low spatial correlation areas. The middle area such as Shanxi, Chongqing, and Guizhou are LH, which means that the middle provinces are low spatial correlation but are surrounded by high spatial correlation provinces.

In Figure 2.6, the results show that the HH provinces are located in the southeast of China, which means that water pollution spatial correlation in the southeast of China is high and clustering. Most LL areas in the west and north of China, this means that the water pollution in west north China spatial correlation is low and clustering.

As we compare economic development and water pollution, we found that economic development is highly correlated with water pollution. The result shows that the southeast of China has HH spatial correlation of both economic development and water pollution, and LL spatial correlation of development and water pollution in the west north of China.

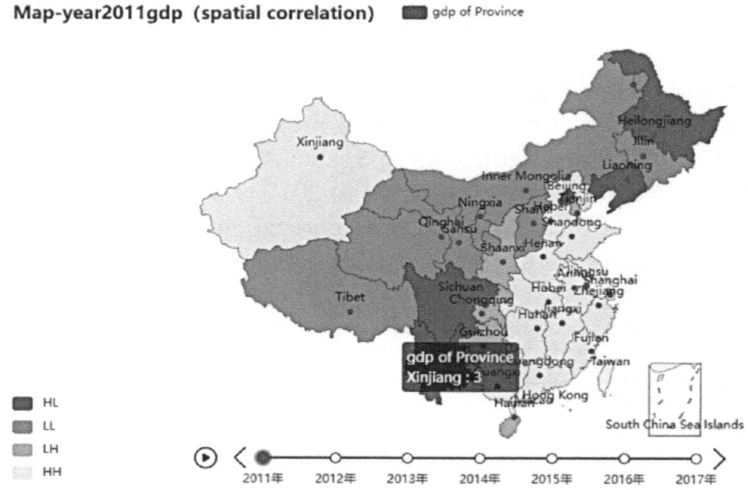

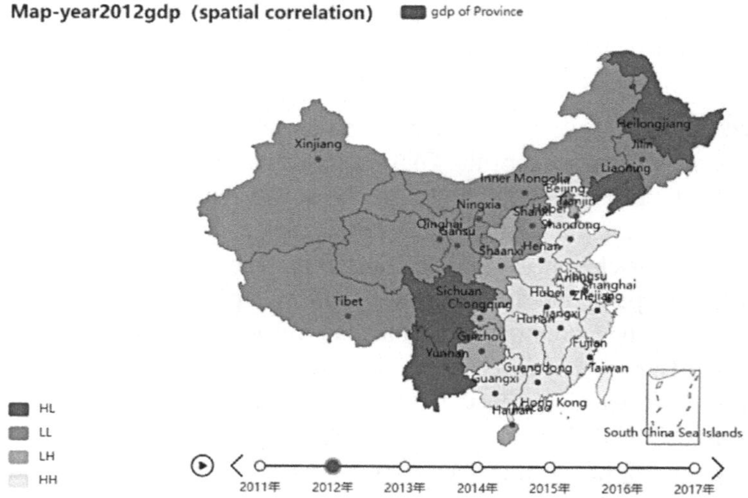

(Continued))

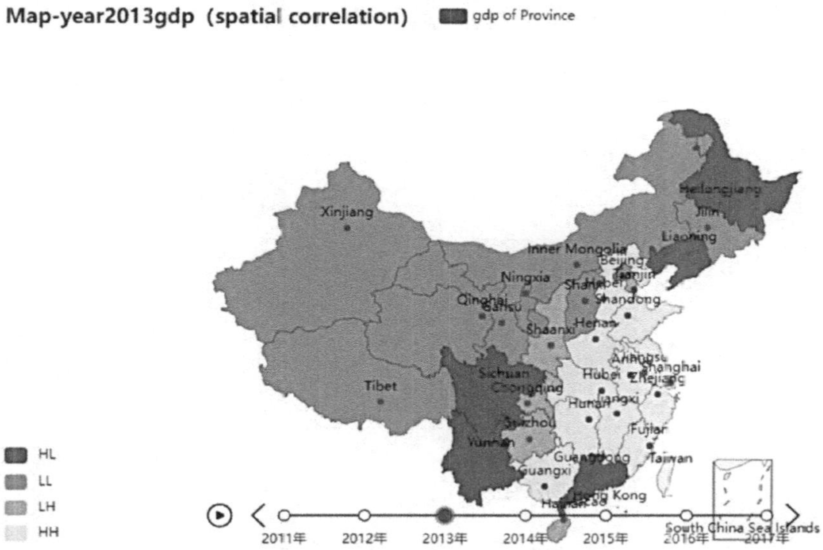
Map-year2013gdp (spatial correlation)
gdp of Province
Xinjiang
Tibet
Inner Mongolia
Ningxia
Qinghai
Gansu
Shaanxi
Henan
Shandong
Sichuan
Yunnan
Guangxi
Hong Kong
Taiwan
Shanghai
Hubei
Fujian
Liaoning
Jilin
South China Sea Islands
HL
LL
LH
HH
2011年
2012年
2013年
2014年
2015年
2016年
2017年

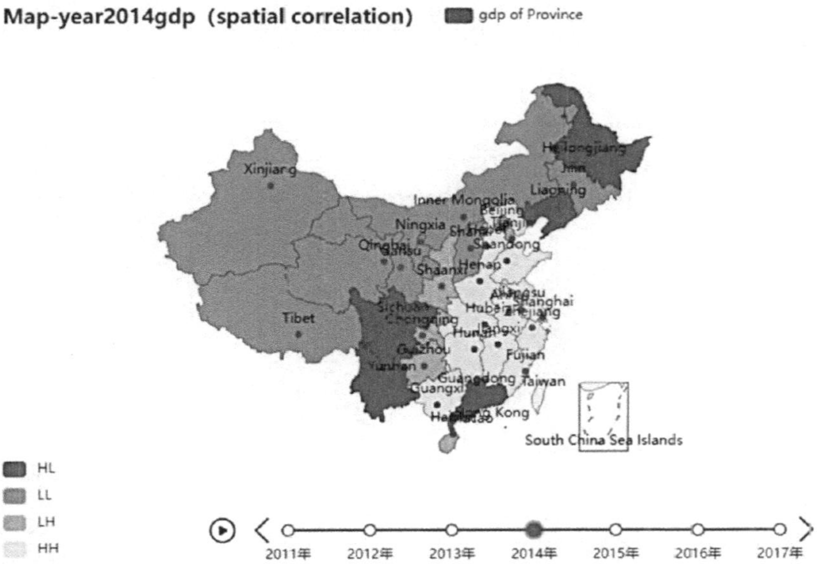
Map-year2014gdp (spatial correlation)
gdp of Province
Xinjiang
Tibet
Inner Mongolia
Liaoning
Ningxia
Qinghai
Gansu
Shaanxi
Shandong
Sichuan
Chongqing
Guizhou
Yunnan
Guangxi
Guangdong
Taiwan
Fujian
Hubei
Shanghai
Hong Kong
South China Sea Islands
HL
LL
LH
HH
2011年
2012年
2013年
2014年
2015年
2016年
2017年

(Continued))

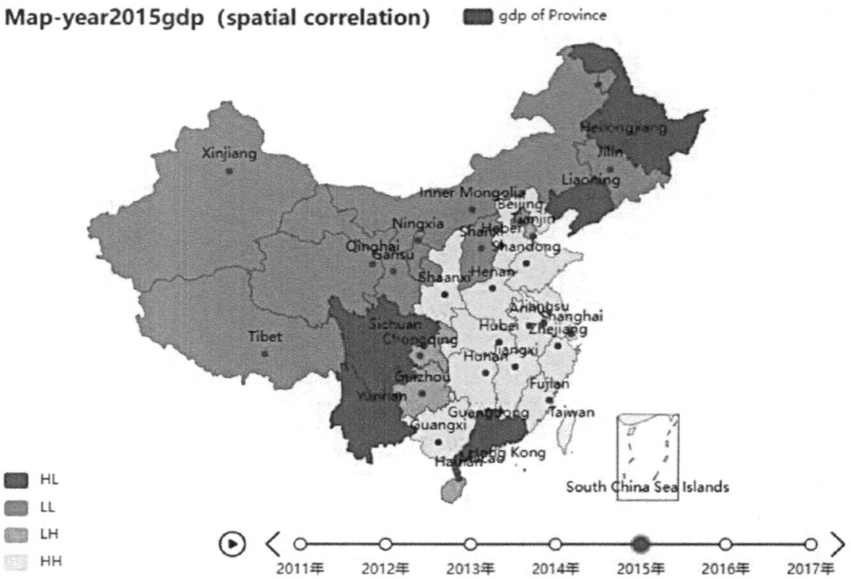
Map-year2015gdp (spatial correlation)
gdp of Province
Xinjiang
Tibet
Inner Mongolia
Heilongjiang
Jilin
Liaoning
Ningxia
Qinghai
Gansu
Shaanxi
Henan
Shandong
Sichuan
Chongqing
Hubei
Hunan
Jiangxi
Zhejiang
Shanghai
Guizhou
Yunnan
Guangxi
Fujian
Taiwan
South China Sea Islands
HL
LL
LH
HH
2011年
2012年
2013年
2014年
2015年
2016年
2017年

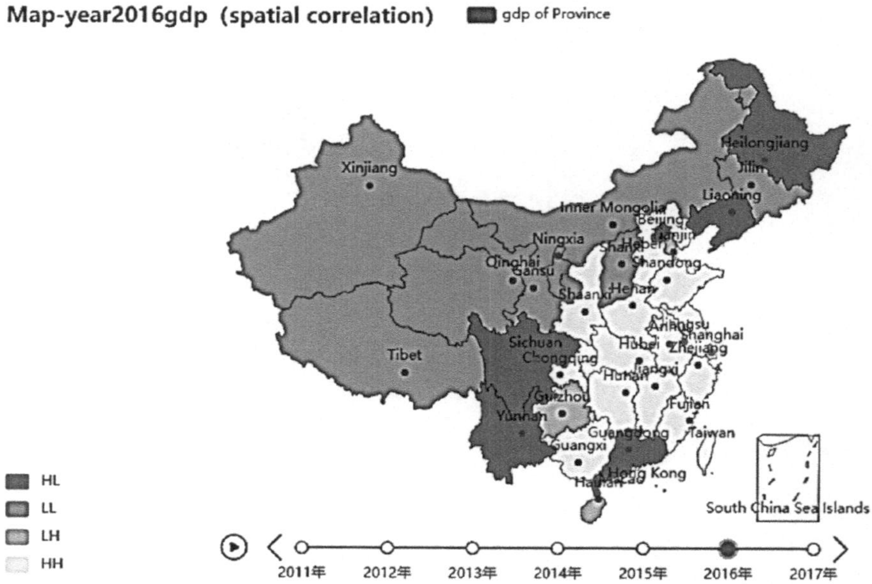
Map-year2016gdp (spatial correlation)
gdp of Province
Xinjiang
Tibet
Inner Mongolia
Heilongjiang
Jilin
Liaoning
Ningxia
Qinghai
Gansu
Shaanxi
Henan
Shandong
Sichuan
Chongqing
Hubei
Hunan
Jiangxi
Zhejiang
Shanghai
Guizhou
Yunnan
Guangxi
Fujian
Taiwan
Hong Kong
South China Sea Islands
HL
LL
LH
HH
2011年
2012年
2013年
2014年
2015年
2016年
2017年

(Continued))

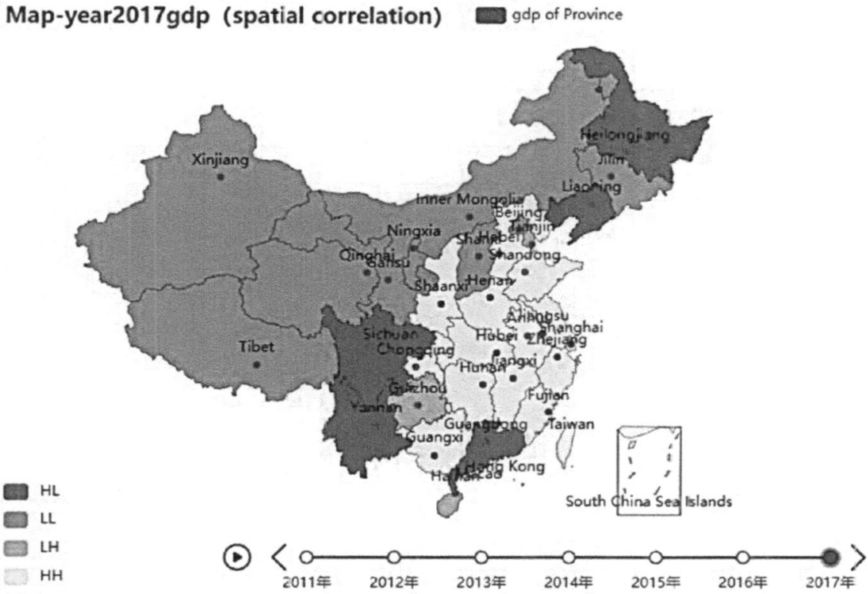

FIGURE 2.8 The spatial correlation of GDP in China's provinces.

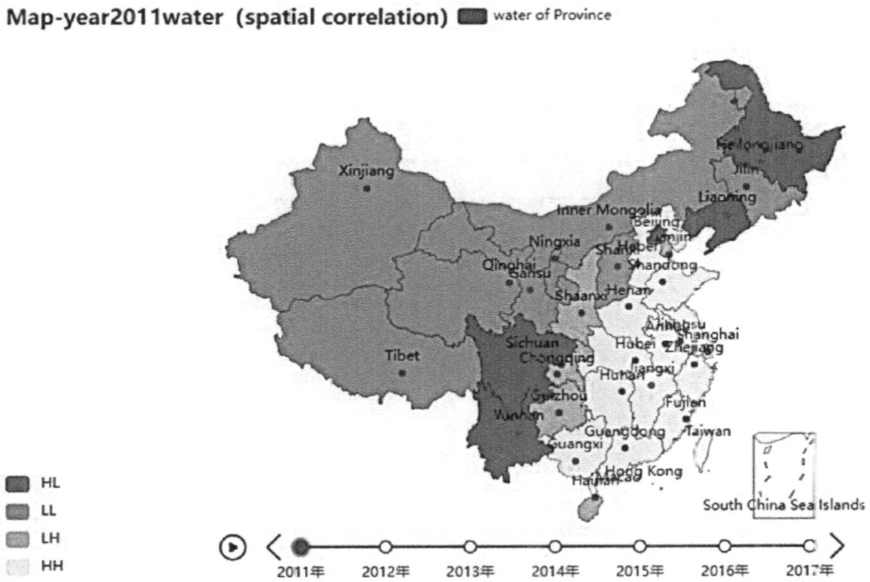

(Continued))

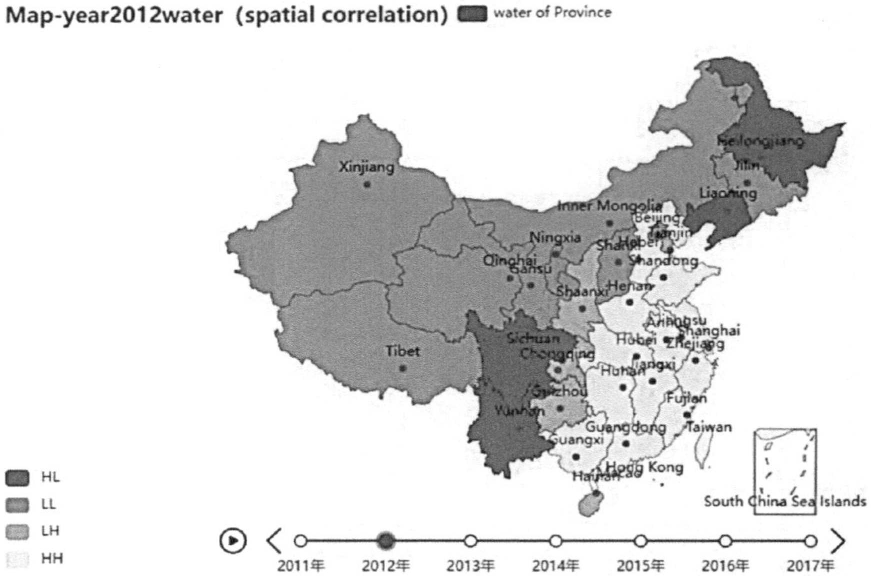
Map-year2012water (spatial correlation)
water of Province
Xinjiang
Tibet
Qinghai
Gansu
Ningxia
Inner Mongolia
Heilongjiang
Jilin
Liaoning
Beijing
Tianjin
Shandong
Henan
Shaanxi
Sichuan
Chongqing
Hubei
Zhejiang
Shanghai
Hunan
Fujian
Guizhou
Yunnan
Guangxi
Guangdong
Taiwan
Hong Kong
South China Sea Islands
HL
LL
LH
HH
2011年
2012年
2013年
2014年
2015年
2016年
2017年

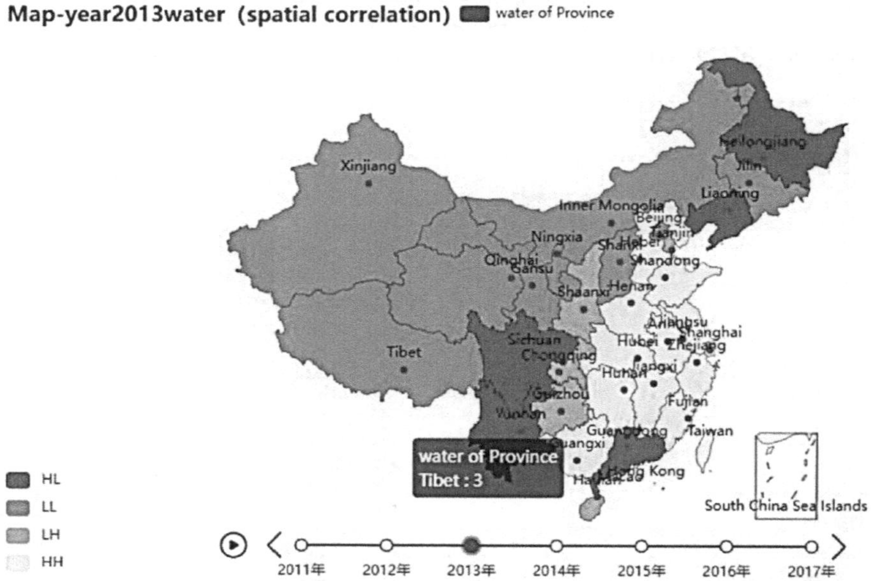
Map-year2013water (spatial correlation)
water of Province
Xinjiang
Tibet
Qinghai
Gansu
Ningxia
Inner Mongolia
Heilongjiang
Jilin
Liaoning
Shandong
Henan
Shaanxi
Sichuan
Chongqing
Hubei
Zhejiang
Shanghai
Hunan
Fujian
Guizhou
Yunnan
Taiwan
Hong Kong
water of Province
Tibet : 3
South China Sea Islands
HL
LL
LH
HH
2011年
2012年
2013年
2014年
2015年
2016年
2017年

(Continued))

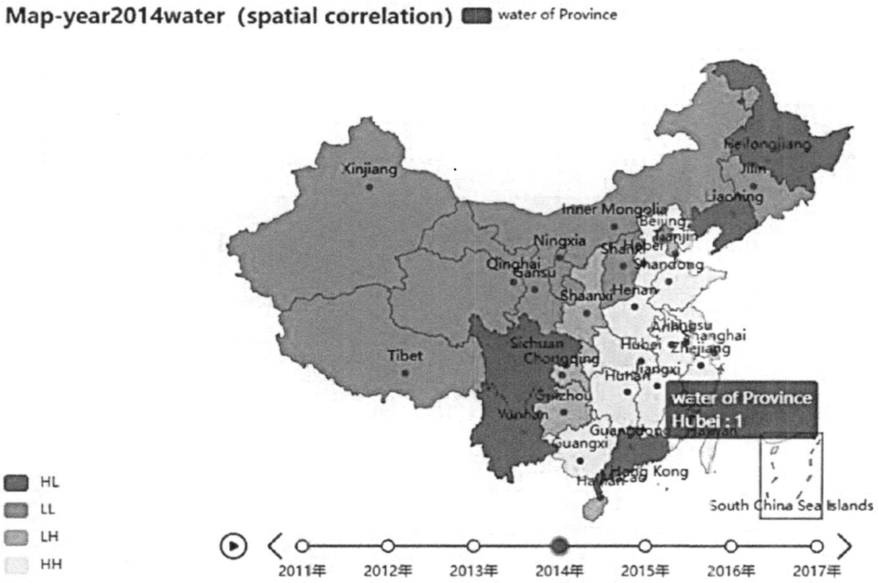
Map-year2014water (spatial correlation)
water of Province
Xinjiang
Heilongjiang
Jilin
Liaoning
Inner Mongolia
Ningxia
Qinghai
Gansu
Shaanxi
Henan
Tibet
Sichuan
Hubei
Hunan
Guizhou
Yunnan
Guangxi
Hong Kong
water of Province
Hubei : 1
South China Sea Islands
HL
LL
LH
HH
2011年
2012年
2013年
2014年
2015年
2016年
2017年

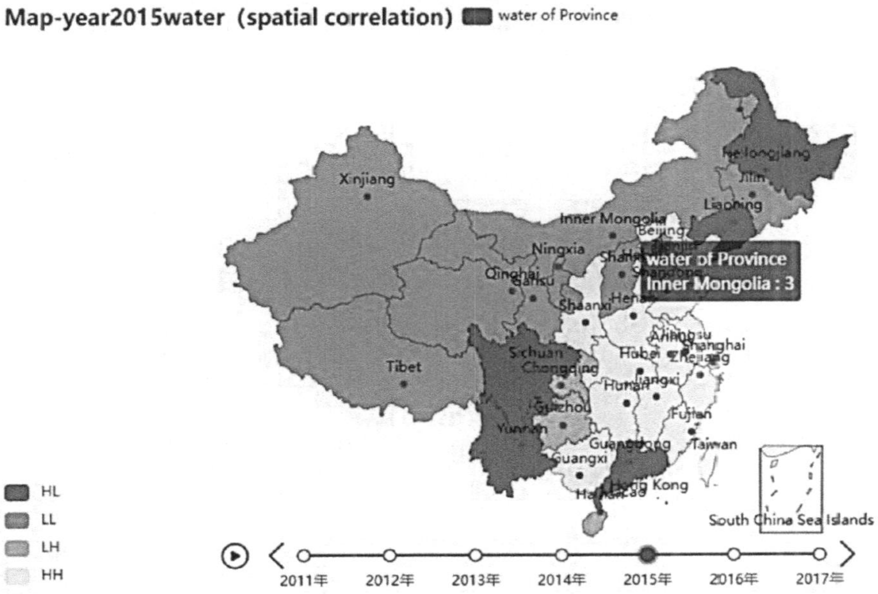
Map-year2015water (spatial correlation)
water of Province
Xinjiang
Heilongjiang
Jilin
Liaoning
Inner Mongolia
Ningxia
Qinghai
Gansu
Shaanxi
water of Province
Inner Mongolia : 3
Tibet
Sichuan
Hubei
Hunan
Guizhou
Yunnan
Fujian
Taiwan
Guangxi
Hong Kong
South China Sea Islands
HL
LL
LH
HH
2011年
2012年
2013年
2014年
2015年
2016年
2017年

(Continued))

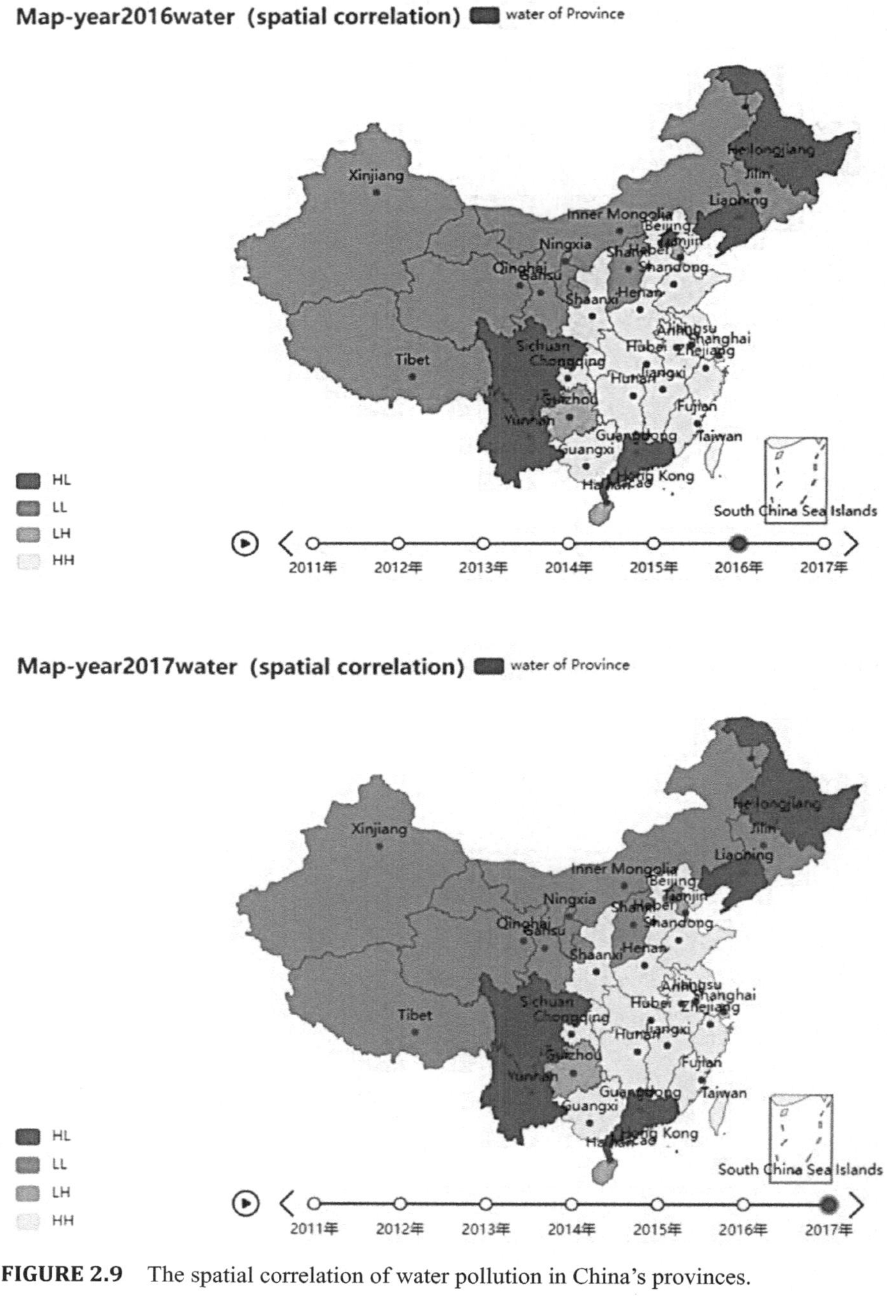

FIGURE 2.9 The spatial correlation of water pollution in China's provinces.

2.5 CONCLUSION

This chapter has attempted to investigate the relationship between energy investment, economics, and water pollution in 31 provinces in China from the period 2011 to 2017. The DEA model and the spatial correlation analysis are both highly popular methods used for empirical research on the efficiency of resource allocation in various studies. Therefore, both of these methods had been applied in this study not only to evaluate the efficiency of decision-making units but also for measuring energy investment-economic-pollution coupling efficiency in spatial correlation analysis.

As discussed in the introduction, climate change would have more serious negative effects on economic growth in poor countries. Yet, the rich countries are often reluctant to increase foreign aid to support the poor countries' coordinated efforts to alleviate the climate change problem. For a country with an income level less than the threshold, economic growth and environmental deterioration are positively associated, in which when the country's economy is growing, the environmental quality is plunging. Meanwhile, for a country with an income level greater than the threshold, there is a positive environment-growth relationship, whereby economic growth would improve the environmental conditions of the country. We tested these theories in the case of 31 provinces in China and found out that as demonstrated in the descriptive analysis, when energy investment and province GRP shows significant growth, water pollution remains the same. Meanwhile, in the section on correlation analysis, the result presented that energy investment, province GDP, and water pollution are positively correlated. It proves that as the province's economy is growing, water pollution is also growing, however, economic growth also improves energy investment.

Further analyzes performed by employing the DEA model and spatial correlation produced result that shows most of the eastern and central provinces are in high efficiency economic, while most low economic efficiency provinces are located in western China. Thus, policies and funding support for the economic dimension are recommended. The eastern area had lower efficiency values than the average in environmental aspects, therefore the government should emphasize pollution treatment more aggressively in this area. To conclude, we came up with the recommendation for sustainable and stable regional development, as well as a regional synergy in energy investment, economic development, and water pollution treatments.

Improvement is certainly required in the central government evaluation indicators and mechanisms regarding the energy investment-economic-pollution nexus relations. This will assist the departments involved in central government as well as regional, for long-term coordinated planning and actions, both internal and external concentrated efforts in preparing efficient solutions for regional imbalance and resource allocations. In the near future, more in-depth studies shall be carried out to understand the relationship between economic growth and environmental conservation as it is one of the major issues in environmental economics, and hopefully, the analysis and results will provide clear insights in aiding policymakers for efficient decision-making.

KEYWORDS

- **correlation analysis**
- **data envelopment analysis**
- **descriptive analysis**
- **economics**
- **energy investment**
- **spatial correlation analysis**
- **water pollution**

REFERENCES

Bai, Y., Deng, X., Jiang, S., Zhang, Q., & Wang, Z., (2018). Exploring the relationship between urbanization and urban eco-efficiency: Evidence from prefecture-level cities in China. *Journal of Cleaner Production, 195*, 1487–1496.

Gao, D., Zhang, B., & Li, S., (2021). Spatial effect analysis of total factor productivity and forestry economic growth. *Forests, 12*(6), 702.

Guo, B., He, D., Zhao, X., Zhang, Z., & Dong, Y., (2020). Analysis on the spatiotemporal patterns and driving mechanisms of China's agricultural production efficiency from 2000 to 2015. *Physics and Chemistry of the Earth, Parts A/B/C, 120*, 102909.

Li, L., & Liu, Z., (2021). Research on efficiency measurement and spatiotemporal disparity of rural public health services in China. *Plos One, 16*(7), e0252871.

Li, Y., & Wang, C., (2013). Spatial correlation analysis on the resource allocation efficiency of regional higher education in China. *Journal of Education and Learning, 7*(2), 101–108.

Liao, J., Yu, C., Feng, Z., Zhao, H., Wu, K., & Ma, X., (2021). Spatial differentiation characteristics and driving factors of agricultural eco-efficiency in Chinese provinces from the perspective of ecosystem services. *Journal of Cleaner Production, 288*, 125466.

Sun, C., Zhao, L., Zou, W., & Zheng, D., (2014). Water resource utilization efficiency and spatial spillover effects in China. *Journal of Geographical Sciences, 24*(5), 771–788.

Tian, D., Zhao, F., Mu, W., Kanianska, R., & Feng, J., (2016). Environmental efficiency of Chinese open-field grape production: An evaluation using data envelopment analysis and spatial autocorrelation. *Sustainability, 8*(12), 1246.

Wang, L., He, Y., & Wang, T., (2021). Analysis of the spatial structure of regional enterprises and its impact on regional economic development based on the data envelopment analysis model (DEA). *Journal of Mathematics, 2021.*

Wang, X., & Du, L., (2017). Carbon emission performance of China's power industry: Regional disparity and spatial analysis. *Journal of Industrial Ecology, 21*(5), 1323–1332.

Xia, B., Dong, S., Ba, D., Li, Y., Li, F., Liu, H., & Zhao, M., (2018). Research on the spatial differentiation and driving factors of tourism enterprises' efficiency: Chinese scenic spots, travel agencies, and hotels. *Sustainability, 10*(4), 901.

Xue, L., Du, Y., Li, W., & Zhao, X., (2019). Space-time effect modeling with machine learning: A scalable approach for assessing efficiency of coal resources. In: *2019 5th International Conference on Big Data Computing and Communications (BIGCOM)* (pp. 178–186).

Zhao, P., Zeng, L., Li, P., Lu, H., Hu, H., Li, C., & Qi, Y., (2022). China's transportation sector carbon dioxide emissions efficiency and its influencing factors based on the EBM DEA model with undesirable outputs and spatial Durbin model. *Energy, 238*, 121934.

Zhou, Y., Tian, G., & Cai, D., (2019). Spatial effects of environmental regulation on regional ecological efficiency. *Ekoloji, 28*(107), 3605–3616.

CHAPTER 3

Green Energy in Malaysia: A Pathway Toward Sustainable Development

WONG SING YUN[1] and JAIN YASSIN[2]

[1]*Faculty of Business, Economics and Accountancy, University Malaysia Sabah (UMS), Kota Kinabalu, Sabah, Malaysia*

[2]*Faculty of Business and Management, Universiti Teknologi Mara (UiTM), Sabah Branch, Kota Kinabalu, Sabah, Malaysia*

ABSTRACT

The transition towards green economy helps the nation in achieving a sustainable development and provides potential solution to the global environmental problems faced over the decades. In heading towards green economy, more investments can be observed in the development of green technology innovation and the increased use of green energy. By definition, green economy has the ability not just to produce clean energy but also allowing the development of technologies that enable cleaner production process of which reduces environmental impact and allow a sustainable use of natural resource (Chapple, 2008). In this chapter, we would like to present a discussion that focus on giving an overview of the development and progress of the transition towards green economy with the focus given especially on green energy consumption and production in Malaysia. A review will be provided on the Malaysian Government's roles in implementing policies and strategies to promote the use of green energy and encouraging the development of green technology innovation. Next, we will attempt to analyze the constraints faced as the nation turn towards green economy development.

Waste Biorefineries: Future Energy, Green Products, and Waste Treatment.
Abu Zahrim Yaser, Siew-Xian Chin, Leonardo Sepúlveda Torre, & A. K. Haghi, (Eds.)

3.1 INTRODUCTION

According to the United Nations Environmental Plan (2011), green economy is considered as improvement of human health and social fairness, while remarkably reduce environmental risks and ecological scarcities. One can also defined green economy as a clean energy economy with emphasis on not just producing clean energy but also green technologies that allow energy saving and improvement of the sustainable ecosystem. Theoretically, Cato (2011) elaborated that green economics differs from neoclassical economics as it seeks to include the perspectives of the marginalized, inherently concerned with social justice, grounded in the process of building a sustainable economy in practice, focus on a steady-state economy, increased role for community-based economic activity and encouraged a shift towards location with the aim to reach self-sufficiency in basic resources.

The development of the green economy was widely seen as offering a potential solution to the multiple global crises. In fact, the green economy has been growing rapidly in the present day through the development of environmentally enhancing goods and services, including cleaner and more efficient green technologies, renewable energy (UNEP, 2011). In the wake of this, the green economy has become the focus of policy development at both the national and international scales. At the operational level, the green economy is seen as one whose growth in income and employment is driven by financing that: low carbon emissions and pollution, energy intensity and resource productivity, and avoid the loss of biodiversity and ecosystem services (Musango et al., 2014). These include both the investments in human and social capital with recognition of the central focus of human well-being and social equity, promotion of income growth and employment in the green economy context. In addition, pursue of green technology in economic and social developments will in certain help to sustain non-renewable fuels, safeguards, and minimizes the environmental degradation.

Meanwhile, researchers in the past have substantially established a clear relationship between energy consumption and economic growth. The economic growth as measured by gross domestic product (GDP) correlates almost exactly with the energy consumption in Malaysia (Shafie et al., 2011). With the rapid economic development, Malaysia will need more and more resources to support its industrial development. This has no doubt raised the demand for energy consumption due to the high economic development rate of Malaysia. A critical challenge lies ahead of the power sector in Malaysia, therefore, will be the issue of sustainability in ensuring the security and reliability of energy supply. The question of security and reliability of energy supply needs to be addressed to ensure smooth implementation of industrial developments that will boost the economic growth in Malaysia. Besides that, the quick depletion of non-renewable resources such as fossil fuels for power generation calls for the adoption of alternative energy. As such, green technology application is seen as the sensible solution to address this main issue of energy and environment (Islam et al., 2009; Shafie et al., 2011; Oh et al., 2010). Similarly, Rahman et al. (2015) supported that green technologies are of a highest necessity with emissions found to be closely related to economic output.

According to Chua & Oh (2011), improvement of energy efficiency, the increase consumption of renewable energy, the use of green technologies in energy, buildings, water and waste management, and transport sectors have helped in unnecessary waste of energy as well as cost and prevents environment degradation. Some of the benefits of green economy presented include: (i) monetary benefits through savings of energy related costs; (ii) save depleting non-renewable fuels using energy efficiently saves overall fuel consumption and extends the lifetime of scarce reserves; (iii) preserve of the environment by adopting renewable energy and energy efficiency to reduce GHG emissions; and (iv) improve foreign reserves. With the benefits as highlighted, the prospects of green economy are certain and the implementation of green policies by the Malaysia government would be highly favorable. This is consistent with the suggestion by Rahman et al. (2015) that the government should design a prudent energy policy that can sustain economic growth while securing energy supplies and protecting the environment.

With the renew interest on green economy development, this chapter will examine the Malaysia's progress within the context of energy consumption and the transition towards renewable energy consumption. Besides that, this chapter will also highlight on the green policies development implemented in Malaysia over the past decades. In many of the existing literatures, the highlight was on one of the important green policies introduced which is the National Green Technology Policy (NGTP, 2009) that has been launched back in 2009. However, the research on the green economy progress and challenges in Malaysia has remained rather sparse over these few years. Hence, the analysis presented from this chapter becomes essential to reveal a better outlook into the green economy development within the present's context.

3.2 ENERGY CONSUMPTION AND PRODUCTION TRENDS IN MALAYSIA

According to Rahim & Liwan (2012), the basic structure of Malaysian economic and productive development is highly dependent on the consumption of fossil fuels, and the main sources of energy supply are petroleum and natural gas which had accounted for more than 88% of the total energy supply. It was projected that the current reservation of petroleum and natural gas in Malaysia would likely last for the next 16 to 36 years, given the current consumption rate. In addition to that, Shafie et al. (2011) shed light on the rise of energy demand in Malaysia with the increase of the population living in the urban areas. Similarly, Oh et al. (2010) stressed that the energy consumption in Malaysia as a rapid developing nation will no doubt keep on increasing. This will definitely raise energy challenges immensely. Figure 3.1 shows the non-renewable energy consumption (% of total energy consumption), while Figure 3.2 indicates the renewable energy consumption (% of total energy consumption)

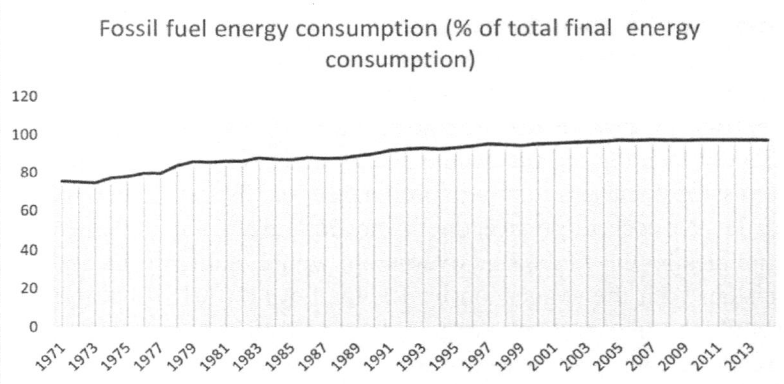

FIGURE 3.1 Non-renewable energy consumption (percentage of total final energy consumption).

Source: Data extracted from: World Bank.

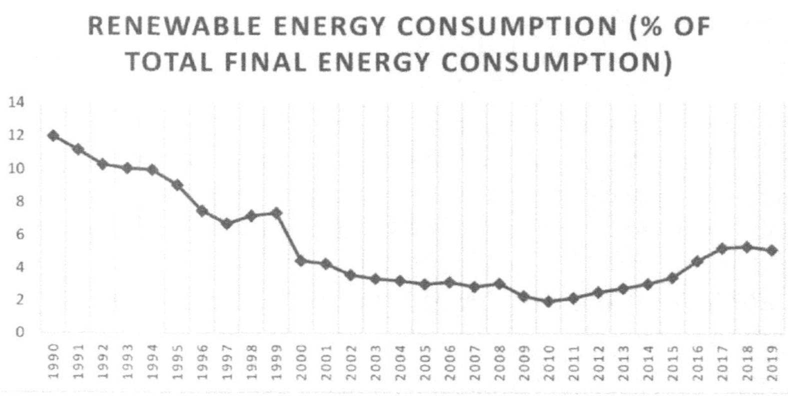

FIGURE 3.2 Renewable energy consumption (percentage of total final energy consumption).

Source: Data extracted from: World Bank.

As illustrated in Figures 3.1 and 3.2, it appeared evident that Malaysia has critically relied on the consumption of non-renewable energy consumption such as fossil fuels over the past decades. The consumption of renewable energy has remained relatively low as compared. Since 1971 to 2019, the consumption of fossil fuels as percentage of the total energy consumption has been found increasing consistently with the highest percentage recorded in the year 2010 (94%). Meanwhile, the consumption of renewable energy was relatively low over the past decades. In between the year 1990 to 1997, the consumption pattern of the renewable energy recorded a consistent decline from 11.98% in 1990 to 6.71% in 1997. Albeit that there was a slight increase of the renewable energy consumption from the year 1998 (7.17%) to 1999 (7.34%), the consumption dropped drastically in the later years from 2000 (4.44%) to 2010 (1.96%). The percentage of renewable energy consumption had remained to be rather sluggish rising to a mere 5.11% recorded in the most recent year of 2019. Figure 3.3 showed the graphical analysis on the data of the energy consumption of data that was obtained from Foo (2015). From Figure 3.3, it was clear that there had been significant increase in the consumption of overall energy demand. Besides that, both the transportation sector and industrial sector were accounted for the highest energy consumers in Malaysia over the decades. In average of the 35 years, the energy supplied to the industry sector was about 39.01% and the transportation sector with 39.15%. Meanwhile, 13.36% of the energy on average was being supplied to residential and commercial sector, 7.82% on average of the energy was supplied to non-energy sector, and the remaining 0.70% on average of the energy to agriculture sector.

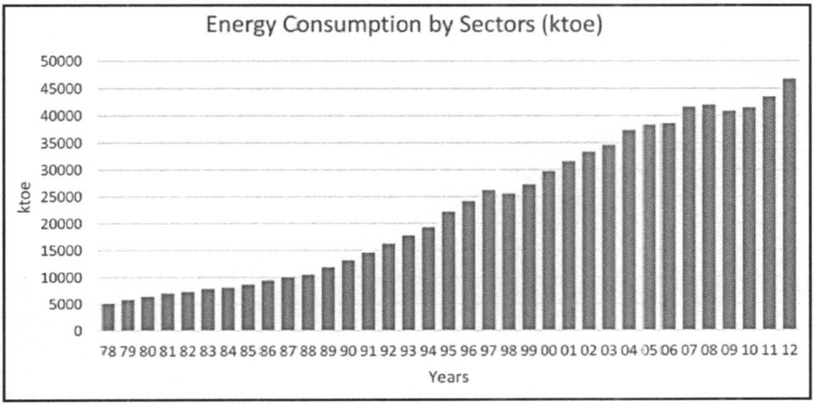

FIGURE 3.3 Energy consumption by sectors (ktoe)
Source: Data extracted from: Foo (2015).

Meanwhile, Figure 3.4 displayed the electricity production trend from non-renewable resources. As observed from Figure 3.4, the percentage of electricity production from non-renewable resources that includes oil, gas, and coal sources had remained relatively high over the past decades. Despite the fluctuations of the electricity produced in a few of the years observed, the trends appeared to be increasing slowly over the years. During the 70's, the electricity production has consistently increase from 72% (1971) to the highest recorded 89% (1977). By the 80's, the trend had lowers slightly from 88% (1979) to the lowest record 71% (1988). However, the electricity production from non-renewable resources again increases during the 90's. This is visibly portrayed from the data with the increase of the production that started in the 1990 with 83% from 76% (1989). This percentage continue to increase in the next seven years with 93% recorded by 1997. Although the trend continues to fluctuate in the remaining observed years, however, it is evident that the percentage of electricity production had remained to be relatively high with the percentage found to lies within 90% for all those years.

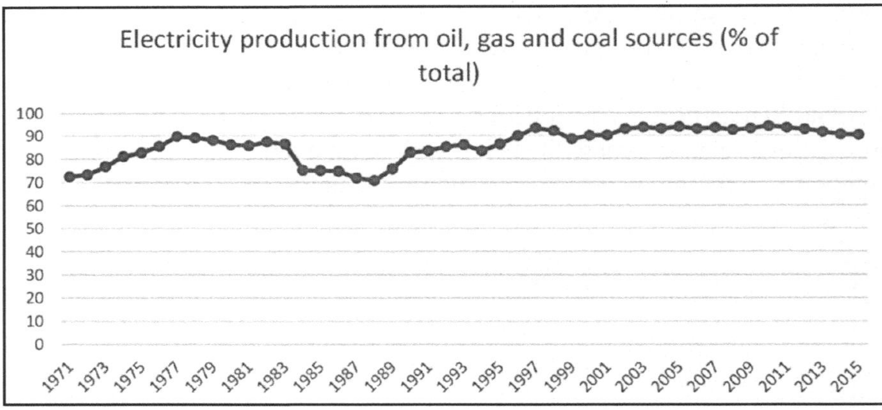

FIGURE 3.4 Electricity production from non-renewable sources (percentage of total).
Source: Data extracted from: World Bank.

It was evident from Figure 3.5 that there was no electricity production from renewable resources from 1971 to 2000, with 0% of energy production being recorded during those years. In between 2001 and 2008, the percentage of energy production using the renewable resources had remained extremely low with less than 0.002% of total energy production. By 2009, the energy production increased to 1%. This increase could be attributed to the formulation of the National Biofuel Policy in April 2009. The National Biofuel Policy was formulated to reflect Malaysia's seriousness in driving the message that 'clean and green' is the way forward towards creating an economy that is based on sustainable solutions (Shafie et al., 2011). The use of renewable energy was being emphasized to satisfy the development of green energy and eventually reduce environmental degradation.

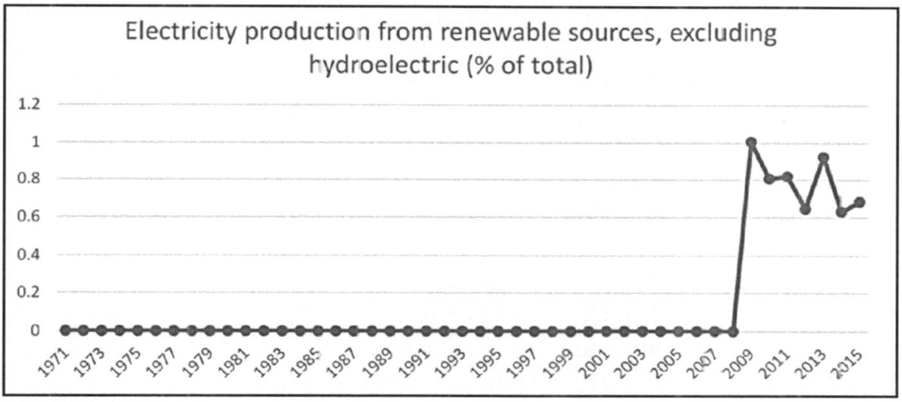

FIGURE 3.5 Electricity production from renewable sources (percentage of total).

Source: Data extracted from: World Bank.

On the 24 July 2009, the National Green Energy Technology Policy was built on four pillars: (i) seek to attain energy independence and promote efficient utilization; (ii) conserve and minimize the impact on the environment; (iii) enhance the national economic development through the use of technology; and (iv) improve the quality of life for all, and five strategic thrusts: (a) strengthen the institutional frameworks; (b) provide conducive environment for green technology development; (c) intensify human capital development in green technology; (d) establish green technology research and innovations; and (e) promotion and public awareness (Foo, 2015). The increased percentage use of renewable sources in electricity production was also in line with the 9th Malaysian Plan (2006–2010), which emphasized the intensification of energy efficiency to address the country's energy challenge.

Between 2010 and 2015, the use of renewable energy for electricity production fluctuated but remained within 1% which was still comparably higher than the years before 2009. In May 2010, the National Renewable Energy Policy and Action Plan was introduced with the goal of increasing renewable energy from 1% to 5.5% (Shafie et al., 2011). Although the comparable energy production using renewable sources was higher than in the past years before 2009, however, arguably the percentage is still rather low. According to Shigeoka (2004), the renewable energy development has been rather slow due to the insufficient commitment of the Malaysian government. This can be seen from the government's failure to achieve the target of the 8th Malaysia Plan (2001–2005) which had earlier aimed to generate 5% of the country's electricity supply using renewable resources by 2005, however, only a mere 0.3% of the energy produced was from renewable resources. Hence, it is pertinent for the government to critically analyze the shortcomings of the poor utilization of the renewable resources in energy production despite the efforts of the Malaysia Plan implemented.

3.3 THE DEVELOPMENT OF GREEN ECONOMY IN MALAYSIA

Since gaining independence from colonial rule in 1957, Malaysia has been a success story in South-East Asia thanks to its abundance of natural resources, quick industrialization, and thriving technology industry. According to Mokthsim & Salleh (2014), Malaysia starts concerned in environmental issues since colonial times. To spearhead a new initiative tackling global challenges on environmental pollution, ozone depletion, global warming, and other related issues, the Malaysian government has accelerated efforts to plan, formulate programs, and coordinate policies to promote green technology. Some of the initiatives of green economy in Malaysia are explored in subsections.

3.3.1 THE MALAYSIA'S PLAN AND GREEN ECONOMY

The Malaysia plan is the planning blueprint of Malaysia to promote the welfare of all citizens and improve the living conditions in rural areas and to end poverty, nonetheless, also execute the sustainable development philosophy's principles of the environment and society consideration. For instance, in accordance with the Ninth Principle of Islam, the government would continue and promote Hadhari environmental stewardship in the Ninth Malaysia Plan to ensure that there is a balance between the needs for development and the environment (Saadatian et al., 2012).

Meanwhile, the 2011–2015 period covered by the tenth Malaysian Plan focuses on 12 National Key Economic Areas or NKEAs which have potential to generate high income. Under the 10th Malaysia. The government's strategy for building a thorough ecosystem for environmental sustainability was laid out in the AFFIRM framework, which stands for awareness, faculty, finance, infrastructure, research, and marketing (Table 3.1).

TABLE 3.1. AFFIRM FRAMEWORK

Awareness	This framework's concept highlights the requirement for all Malaysians to embrace a joint responsibility for environmental preservation and protection. This regulation requires businesses, homes, and other institutions to separate their garbage and to cut back on expenses for utilities like electricity. To promote awareness of this need, collaborative efforts will also be made with the commercial sector and civil society organizations.
Faculty	The government will work to incorporate green practices and technologies into the curriculum of universities and other institutions of higher learning.
Finance	Businesses who incorporate green technology into their operations will be given discounts and incentives. A RM 1.5 million soft loan grant will be made available to kickstart the adoption of sustainable development techniques. Additionally, building contractors and designers who choose to design environmentally friendly structures and infrastructure will be given tax discounts and incentives.
Infrastructure	Part of the government's long-term plan is to create green communities in Putrajaya and Cyberjaya. These communities will spearhead the adoption of guidelines aimed at reducing the carbon footprint as compared to other townships.
Research	The government will work in collaboration with regional research centers and institutes to advance research, development, and commercialization efforts on green technology as part of its overall development plan. To support local efforts, foreign expertise will also be enlisted, keeping in mind the acceptance of the international norms for sustainable development.
Marketing	The creation of a nationwide eco-labeling scheme and standards will be accelerated by the government. The government's green technology program will be aided by these standards, which embrace worldwide standards for labeling and recognizing items and services utilizing green technology. Additionally, it will help local firms grow their regional economies while simultaneously considering any potential environmental impacts.

In the 11th Malaysia Plan, pursuing green growth is one of the six strategic thrusts. It represents a commitment to pursue development in a more environmentally friendly way from the outset as opposed to the more traditional and expensive paradigm of grow first, clean up later. According to EPU

(2020), successful green growth trajectory requires fundamental changes across every major dimension as in Figure 3.6.

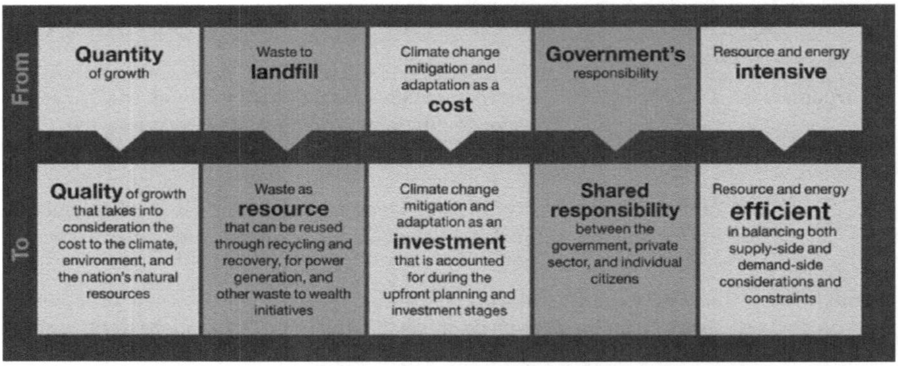

FIGURE 3.6 Fundamental for pursuing green growth under 11th MP.

Source: EPU (2020).

As a continuation, the 12th Malaysia Plan (12 MP) encompassed three themes, namely economic empowerment, environmental sustainability, and social reengineering which aims to maintain Malaysia's commitment to sustain the environment while ensuring ongoing economic growth. The 12 MP prioritizes green growth drivers such as improving resource management and protection, embracing the circular economy, transitioning to a low-carbon nation, and boosting climate change resilience. According to the current Malaysia Prime Minister Datuk Seri Ismail Sabri Yaakob, accelerating the transition to the circular economy, encouraging environmentally friendly and disaster-resilient towns and cities, enhancing green mobility, boosting low-carbon energy, expanding producer responsibility, and implementing evidence-based and risk-informed actions were some of the strategies to be put into action (The Star, 27 September 2021).

Under 12th Malaysia Plan, there are a few targets related to green economy:

i. Malaysia's to move to low-carbon, clean and resilient development with up to 45% cutback in greenhouse gas (GHG) release from 2005 measurements by 2030.
ii. To achieve a 2025 goal for a 40% rate of domestic waste recycling, a 67% reduction of hydrochlorofluorocarbon consumption and 25% government green compass.

iii. To control natural resources productivity, the selected goals included having 100% of the tiger population stabilized by 2025 and at least 10% of coastal and marine areas conserved through various conservation measures.
iv. To have at least 20% of earthbound and inland water areas conserved through various conservation measures by 2025, as well as performance of 10 Integrated River Basin Management plans by then.

3.3.2 MINISTRY OF ENERGY, GREEN TECHNOLOGY, AND WATER MALAYSIA (KETTHA)

The establishment of KeTTHA has resulted in significant accomplishments in various green technology initiatives and programs in Malaysia. Ministry of Energy, Green Technology and Water Malaysia (KeTTHA) was established on 9 April 2009 and lead a new initiative addressing global issues such as environmental pollution, ozone depletion, 'global warming' and issues related thereto. The objectives of the Ministry are:

i. To provide the fulfillment of development policies in the water, power industry, alongside with green technology adequately.
ii. To arrange comprehensive infrastructure alongside with integrated standards and quality.
iii. To provide a conducive ecosystem for development of industrial sectors and technology innovations.
iv. To improve the application of innovative technologies by expansion of research and development (R&D).
v. To provide a very productive, efficient, and affordable service delivery system.
vi. To carry out the regulatory mechanisms based on the provisions of existing legislation.
vii. To improve the capabilities of related organizations on a current basis to achieve the green industry and innovative technologies.

3.3.3 NATIONAL GREEN TECHNOLOGY POLICY (NGTP)

The National Green Technology Policy (NGTP) was introduced in Malaysia in July 2009 in line with the general trend toward the adoption of green technology. It aims to give Malaysians guidance and inspiration so they can continue to enjoy a high standard of living and a healthy environment while reducing their carbon footprint without jeopardizing economic progress.

This policy intends to, in accordance with the Economic Planning Unit of the Prime Minister's Office of the Government of Malaysia (2009):

i. To reduce the energy consumption rate and at the same time enhance economic advancement.
ii. To ease the growth of the Green innovative Technologies in industrial sectors and enhance its contribution to the national economy.
iii. To increase national potentiality and capability for innovation in Green Technology development and enhance Malaysia's competitiveness in innovative Green Technology worldwide.
iv. To manage sustainable development and preserve the ecosystem for future generations.
v. To strengthen public education and recognition on Green innovative Technology and encourage its broad application.

3.3.4 NATIONAL GREEN TECHNOLOGY COUNCIL AND CLIMATE CHANGE (2009)

In line with the NGTP, to promote coordination among government ministries, agencies, the corporate sector, and important stakeholders on green initiatives, the National Green Technology Council and Climate Change at the ministry level were established in 2009.

The Council is supported by a steering committee to assists the Council in developing policies, identifying strategic issues, coordinating, monitoring, and evaluating the NGTP and Green Technology programs at the federal level. The National Green Technology Council and another organization, the Malaysian Green Tech Agency (GTAC), were both responsible for coordinating and implementing the Malaysian government's green technology initiatives and programs (Chua & Oh, 2021).

3.3.5 GREEN TECHNOLOGY FINANCING SCHEME (2010)

The green technology financing scheme (GTFS) was introduced in Malaysia in 2010 to persuade the business sector to embrace environmentally friendly operations and practices and to use green technology in their creations. As announce in budget 2010, the initiative received an RM 1.5 billion budgetary allotment for the producer and consumers of green technology. The largest loan that can be given is RM50 million for producers and RM10 million for user enterprises.

From this initiative, there are 219 projects in total received green certification in 2010. Out of the total projects, 76 projects had received financing offers totaling RM1.016 billion from financial institutions. The Malaysian government has increased financing by an additional RM 2 billion as a demonstration of its dedication to fostering an atmosphere that will foster the development of green technologies.

3.3.6 THE GREEN TECHNOLOGY MASTER PLAN (GTMP)

The GTMP established in 2017 is fundamentally an outcome of the 11th Malaysia Plan (2016–2020) which has earmarked green growth as one of six game changers altering the trajectory of the nation's growth. The GTMP introduced framework in which aims to facilitates the mainstreaming of green technology into the planned developments of Malaysia while encompassing the four pillars set in the NGTP, namely energy, environment, economy, and social (Abdul & Abbas, 2017).

The GTMP's first edition focuses on six main industries: energy, manufacturing, transportation, buildings, waste, and water. It aims to harmonize the policy directions of each industry toward the shared objective of sustainable resource use. Each of these sectors' green technology targets will be gradually realized and adjusted through the policies and initiatives created during each five-year National Development Plan period.

3.3.7 MYHIJAU PROGRAM

To enhance and develop green products and services in Malaysia, Malaysian government has introduced MyHIJAU Program. The program consists of four sub-programs which will include the production stage up to the stage it is marketed, the MyHIJAU SME and entrepreneurs, MyHIJAU Labeling, MyHIJAU Directory and MyHIJAU Acquisition (Abdul, 2015).

MyHIJAU SME and entrepreneurs Program seeks to support the growth of local SME including Bumiputera and women businesses, to adopt green practices and produce local green products and services. SMEs enrolled in this program will undergo a training, business clinic and advisory sessions so they can get green label certification for their products and services.

Under the MyHIJAU labeling, official green recognition scheme endorsed by the Government of Malaysia, will bringing together certified products and services that meet local and international environmental standards under one single mark. Meanwhile, MyHIJAU Directory provided comprehensive guide to a wide range of green products and services, systematically categorized with detailed information and description for consumer reference.

Next, companies that acquire qualifying green technology assets, projects for business or own consumption and listed under the MyHIJAU Directory may be eligible for green investment tax allowance (GITA). Furthermore, new green service providers registered with MyHijau Mark may be eligible for green income tax exemption (GITE).

3.3.8 GREEN JOBS AND EDUCATION

The target to shift towards the green jobs' initiative is consistent with 12 Malaysia Plan. The goal of the Green Jobs Malaysia initiative is to increase the ability of the government and social partners to comprehend the prevalence of green jobs throughout the economy and to pinpoint opportunities for additional green job creation. By 2030, the Malaysian government hopes to generate RM 180 billion in income and create more than 2,00,000 green jobs (Entrepreneur Asia Pasific, 2018).

Under Green Jobs program, the energy management and climate change program (EMCP) introduced to offer certification and professional program. Organizations looking to embrace new green technologies must first equip their workforce adequately with both knowledge and skills necessary to capitalize on such innovation.

Meanwhile, the Green Jobs Malaysia Project, the partnership between the Institute of Labor Market Information and Analysis (ILMIA) of the Ministry of Human Resources (MoHR), the Ministry of Energy, Green Technology and Water (KeTTHA), the Malaysian Employers Federation (MEF), the Malaysian Trade Union Congress (MTUC) aims to promotes green jobs. This project consists of four phases:

- **Phase 1:** Collecting data, outlook, and design green jobs in Malaysia.
- **Phase 2:** Building a dynamic social accounting matrix (DySAM) that will resolve ecosystem activities (green DySAM).
- **Phase 3:** Expanding the capability of future DySAM experts in Malaysia.
- **Phase 4:** Managing green abilities standards development program by building on the results of the DySAM analysis to pilot test the performance of national occupational standards on green skills in key sectors (waste and water management, building energy, transportation).

Additionally, it promotes a bigger role for TVET organizations to improve the employability of people with green technology skills. For instance, Vocational Colleges is one of the well-known TVET institutes in Malaysia that offers instruction and work-based training incorporating green technology in the curriculum. Vocational College is an institution that is placed under the Ministry of Education Malaysia (MOE) and offers variety of courses that fit the industry's need in providing secondary and tertiary TVET education at certificate and diploma level (Kaliappan & Hamid, 2021).

Since then, to promote awareness on green jobs and technology, the educational strategy for education for sustainable development (ESD) has developed into one of the useful strategies for educating students at the school level to raise awareness of the environment. For instance, the Sustainable Schools Program Environmental Award (SLAAS). The goal of SLAAS was to create school culture that promotes environmental preservation and conservation in the areas of management, curriculum, co-curriculum, and ongoing green activities to establish a way of life consistent with the idea of sustainable development. The application of ESD has five goals: (i) cultivate the environmental good virtue in the school community; (ii) raise awareness of the significance of environmental conservation and preservation in the school community; (iii) encourage the school community to engage in effective environmental-friendly activities; (iv) increase cooperation between schools and communities; and (v) to establish a supportive learning environment to develop a community of students who practice sustainable living (Hanifah et al., 2015).

3.3.9 LOW CARBON INITIATIVES

Under Ministry of Energy, Green Technology and Water Malaysia (KeTTHA) low carbon cities framework (LCCF) has been developed with collaboration with the GreenTech Malaysia and the Malaysian Institute of Planners in 2010–2011. LCCF designed to can help local governments, developer, and city planners create townships that have a 40% reduction in emissions and used to promote application and development of the green technology in Malaysia and help strengthen and improve sustainability of the community towards creating green and sustainable practice.

To reduce the consumption of fossil fuels, such as diesel, Malaysia government has promoted the usage of electric vehicles. Electric Vehicle Infrastructure Roadmap Masterplan initiated in 2013 to get the technical data and the views of the public in preparing the Master Plan Electric Vehicles.

3.4 CHALLENGES TO GREEN ENERGY DEVELOPMENT

According to Oh et al. (2010), the presence of the economic and political arguments posed as a major difficulty to the government's aspirations of switching to renewable energy production. For example, the promotion of renewable energy through Five-Fuel policy that was seen as ad hoc and failed to achieve its desired results. It is apparent that the current progress in bringing renewable energy generation into the selected target has been slow, and financing has been one of the main issues. It was postulated that the massive support for non-renewable energy sources provided in the forms of incentive, subsidies, tax reduction and export credits by regulatory bodies has become a major obstacle for the efforts to switch towards renewable energy use. Hence, similar incentives will need to be offered for renewable energy to be competently economic.

Similarly, Foo (2015) agreed that the country's transformation effort towards green energy use is hindered by the shortage of huge investment and inconsistent financial support for renewable projects, due to the unattractive tariffs, utilities, and poor economic efficiency. Mustapa et al. (2010) pointed out that the seasonal fluctuations and weather changes had largely limited the adoption of the emerging green technologies. Besides that, lack of technical support and training being provided would also become a hindrance to the feasibility of the green project implementation in Malaysia. Indeed, the success of the green technologies industry would highly be dependable upon the right support mechanism to be in place (Oh et al., 2010).

Jaafar et al. (2003) highlighted the lack of economies of scale in renewable energy projects which resulted extremely high costs of such project. No doubt, this created another key barrier to the use of renewable energy. Besides that, the poor perception of the potential and commercial viability of renewable energy resources also discourages any potential investment on renewable energy. The high financing costs of green energy projects will indirectly raise the price of the alternative renewable energy which further making it difficult to create a green market. In other words, the use of green energy become economically unattractive due to its high cost.

Interestingly, Zhou (2009) shed light on the environmental issue that appeared to impede the green energy progress. Some of these environmental issues highlighted include the extensive clearance of pristine rainforest, erosion, and loss of biodiversity and habitat for wildlife that arises due to the installation of renewable energy plants. Lastly, the lack of regulatory framework was also identified as one of the key challenges to green energy progress. The lack of functional institution network on green energy and the lack of join-effort between government agencies and private institutions will further impede the advancement of green technologies and clean energy innovation.

3.5 CONCLUSION

In conclusion, it was evident that Malaysia faces critical issues of energy consumption and production that negatively affected the environment since in the past. The energy production using non-renewable resources such as fossil fuels releases emissions that become a great environmental concern. Despite those efforts by the governments such as the development of green energy policies, however, the improvement in terms of the renewable energy consumption and production appears to be still lacking. Hence, there is an urgent need to press for the development of a more effective green energy policies and to address the challenges or barriers that harms the progress of sustainable energy development. Besides the government, all the main stakeholders such as institutions, non-government organizations (NGOs), industry and society should also be involved in the efforts for the nation to make a smooth transition in becoming a green country.

KEYWORDS

- **carbon**
- **green economy**
- **green energy development**
- **green jobs**
- **green technology**
- **green technology financing scheme**
- **MyHIJAU program**

REFERENCES

Abdul, M. A., (2015). *MyHIJAU Program and Green Technology Financing Scheme*. Retrieved from: https://inis.iaea.org/collection/NCLCollectionStore/_Public/47/122/47122491.pdf (accessed on 15 November 2023).

Abdul, M. N., & Abbas, A., (2019). Economic and environmental sustainability of low-carbon power generation: Relevancy in the Malaysia Green Technology Master Plan (GTMP). *Journal of Chemical Technology & Biotechnology, 94*(5), 1425–1432.

Cato, M. S., (2011). *Environment and Economy*. Routledge, London.

Chapple, A., (2008). *Making the Voluntary Carbon Market Work for the Poor, Forum for the Future*. London.

Chua, S. C., & Oh, T. H., (2011). Green progress and prospect in Malaysia. *Renewable and Sustainable Energy Reviews, 15*(6), 2850–2861.

Fatimah, Z., (2021). *Green Growth a Priority Under 12MP.* The Star. Retrieved from: https://www.thestar.com.my/news/nation/2021/09/27/green-growth-a-priority-under-12mp (accessed on 15 November 2023).

Foo, K. Y., (2015). A vision on the opportunities, policies and coping strategies for the energy security and green energy development in Malaysia. *Renewable and Sustainable Energy Reviews, 51*, 1477–1498.

Hanifah, M., Shaharudin, I., Mohmadisa, H., Nasir, N., & Yazid, S., (2015). Transforming sustainability development education in Malaysian schools through greening activities. *Review of International Geographical Education Online, 5*(1), 77–94.

Islam, M. R., Saidur, R., Rahim, N. A., & Solangi, K. H., (2009). Renewable energy research in Malaysia. *Engineering e-Transaction, 4*(2), 69–72.

Jaafar, M. Z., Kheng, W. H., & Kamaruddin, N., (2003). Greener energy solutions for a sustainable future: Issues and challenges for Malaysia. *Energy Policy, 31*(11), 1061–1072.

Kaliappan, A., & Hamid, H., (2021). Green technology: A must or a need in TVET education in Malaysia? *Journal of Technical Education and Training, 13*(1), 86–96.

KeTTHA, (2017). *Green Technology Master Plan Malaysia 2017–2030 (GTMP).* Ministry of Energy, Green Technology and Water (KeTTHA).

Komal, N., (2018). *Malaysia Aims to Create 200,000 Green Jobs by 2023 in ASEAN.* Entrepreneur Asia Pasific. Retrieved from: https://www.entrepreneur.com/article/324085 (accessed on 15 November 2023).

Mokthsim, N., & Salleh, K. O., (2014). Malaysia's efforts toward achieving a sustainable development: Issues, challenges, and prospects. *Procedia-Social and Behavioral Sciences, 120*, 299–307.

Musango, J. K., Brent, A. C., & Bassi, A. M., (2014). Modelling the transition towards a green economy in South Africa. *Technological Forecasting and Social Change, 87*, 257–273.

Mustapa, S. I., Peng, L. Y., & Hashim, A. H., (2010). Issues and challenges of renewable energy development: A Malaysian experience. In: *Proceedings of the International Conference on Energy and Sustainable Development: Issues and Strategies* (pp. 1–6). IEEE.

Oh, T. H., Pang, S. Y., & Chua, S. C., (2010). Energy policy and alternative energy in Malaysia: Issues and challenges for sustainable growth. *Renewable and Sustainable Energy Reviews, 14*(4), 1241–1252.

Rahim, K. A., & Liwan, A., (2012). Oil and gas trends and implications in Malaysia. *Energy Policy 50*, 262–271.

Rahman, M. S., Junsheng, H., Shahari, F., Aslam, M., Masud, M. M., Banna, H., & Liya, M., (2015). Long-run relationship between sectoral productivity and energy consumption in Malaysia: An aggregated and disaggregated viewpoint. *Energy, 86*, 436–445.

Saadatian, O., Haw, L. C., Mat, S. B., & Sopian, K., (2012). Perspective of sustainable development in Malaysia. *International Journal of Energy and Environment, 6*(2), 260–267.

Shafie, S. M., Mahlia, T. M. I., Masjuki, H. H., & Andriyana, A., (2011). Current energy usage and sustainable energy in Malaysia: A review. *Renewable and Sustainable Energy Reviews, 15*(9), 4370–4377.

Shigeoka, H., (2004). *Overview of International Renewable Energy Policies and Comparison with Malaysian's Domestic Policy.* School of Int'l & Public Affairs Columbia University.

UNEP, (2011). *Towards a Green Economy: Pathways to Sustainable Development and Poverty Eradication*. UNEP, Geneva.

Unit Perancangan Ekonomi, (2020). *Eleventh Malaysia Plan: Anchoring Growth on People*. Rancangan Malaysia Kesebelas (11th Malaysia Plan) – 2016.

World Bank, (2022). *World Development Indicators Database*. Retrieved from: https://databank.worldbank.org/source/world-development-indicators (accessed on 15 November 2023).

Zhou, A., & Thomson, E., (2009). The development of biofuels in Asia. *Applied Energy, 86*, S11–S20.

PART II

EMERGING TECHNOLOGIES FOR WASTE VALORIZATION

CHAPTER 4

Emerging and Eco-Friendly Approaches for Green Contaminated Water Effluent Remediation Alternatives: Compilations of Case Base Analyzes to Sustainability

SITI IRBAH ANUAR,[1] HAZLINDA IBNO,[1,2] CAROLYN PAYUS,[1,2,3] OLGA DAVID,[2] NG CHING WEI,[2] THOO KHYE SEAN,[2] MOHD. ASYRAF,[2] HARRY CHONG LYE HIN,[2] SITI AISHAH MOHD. ALI,[2] JUSTIN SENTIAN,[1,2] and MOH PAK YAN[2,3]

[1]*Natural Disaster Research Center (NDRC), University Malaysia Sabah (UMS), Jalan UMS, Kota Kinabalu, Sabah, Malaysia*

[2]*Faculty of Science and Natural Resources, University Malaysia Sabah (UMS), Jalan UMS, Kota Kinabalu, Sabah, Malaysia*

[3]*Water Research Unit, Faculty of Science and Natural Resources, University Malaysia Sabah (UMS), Jalan UMS, Kota Kinabalu, Sabah, Malaysia*

ABSTRACT

Bioremediation treatment of contaminated water effluent, especially from heavy metals such as acid mine drainage (AMD), pollutants from nutrients, microbiological, oil, and grease in water bodies, is crucial due to its adverse impacts on humans and the environment alongside water scarcity issue. AMD is a high acidic runoff with high content of heavy metals, usually contaminating surface and groundwater. Therefore, it is crucial as its impact persists during and even after abandoned mine operations. Typically, water treatment centers around civil engineering methods or process-based technologies, which are considered not eco-friendly, expensive, unsustainable,

Waste Biorefineries: Future Energy, Green Products, and Waste Treatment.
Abu Zahrim Yaser, Siew-Xian Chin, Leonardo Sepúlveda Torre, & A. K. Haghi, (Eds.)

and environmentally damaging. As a result, there is a rising demand for an alternative remediation approach, a more sustainable solution to prevent further environmental harm and potential pollution. Thus, this chapter will compile and discuss the efficiency of the few options of alternative methods for a more sustainable solution for water and effluent treatment utilizing solid waste, namely bone meal, eggshells, durian husk, and peanut hull. Specifically, both bone meal and eggshells centers around heavy metal removal for treatment of effluent contaminated by AMD, while durian husk and peanut hull centers around removal efficiency on total hardness (TH), electrical conductivity (EC), and total dissolved solids (TDS). Bioremediation is a more organic, biodegradable, and inexpensive value-added technique to remediate the contaminated water resources presented in this chapter. Furthermore, because they include proteins, minerals, and sugars, the aforementioned materials have the potential to buffer and recover polluted water, resulting in reduced environmental complications. Therefore, it is essential to reduce organic waste by reusing and utilizing them as a low-cost compound, substrate, or raw material for water pollution remediation and treatment, as will be explored by completing case base analyzes in this chapter.

4.1 INTRODUCTION

Water is essential to humankind. Yet water pollution, attributed to anthropogenic activities or natural factors, is a fundamental concern and a present threat (Pachecho et al., 2020). Water pollution occurs when harmful substances such as organic compounds, inorganic compounds or microorganisms contaminate the water source, negatively impacting humans and the environment (Ojha et al., 2021). The increasing human population and their activities damage the environment, including water pollution through exposure to dangerous pollutants. For instance, mining activities usually will lead to acid mine drainage (AMD), which might pollute the surface and groundwater. AMD can occur naturally and through mining activities, either operating or abandoned. Generally, it occurs through the oxidation of sulfide minerals due to exposure to oxygen, water, and microorganisms, which are further accelerated by mining activities due to increased exposure of the sulfide minerals to the precursors (Akcil & Koldas, 2006). AMD is a strong acidic effluent with high concentrations of metal, sulfate, and high number of suspended solids and is a concerning water pollutant (Muhammad et al., 2015). Therefore, treatment of polluted water in general, specifically contaminated by AMD, is crucial due to its potential to negatively affect human health and the environment, with the rising water scarcity issue alongside an effort to achieve Sustainable Development Goal 6, which targets clean water as one of its goals.

There are a lot of methods and technologies to treat polluted water and effluent techniques such as membrane filtration, reverse osmosis, chemical precipitation, solvent extraction, oxidation, and adsorption (Younas et al., 2021). Among all the techniques, adsorption, a part of bioremediation, is the most sustainable and friendly approach. Immobilization of microbial cells and enzymes by adsorption is a type of bioremediation process. The method takes place through their physical interaction with the surface of water-insoluble carriers, and the adsorption is achieved through the formation of weak bonds. It is mostly utilized for its simplicity, quickness, ease, reasonably accessible, eco-friendly, and cost-effective. The development of effective adsorbents from solid waste is emerging research where they are converted into cost-effective adsorbents through chemical or physical modifications (Younas et al., 2021). Few studies have shown the efficiency of this method as this adsorption application is frequently utilized in the removal of heavy metals for industrial effluents (El-Haddad et al., 2016; Buksh et al., 2018). This is also in line with the increasing organic solid waste problem in both developed and developing countries, namely agricultural waste, household food waste, and human and animal waste.

Therefore, this chapter aims to point out the potential of the respective solid waste, namely durian husk, peanut hull, bone meal and eggshells, as a low-cost sorbent for water treatment. It will be a case-based analysis.

4.2 UTILIZATION OF SOLID WASTES AS ADSORBENT FOR WATER EFFLUENT REMEDIATION

4.2.1 BONE MEAL FOR AMD TREATMENT

4.2.1.1 INTRODUCTION

In Malaysia, the mining industry, such as the production of tin, copper, bauxite, gold, and iron ore, has contributed a lot to the country's economic development. Sabah, which is a part of Malaysia, is located on Borneo Island, and it is abundant in these types of minerals. However, there are few known places for its mineral deposits in Sabah, such as coal in Meliau and Maliau Basins, gold in Bukit Mantri, Semporna, and copper at Mamut copper mine, Ranau (Jopony & Tongkul, 2009). According to Jopony & Tongkul (2009), copper production operated from 1975 to 1999. The mine aforementioned was Malaysia's only copper mine site, located in East Malaysia, Sabah. The water at the mining pit has a very low pH of 2.90 to 3.75, a high sulfate of contained and elevated concentrations of dissolved metals such as Al, Mn, Fe, Cu, and Zn. Various health problems and aquatic life disturbances resulted from the acid mine drainage (AMD). It causes the death of fish in the contaminated river, crops to be decreasing and skin rashes to exposed people living nearby the mining pit. The mining pit contained few minerals with an area of 4,800 hectares, depth of 500 m and diameter of 1,200 m. Among those minerals are chalcopyrite ($CuFeS_2$) co-existed with other sulfide minerals such as sphalerite (ZnS), galena (PbS) and different non-significant pyrite (Chan, 2000). With reference to Mohd Azizli et al. (1995), the open pit and its facilities are located on the southeastern flank of Mount Kinabalu, at 1,300 to 1,600 m above sea level.

A study at Mamut ex-Copper mine showed that the water contaminated by AMD is traveled to the connected rivers that act as the primary water source for the surrounding villagers. Since the polluted water with elevated concentrations of heavy metal is used by the villagers for household purposes such as cooking and personal hygiene purposes such as showering, they are at risk of suffering heavy metal poisoning. The symptoms, however, depend highly on the type of heavy metals they are exposed to. Examples of heavy metals typically found in AMD are sulfate, chloride, aluminum, iron, manganese, arsenic, antimony, cadmium, and lead. Therefore, the treatment of effluent-containing heavy metals such as AMD is crucial due to the negative impacts on humans and the environment and high-water demand. The adverse effects of AMD are not limited only to aquatic life and human health; it can also cause vegetation and rehabilitation difficulties. Soils contaminated by AMD are, at best, a significant limitation on vegetation types that can be used for rehabilitation and, at worst responsible for a failed rehabilitation plan (Jeff et al., 2005). AMD will lead to high acidity and metal contamination of streams or rivers, which can adversely impact humans, wildlife, and the environment (Robinson-Lora & Brennan, 2009). Water in the main pit areas has low pH between 2.90 and 3.75 (Jopony & Tongkul, 2009), which is very acidic and dangerous to human health. The effluent from the MCM, even though treated, will continue to flow into each of nearby rivers, causing the death of fish in the river or accumulation in the fish body and disturbing the domestic use of the villagers as they depend mainly on the river for the water source. When the human body consumes the contaminated water, it will cause skin rashes and leukemia (Weissmen et al., 1994). Nearly 2,000 acres of paddies and over 500 households will be affected when the contaminated water from the Mamut mine pit flows into the irrigation canals. Villagers near the contaminated fields or downstream are exposed to possible health hazards (Rajendran & Reich, 1981).

Previous research had been done on using bone meal for contaminated soil treatment (Hodson et al., 2001; Deydier et al., 2003; Sneddon et al., 2008). However, this study was conducted differently by treating contaminated effluent. Specifically, this study was conducted to analyze the potential of bone meal for neutralization and for heavy metal removal, specifically from AMD-contaminated water samples. This is mainly due to the reaction of the bone meal with AMD, which generally will lead to salt production.

According to Jung et al. (2007), the major component of bone is calcium phosphate. Various industries utilize calcium from dolomite, bone meal, and oyster shell. For instance, the rubber industry uses calcium to produce additives in the pulp and paper industry and as early strengthening agents in the construction industry. In addition, calcium is used as a foodstuff antiseptic to prevent fruits and vegetables from getting rotten too fast and getting involved in cheese-making. Rey (1998) stated that apatite is a principal mineral constituent of bone. Past research has proved the source of phosphate ions from the formation of phosphate metals released by the synthetic hydroxylapatite. Lambert et al. (1997) and Hodson et al. (2000) conducted experiments to treat metal-contaminated soils.

Reduction in metal availability in soils has been proved after the addition of a phosphate (Hettiarachchi & Peirzynski, 1999). Another previous research by Sneddon et al. (2005) has shown that incinerated bone meal is very effective as a sorbing agent for As(V). On the other hand, Valsami-Jones et al. (1998) suggested that due to the poorly crystalline structure of bone meal and its intermediate solubility, it would be a potential source of phosphate for remediation of contaminated soils. Sneddon et al. (2005) proved the reduction of Cd and Pb availability in soil with bone meal additions. In 2001, Hodson et al. showed an increase in soil and leachate pH (3.6 to 4.65) and metal immobilization in the soils. However, these are all about the past research on the potential of bone meal as remediation for toxic metals contamination in soil.

Nevertheless, in wastewater, only a few significant international research has been done involving synthetic bone meal. Generally, the bone meal chosen will be comprised of chicken bones. The main reason would be that the sources of these materials are available at food restaurants and even household food waste. They are a sustainable and commercially available waste product of the food industry. Besides, they are very cost-effective, reducing the cost of buying chemical compounds, and thus are economically viable compared to chemical compounds or other materials that have been used before. Since they are organic waste, they are biodegradable and thus are environmentally friendly. In other words, they give fewer complications or interference with the environment.

4.2.1.2 FINDINGS

Scanning electron microscope (SEM) images of bone meal samples before and after treatment were taken under three different magnifications and illustrated by the figures below. Visible pores are much more observed after treatment compared to before treatment. Under higher magnification, an uneven surface was observed prior to treatment, and a more smooth and scattered bone meal surface was observed after treatment. These alterations are related to AMD's reactivity with the surface of the bone meal. Heavy metals bind to the bone meal's wall, blocking and enlarging pores, while the scattered surface of the bone meal precipitates out of the solution following treatment.

In Figure 4.1(a), no pores were observed before treatment, as compared to after treatment in Figure 4.1(b) where pores are visible on the surface of the bone meal. The morphology is represented by a fine grain size of 75 μm bone meal. The porosity nature of bone meal makes it a suitable permeable reactive medium, an excellent condition and competent to support a high flow of solution (Conca & Wright, 2006)

FIGURE 4.1(a) Bone meal before treatment (magnification of 500x).

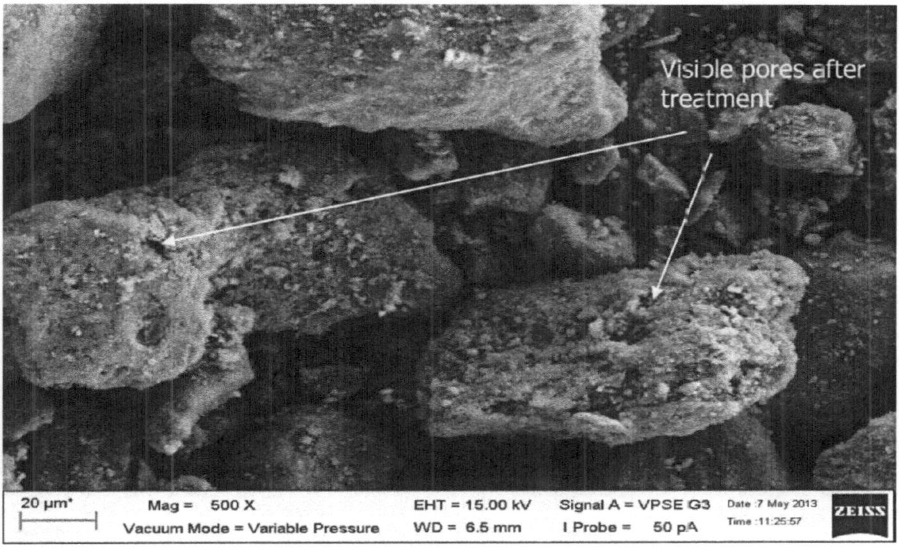

FIGURE 4.1(b) Bone meal after treatment (magnification of 500x).

Figures 4.2(a) and 4.3(a) show both irregular surfaces of the bone meal before the treatment with AMD. In Figure 4.2(a) is seen under 10 µm resolution and is focused more under the resolution of 2 µm for a closer look, as in Figure 4.3. A few lumps can be observed on the bone meal surface in both resolutions in Figures 4.2(b) and 4.3(b). After the neutralization treatment of AMD with bone meal, it was observed that the surface of the bone meal has a smooth texture and is scattered compared to the surface image before the treatment. This could be due to the reaction during the treatment; as the acid reacts with bone meal, it erodes the surface of bone meal, causing the lump to be eroded. In past research by Rios et al. (2008), other than the reactivity and mode of metal uptake into the sorbents, the crystal size and morphology of the sorbents are essential to trigger their sorption properties. The crystallinity of bone meal is low, but according to Hodson et al. (2001), after the bone meal is incinerated, it increases the crystallinity of bone meal. As the crystallinity of bone meal increases, it will also increase the surface area, as more pores are exposed because the organic matter was removed, which will increase the metal immobilization. In this experiment, the bone meal was incinerated in a furnace under 500°C until a light gray powder was obtained. However, the images of bone meal shown under SEM are not too clear because it is in powder form.

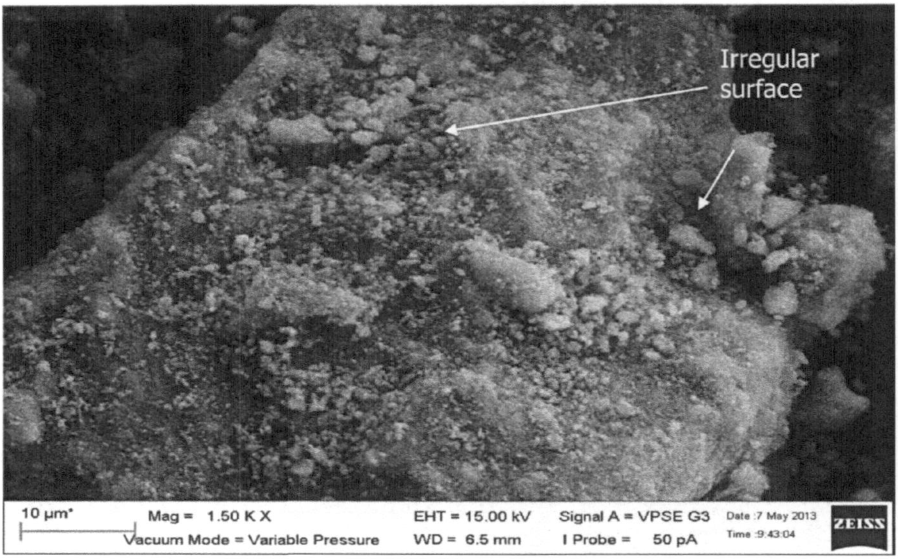

FIGURE 4.2(a) Bone meal before treatment (magnification of 1500x).

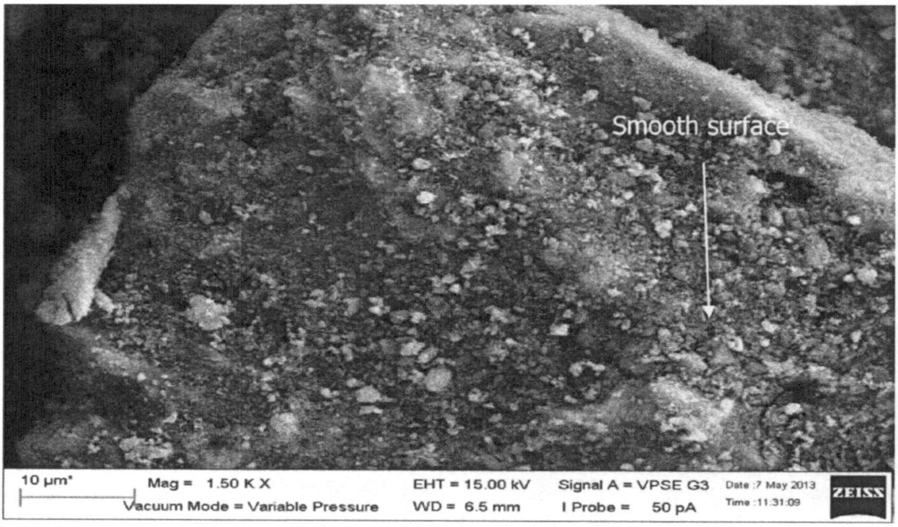

FIGURE 4.2(b) Bone meal after treatment (magnification of 1500x).

FIGURE 4.3(a) Bone meal before treatment (magnification of 3000x).

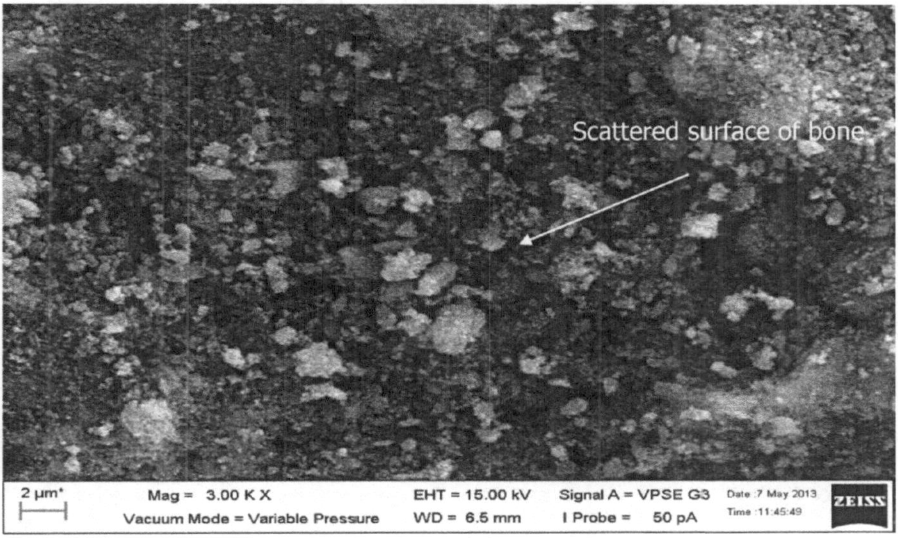

FIGURE 4.3(b) Bone meal after treatment (magnification of 3000x).

The AMD is very acidic and can be attributed to the oxidation of metal sulfides in the mine wastes (Jopony & Tongkul, 2009). Rios et al. (2008) also supported this when they stated that pyrite, an iron sulfide mineral, caused the acidic characterization of AMD when after being exposed to air and water, it oxidized and dissociates, releasing Fe^{2+}. This study also includes neutralization of AMD after treatment with bone meal. Essentially, the release of hydroxide ions leads to a slow increase in the pH of AMD with an increased amount of bone meal used over a longer contact time, as shown in Figure 4.4, and reduced total acidity, as shown in Figure 4.5.

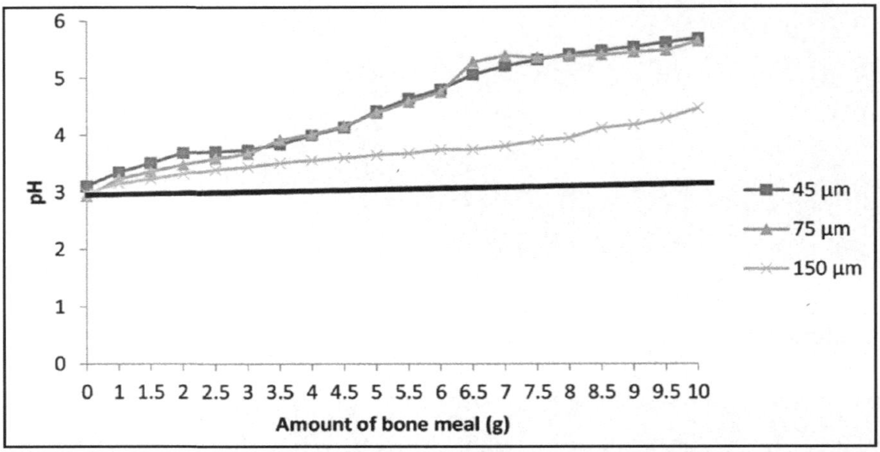

FIGURE 4.4 pH of AMD after neutralization of different selected sizes of bone meal over amount of bone meal used (g).

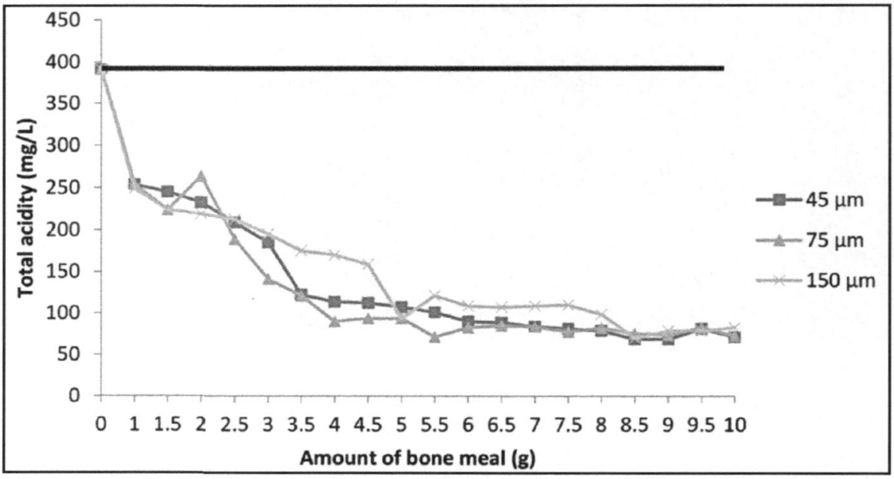

FIGURE 4.5 Total acidity (mg/L) of different selected sizes of bone meal over amount of bone meal used (g).

Different grain sizes of the bone meal were used, such as 45 μm, 75 μm and 150 μm, with the addition of 1 g for the first minute (20 min) and 0.5 g for the next minute until it reached 10 g for a 6 hour and 30 minutes duration

Based on the result above, it is observed that the capacity of bone meal in neutralizing the acid mine drainage (AMD) water sample is dependable on the surface area and amount of bone meal used. For the finest grain, 45 μm, the pH increased from 3.11 to 5.69, for grain size 75 μm the pH increased from 2.93 to 5.64, for grain size 150 μm the pH increased from 2.98 to 4.47. The crystallinity of bone structure, which refers to the porous structure of bones, is likely to have increased along with the surface area of bone meal (Hodson et al., 2001). The removal of organic material from the bone through incineration during the pre-treatment process allowed larger pores of the bone to be exposed. The solubility of the bone meal phosphate mineral is low in water (Valsami-Jones et al., 1998). Therefore, it requires a longer period of contact time for the neutralization reaction to be more efficient. In this experiment, the pH of the AMD during treatment is observed every 20 minutes time interval for 6 hours and 30 minutes.

The concentration of magnesium shows a significant inconsistent trend for different sizes of the bone meal. Fluctuations of heavy metals concentration with the finest grain size occurred throughout the time. From the graph in Figure 4.6, for size 45 μm bone meal, there was a significant rise in magnesium concentration further away from the initial concentration. In contrast, 75 μm and 150 μm bone meal shows a constant decreasing trend. It could be associated with the magnesium element from the bone meal leaching out to the AMD water. Another reason would be because metals have different affinity for adsorption sites (Stella, 2008). Since AMD consists of multiple metals such as Fe, Zn Mg and Cu, they may compete for binding sites during the adsorption. Therefore, adsorption of other metals might lower the removal efficiency of magnesium concentration from the AMD. Thus, it can be concluded that magnesium removal is not efficient with bone meal, and it might also have been desorption instead of adsorption (Sneddon et al., 2006). Adsorption or precipitation works when optimum pH is reached and minimum solubility is achieved. Rios et al. (2008) stated that the pH required to precipitate most metals from water ranges from pH 6 to 9. No significant adsorption or precipitation occurred for this element. For adsorption to occur with Mg, it has to reach optimum pH solubility of 10. From this experiment, the maximum pH after the treatment only reaches 5.69, and the pH range is not favorable for any adsorption process of Mg to occur, so in this experiment, magnesium is not adsorbed or precipitated with the bone meal reaction.

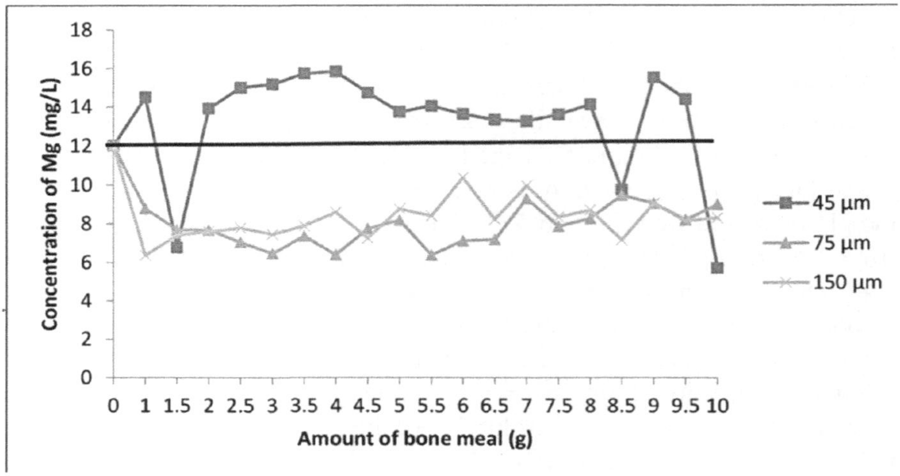

FIGURE 4.6 Mg concentration (mg/L) of different selected sizes of bone meal over amount of bone meal used (g).

Figure 4.7 shows the concentration of Zn after the treatment with bone meal. A sudden increase was observed during the first interval, and a stable decreasing trend was detected for the rest of the intervals.

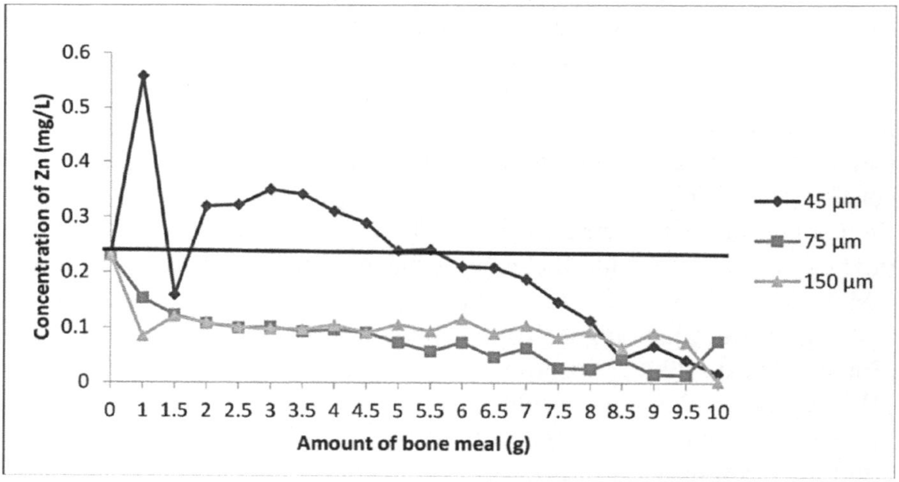

FIGURE 4.7 Zn concentration (mg/L) of different selected sizes of bone meal over amount of bone meal used (g).

Figure 4.8 generally shows a progressive decrease in Al concentrations after the treatment, although there is a slight increase within the first dosage.

The initial concentration of Al decreased from 3.028 mg/L and then dropped to – 0.030 mg/L at or below the detection limit. This is because, at a pH of more than 5.0, the Al^{3+} starts to decrease and form a solid precipitate aluminum hydroxide ($Al(OH)_3$) (Daubert & Brennan, 2007), it would precipitate out of the solution based on their solubility product constants (Ksp) (McQuarrie & Rock, 1984) and therefore would lower the concentration of Al.

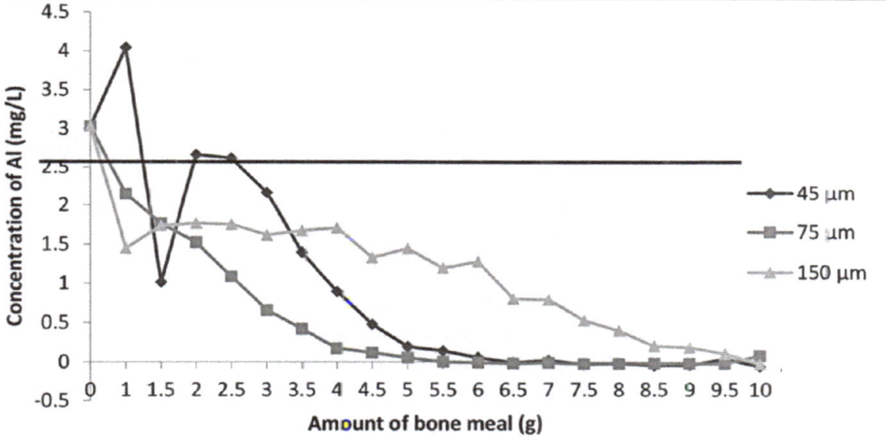

FIGURE 4.8 Al concentration (mg/L) of different selected sizes of bone meal over amount of bone meal used (g).

Figure 4. shows the concentration of Cu after the treatment of AMD with bone meal. An abrupt increase was observed for the first dosage of bone meal addition but somehow proceeded with a steady decrease of Cu concentration until the dosage of a bone meal increased to 10 g. Since the minimum solubility for Cu to precipitates is at pH 4.0 (Lee et al., 2002), it can be seen from this experiment that as the pH increased up to 4.0 with an increasing amount of bone meal used, the removal of Cu occurred at this phase and reduced the concentration of Cu at the end of the experiment. Figure 4.10 shows the concentration of Fe after the treatment of AMD with bone meal. The concentration of Fe decreased linearly after the treatment with bone meal. The concentration dropped from 0.392 mg/L to 0.01 mg/L, where it almost reached a complete removal of Fe metal in AMD. This result indicates that bone meal has high efficiency in removing Fe with a progressive decrease of Fe concentration for the rest of the intervals. Evangelou (1955) stated Fe^{3+} could be precipitated as $Fe(OH)_3$ at a pH of more than 3.5, and a proportion of the dissolved Fe may be precipitated out from the solution. The remaining Fe, therefore, is likely to be dominated by Fe^{2+}.

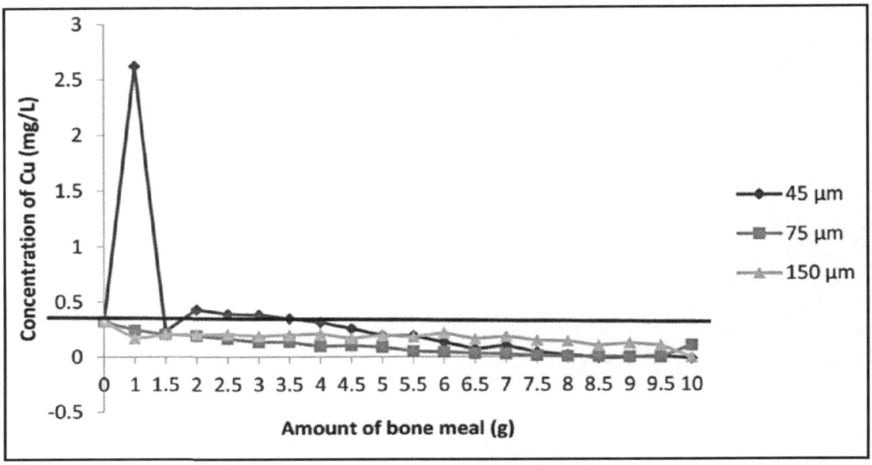

FIGURE 4.9 Cu concentration (mg/L) of different selected sizes of bone meal over amount of bone meal used (g).

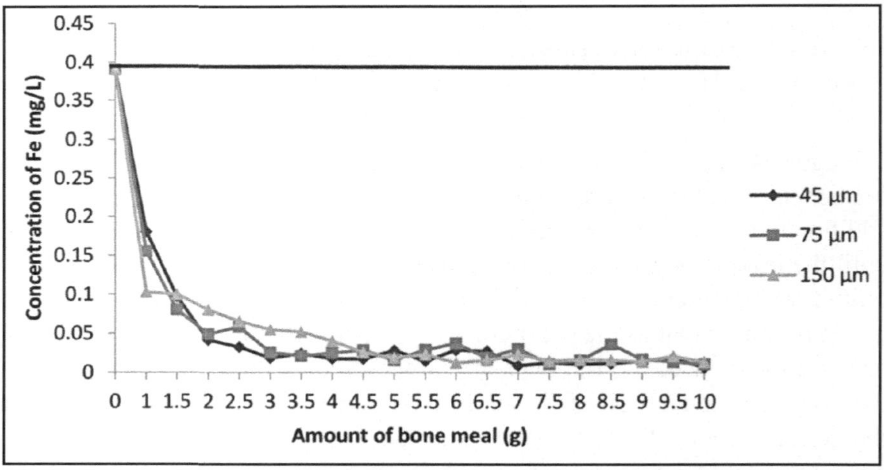

FIGURE 4.10 Fe concentration (mg/L) of different selected sizes of bone meal over amount of bone meal used (g).

Fe has the highest removal efficiency of almost 99%, while Mg has the least with 50% to 20% removal efficiency. The high removal efficiency of Fe might be attributed to Fe that has reached the minimum solubility at a pH of more than 3, which is enough to precipitate the metal to become metal hydroxide. As for Mg, it might be attributed to the pH solubility at pH 10, which is not favorable for this experiment. Based on Lee et al. (2002), the pH range of the drainage system, the relative abundance of dissolved metals, and the solubilities of different Fe, Al, and Mn hydroxides and hydroxysulfates are the key parameters influencing the removal of trace metals in AMD polluted waters. Al and Cu showed less significant removal efficiency among other metals because it falls below the detection limit at a specific mesh size. Therefore, the order of removal efficiency of metals in decreasing order based on the results are Fe>Zn>Al>Cu>Mg. A rapid removal of Fe and Zn in this result shows that bone meal has the potential for metal removal. However, further research is needed to completely comprehend the metal removal processes done by bone meal.

4.2.2 EGGSHELLS AND LIMESTONE FOR AMD TREATMENT

4.2.2.1 INTRODUCTION

Eggshell is largely crystalline calcium carbonate. The calcium comes partly from the hen's bones; when necessary, the hen can mobilize 10% of her bone. Eggshell has the potential to be a suitable sorbent due to its good adsorptive properties, such as its porosity, availability of calcium carbonate and protein acid mucopolysaccharide. The carboxyl, amine, and sulfate functional groups of protein acid mucopolysaccharide may create ionic bonds with heavy metal ions (Surasen, 2002). Furthermore, the eggshell acts as a neutralizing agent. Because any aqueous solution that has been equilibrated with eggshell becomes more basic (Brown & Lemay, 1985), heavy metals can precipitate and deposit on eggshell particles. Remediation of AMD, for instance, requires three different processes: the addition of a neutralizing agent, the reduction of sulfate concentrations, and the removal of dissolved heavy metals. There are usually two broad classes of methodologies used to treat AMD, which are passive treatment and active treatment. Passive treatment involves natural chemical and biological reactions occurring in a controlled microbiological-chemical reactor without powered mechanical assistance. Active treatment, on the other hand, involves the mechanical addition of alkaline chemicals to raise pH and precipitate metals (Neculita et al., 2007). Limestone is a low-cost and efficient chemical for neutralizing waste acid streams that are low in dissolved iron and manganese. Although AMD can contain significant amounts of iron and manganese, limestone is the preferred neutralizing agent for AMD effluent remediation. The magnitude, extent, and longevity of AMD pollution, as well as the lack of a financially accountable party for many AMD pollutants discharges, make cost the most important consideration when weighing remediation options.

The neutralization of AMD is conducted using a treatment reactor that was made up of a wooden gadget. Four containers were placed on the gadget, and one container was placed under the gadget, as shown in Figure 4.11.

FIGURE 4.11 Reactor used for the neutralizing treatment in real scale.

The flow of AMD is downward vertically from the container at the top (Container A) to the container at the bottom (Container E). Vertical AMD flow is preferred over horizontal AMD flow to ensure the entirety of the neutralizing agents is utilized, in contrast to horizontal AMD flow, in which only the surface of the agents is utilized for the AMD treatment. Therefore, two sets of arrangements of neutralizing agents are used in this experiment, as shown in Table 4.1.

TABLE 4.1 Different Arrangement of Neutralizing Agents in Two Sets of Experiment

Containers	Set	
	A	**B**
B	Limestone	Eggshells
C	Eggshells	Limestone
D	Limestone and eggshells	Limestone and eggshells

4.2.2.2 FINDINGS

The result of the study showed that the neutralization process was more significant in Set A than in Set B. By then, the AMD pH for Set A and Set B was pH 4.48 and 5.08, respectively. This climb of AMD pH is continued throughout the experiment. However, the performance of limestone in Set A is poorer than the eggshells in Set B. For example, the time taken for Set A and Set B to reach pH 6 is 1,140 minutes and 180 minutes, respectively, after the experiment is started. Interestingly, given the sufficient length of experiment time, both sets ended in similar readings, pH 6.77 in Set A and pH 6.79 in Set B. A similar trend in the change of pH is observed by comparing Pail 2 between Set A and Set B. The neutralizing agents used in Pail 2 in Set A and Set B are eggshells and limestone, respectively. During the early stage of the experiment, Set B displayed a faster climb of AMD pH than Set A. However, the AMD pH in Set A is higher than that of Set B two hours after the experiment is started. The climb is continued through the experiment and also ends in similar readings, 7.47 in Set A and pH 7.19 in Set B, after 3 days. While the dissolution of eggshells and limestone has been produced that contributed to the increase in AMD pH, the relatively poorer performance of limestone is due to the armoring process that hinders limestone dissolution and compromises the effectiveness of the treatment system. Armoring is the process in which the dissolved iron and manganese form insoluble precipitates on the limestone surface that inhibit limestone dissolution (Loeppert & Hossner, 1984).

The precipitated from the chemical reaction between pyrite and water form on the surface of limestone through adsorption of Fe(II), followed by oxidation to Fe(III) (Evangelou, 1995). Armoring of limestone causes two setbacks in AMD treatment. Firstly, it causes the limestone to have a lesser surface area available for limestone dissolution. Secondly, the iron precipitation reactions take time, that is, there is no immediate and significant increase in pH (Dousma & de Bruyn, 1976). This is in corresponds to the statement made by Ziemkiewicz et al. (1997) that armoring of limestone will cause a wide range of inhibition but still partly effective in acid neutralization. Armoring of limestone explains the presence of lag time in the AMD treatment, namely 15 minutes to 240 minutes in Pail 1 (Figure 4.12) and 15 minutes to 60 minutes in Pail 2 (Figure 4.13), where the climb of pH is relatively insignificant compared to pails containing eggshells as neutralizing agents.

On the other hand, eggshells offered more stable and effective performance in increasing pH for AMD in comparison to that limestone. Since the content of Pail 3 is the same in Set A and Set B, which is a mixture of eggshells and limestone, the AMD pH in both sets showed similar readings. The AMD pH in both sets climbed steadily and reached similar after 3 days, with pH 7.57 in Set A and pH 7.37 in Set B.

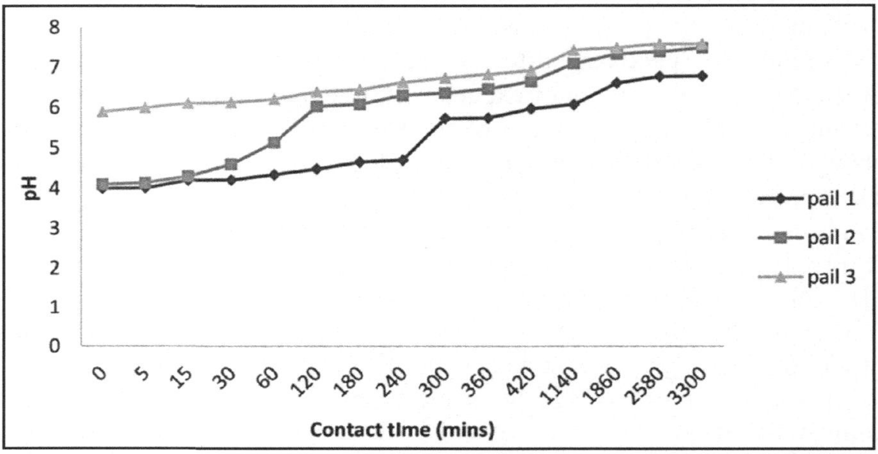

FIGURE 4.12 Changes of pH over time in Set A (Note: first sample is collected after 30 minutes of retention time).

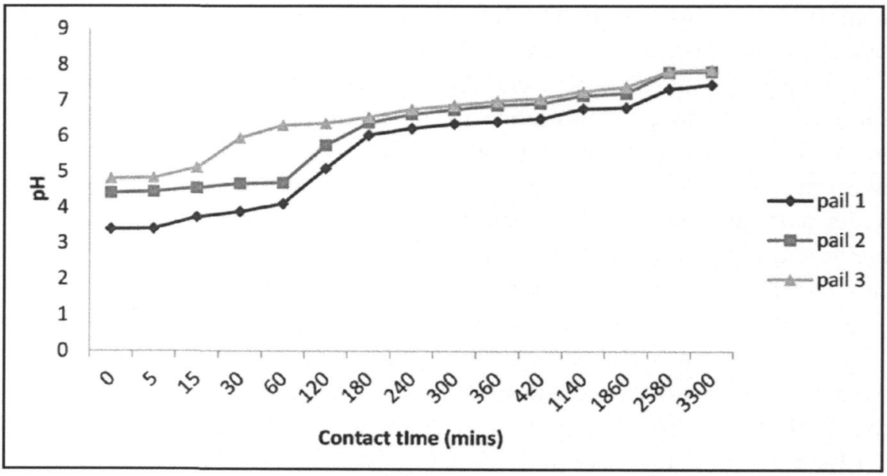

FIGURE 4.13 Changes of pH over time in Set B (Note: first sample is collected after 30 minutes of retention time).

While the acidity of the treated AMD is indirectly proportional to the pH in the early experiments, it showed that Set A showed lower acidity than that of Set B, with 142.5 mg $CaCO_3$/L in Set A and 147.5 mg $CaCO_3$/L in Set B an hour after the experiment is started, as shown in Figures 4.14 and 4.15. However, the acidity in Set B is significantly lower at 40 mg $CaCO_3$/L as opposed to the 97.5 mg $CaCO_3$/L in Set A two hours after the experiment started. The trend

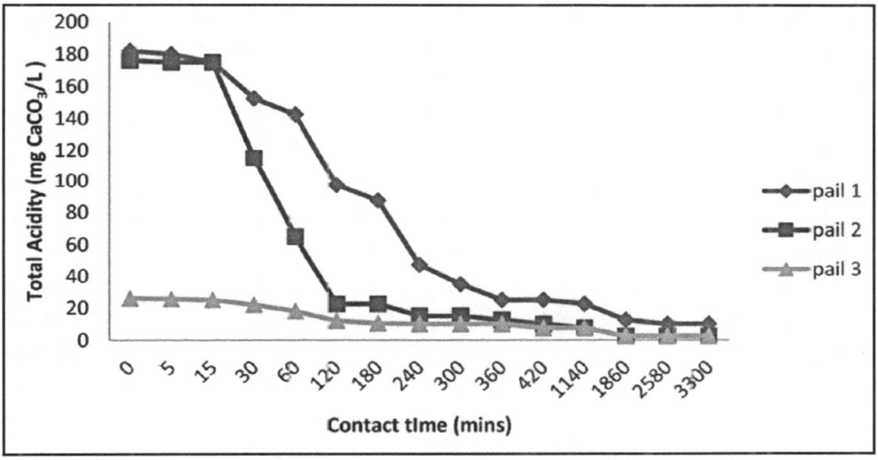

FIGURE 4.14 Changes of total acidity (mg $CaCO_3$/L) over time in Set A (Note: first sample is collected after 30 minutes of retention time).

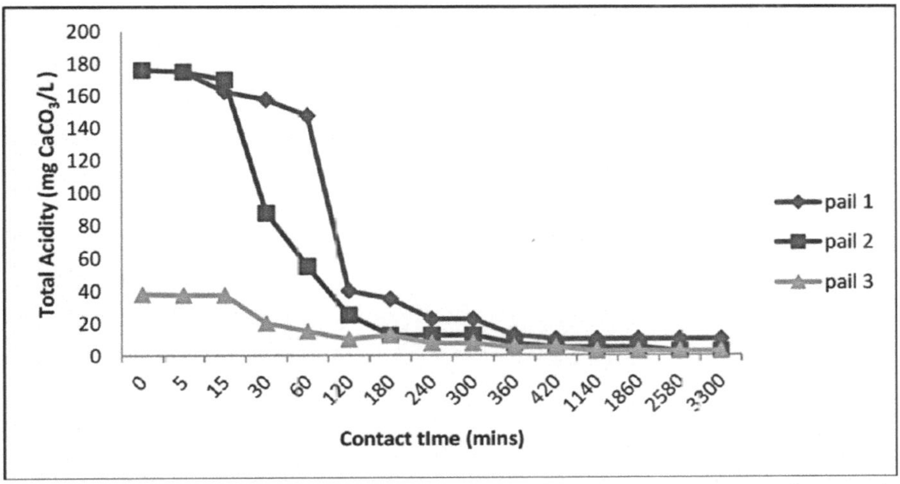

FIGURE 4.15 Changes of total acidity (mg $CaCO_3$/L) over time in Set B (Note: first sample is collected after 30 minutes of retention time).

of decreasing acidity in both sets is consistent throughout the experiment after 2 hours. Similar to the change in pH, Set A showed an apparent lag phase. The total acidity is lowered from 97.5 mg $CaCO_3$/L to only 47.5 mg $CaCO_3$/L after 120 minutes to 240 minutes after the experiment is started. This reduction in total acidity is significantly lower than the yield in Set B, having 40 mg $CaCO_3$/L reduced to as low as 22.5 mg $CaCO_3$/L in the same period of time. However, given sufficient experiment time, the final acidity of Pail 1 in Set A and Set B reached the same reading, 10 mg $CaCO_3$/L, after 3 days. A similar phenomenon is observed in Pail 2 for both sets. The neutralizing agents for Pail 2 in Set A and Set B are eggshells and limestone, respectively. An hour after the experiment started, the total acidity of AMD was lower in Set B than in Set A. The acidity was 65 mg $CaCO_3$/L in Set A and 55 mg $CaCO_3$/L in Set B. However, the acidity in Set A is lower (22.5 mg $CaCO_3$/L) than in Set B (25 mg $CaCO_3$/L) two hours after the experiment started. This trend of decrease in total acidity is then consistent, thorough experiment, yielding 2.5 mg $CaCO_3$/L in Set A and 5 mg $CaCO_3$/L in Set B. The change in acidity for Pail 1 and Pail 2 in both sets have shown a similar trend to the change in pH. This is evidence to support further the speculation that armoring of limestone has occurred in the AMD treatment process. Since Pail 3 in both sets contains a mixture of eggshells and limestone as neutralizing agents, the trend in both sets showed a similar decrease in total acidity at any given time. The final acidity is 2.5 mg $CaCO_3$/L for both sets after 3 days of experiment time. As shown in Figures 4.16 and 4.17, it is evident that the total acidity of the AMD decreases

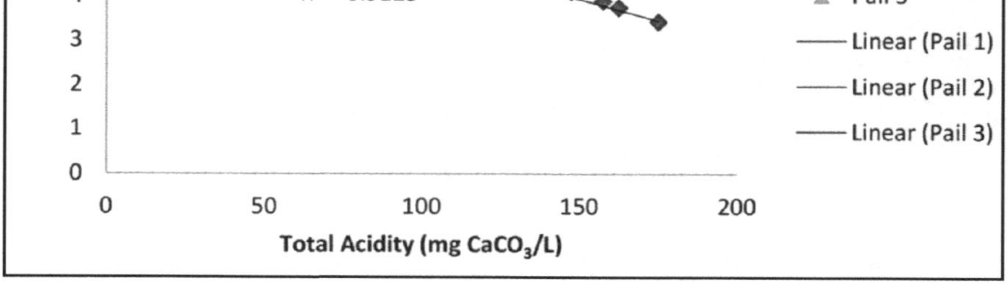

FIGURE 4.16 pH against total acidity (mg $CaCO_3$/L) in Set A.

as the pH climbs. While the decrease in the concentration of hydrogen ions has contributed to the increase in pH and decrease in total acidity, it should be noted that in this study, however, there is a significant linear relationship between pH and total acidity from all outputs in both sets.

Figures 4.18–4.31 show that after conducting the experiment for 3,300 minutes (3 days), it is found that most heavy metals (i.e., iron, copper, aluminum, lead, and cadmium) can be reduced to below the maximum admissible limit set in MNDWQS and WHO, except nickel and manganese contents. Iron, copper, aluminum, lead, and cadmium are reduced to a minimal 0.0881 ppm, 0.5117 ppm, 0.4980 ppm, 0.007051 ppm and 0.001607 ppm, respectively. Limestone, eggshells, and the mixture of these two cannot treat nickel and manganese effectively. The final concentration of nickel and manganese are as high as 0.4041 ppm and 7.1586 ppm. The reduction in the heavy metals concentration by limestone and eggshells can be attributed by two factors. Firstly, the surface of the limestone and eggshells provides solid contact resulting in chemisorptions of metal ions at low concentrations (Sturchio et al., 1997). Secondly, the presence of dissolved calcium carbonate from the neutralizing agents increased the pH of the solution above the solubility point, which caused metals to precipitate as metal oxide (Xu et al., 1996). Eggshells and limestone are known to contain metal oxides. In order to reach chemical equilibrium, both leaching (also desorption) and adsorption of metal oxides from both neutralizing agents in AMD are speculated to occur. While the input of metal oxides from eggshells is negligible, as

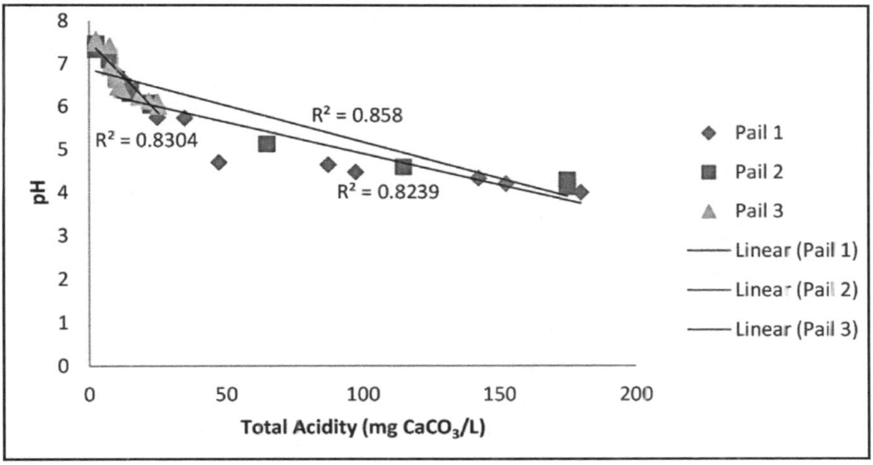

FIGURE 4.17 pH against total acidity (mg $CaCO_3$/L) in Set B.

evidenced by the lack of increase in heavy metal concentration observed, the leaches of metal oxide from limestone are more significant as the concentration of heavy metals (i.e., iron, copper, aluminum, cadmium, cadmium, nickel, and manganese) are increased in AMD after 420 minutes (7 hours) of AMD treatment. However, nickel and manganese aside, the concentration of the heavy metals are successfully treated to below maximum admissible limits after 3,300 minutes (3 days) of AMD treatment.

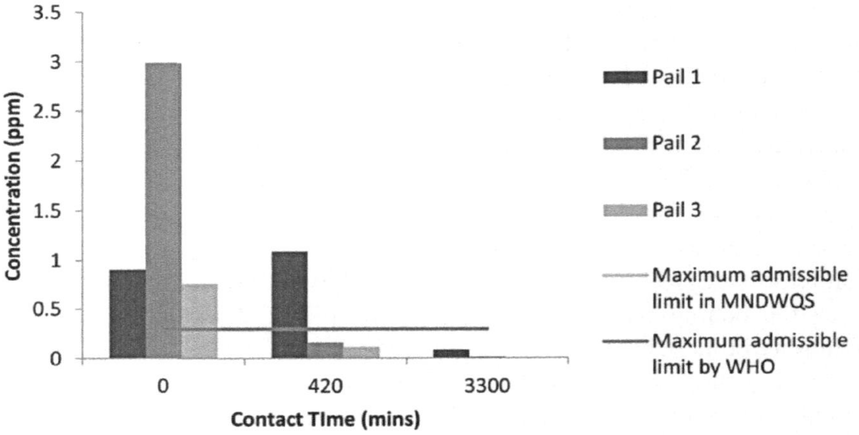

FIGURE 4.18 Concentration of iron in Set A compared to maximum admissible limits.

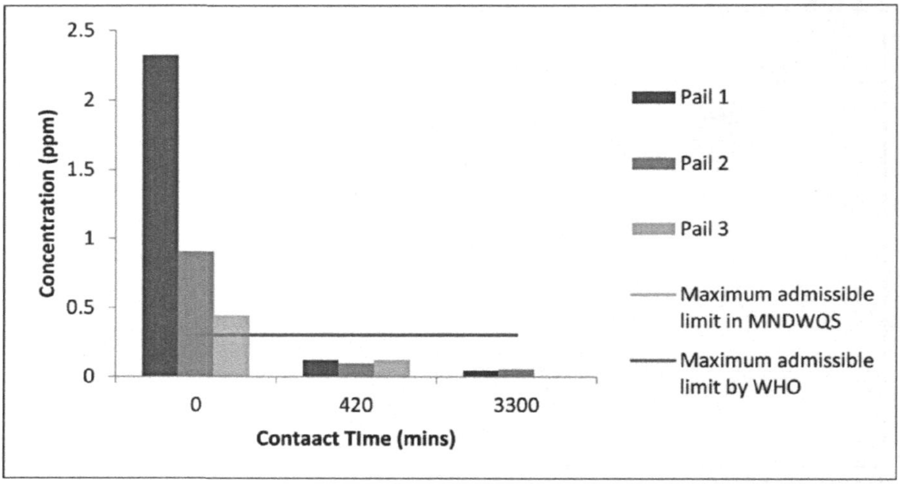

FIGURE 4.19 Concentration of iron in Set B compared to maximum admissible limits.

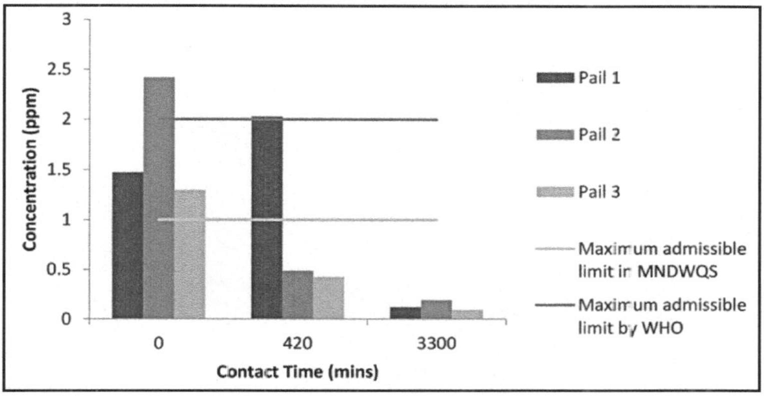

FIGURE 4.20 Concentration of copper in Set A compared to maximum admissible limits.

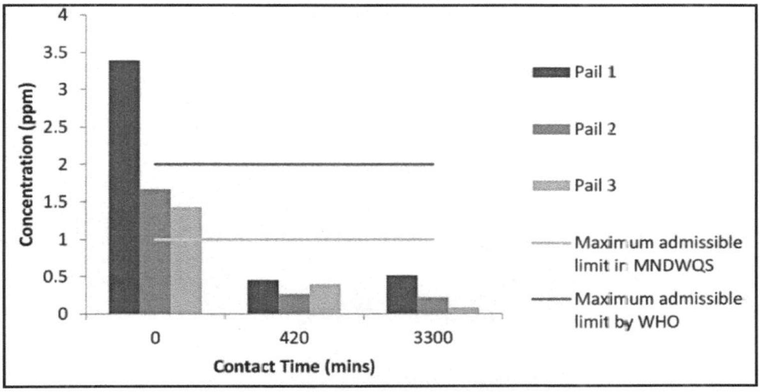

FIGURE 4.21 Concentration of copper in Set B compared to maximum admissible limits.

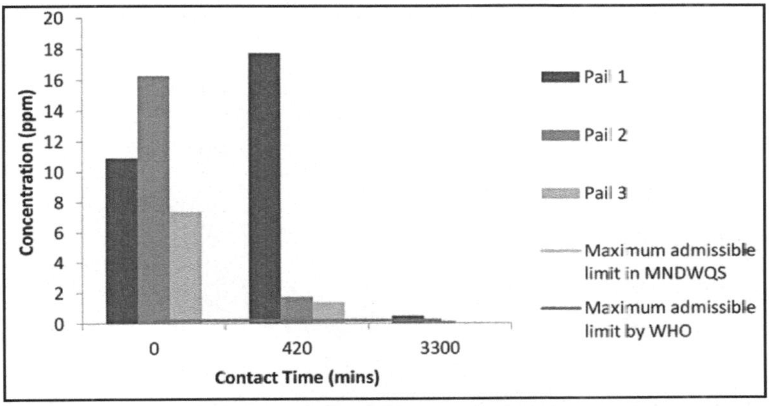

FIGURE 4.22 Concentration of aluminum in Set A compared to maximum admissible limits.

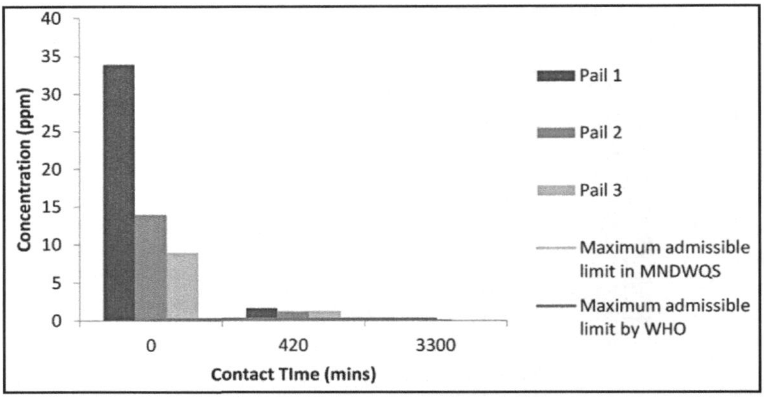

FIGURE 4.23 Concentration of aluminum in Set B compared to maximum admissible limits.

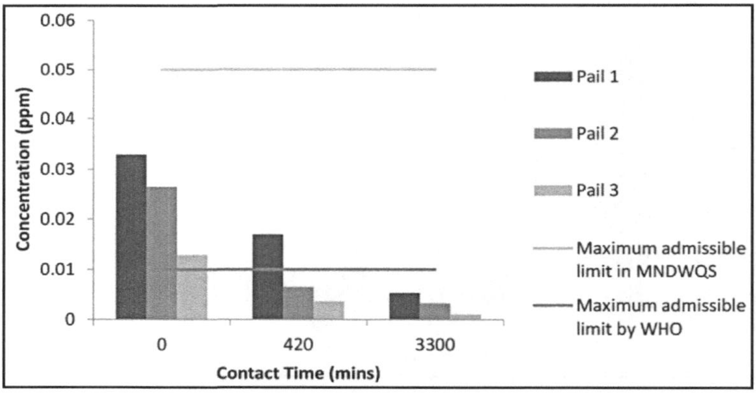

FIGURE 4.24 Concentration of lead in Set A compared to maximum admissible limits.

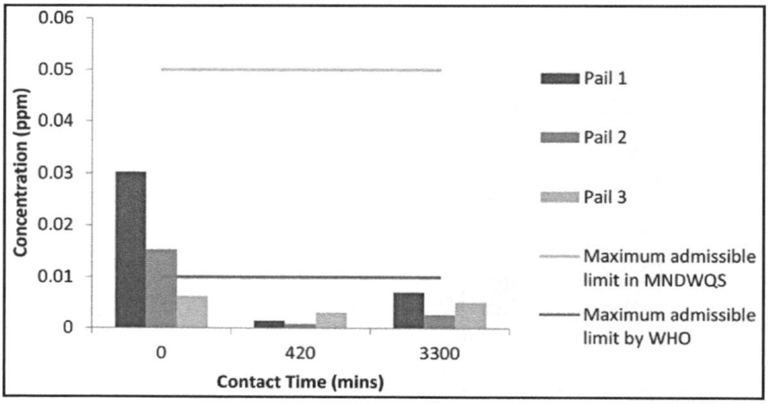

FIGURE 4.25 Concentration of lead in Set B compared to maximum admissible limits.

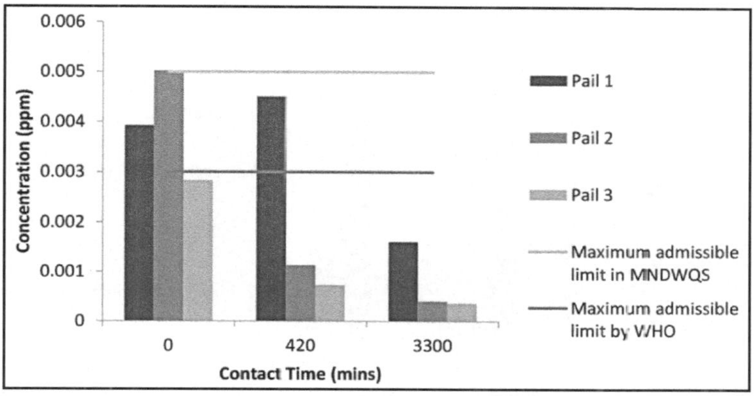

FIGURE 4.26 Concentration of cadmium in Set A compared to maximum admissible limits

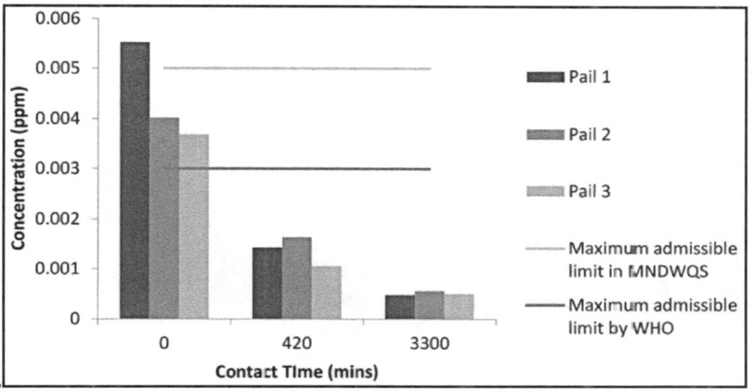

FIGURE 4.27 Concentration of cadmium in Set B compared to maximum admissible limits.

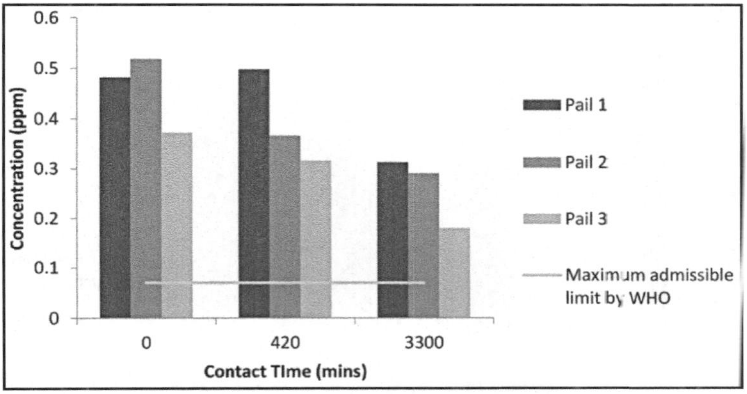

FIGURE 4.28 Concentration of nickel in Set B compared to maximum admissible limits.

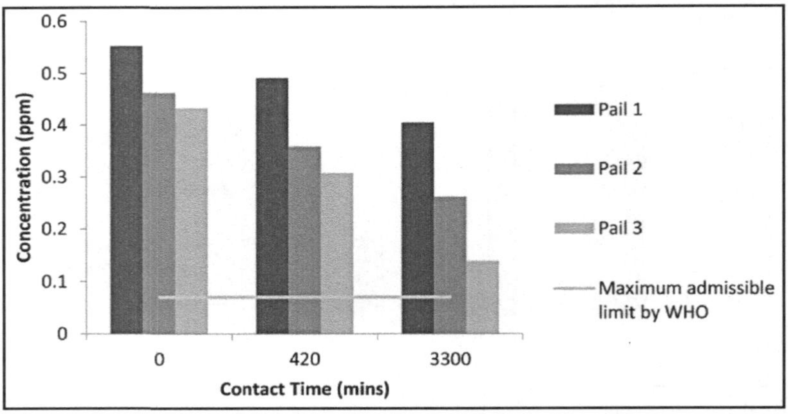

FIGURE 4.29 Concentration of nickel in Set A compared to maximum admissible limits.

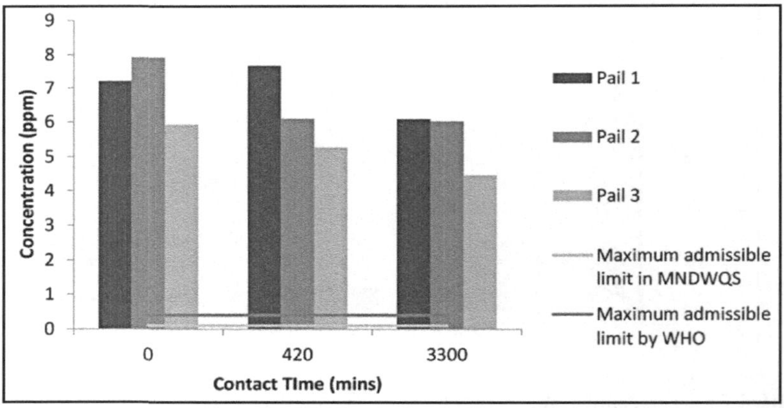

FIGURE 4.30 Concentration of manganese in Set A compared to maximum admissible limits.

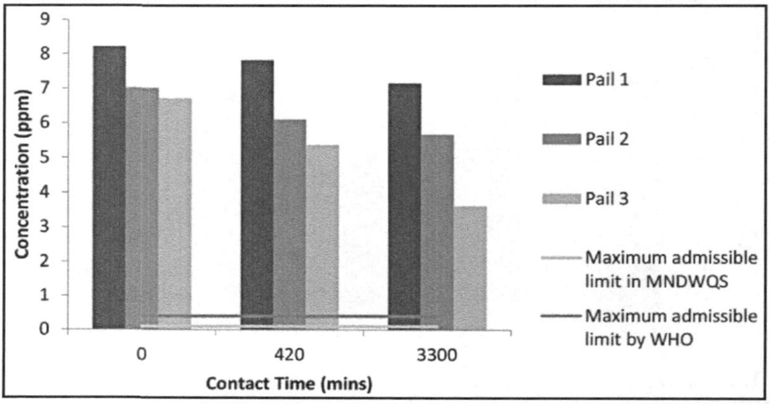

FIGURE 4.31 Concentration of manganese in Set B compared to maximum admissible limits.

The surface morphology of eggshells and limestone was observed under 500x magnifications with a scale of 1 cm: 10 μm. Before the eggshells are used as neutralizing agents, it is observed that their surfaces are smooth and do not have cracks on the surface (Figure 4.32). However, after AMD treatment, the surface has become rough. Visible cracks can be seen as well. The porosity is also notably bigger, and metal oxides can be deposited on the surface (Figure 4.33). This finding suggests that while eggshells effectively remove heavy metals and increase pH, their lifespan as neutralizing agents is limited as cracks are visible on most surface parts.

On the other hand, the limestone surface appeared to be smooth with no visible cracks before AMD treatment, as shown in Figure 4.34. However, dissolution has occurred as some surface parts are rugged after AMD treatment. The visible hole can be seen as well in Figure 4.35. In contrast to the numerous cracks observed on the eggshell surface, the holes on the limestone surface are relatively fewer. Hence it is speculated that limestone can be used for more than 3,300 minutes (3 days), yielding a longer lifespan than eggshells. However, no visible metal oxides are observed on the limestone surface.

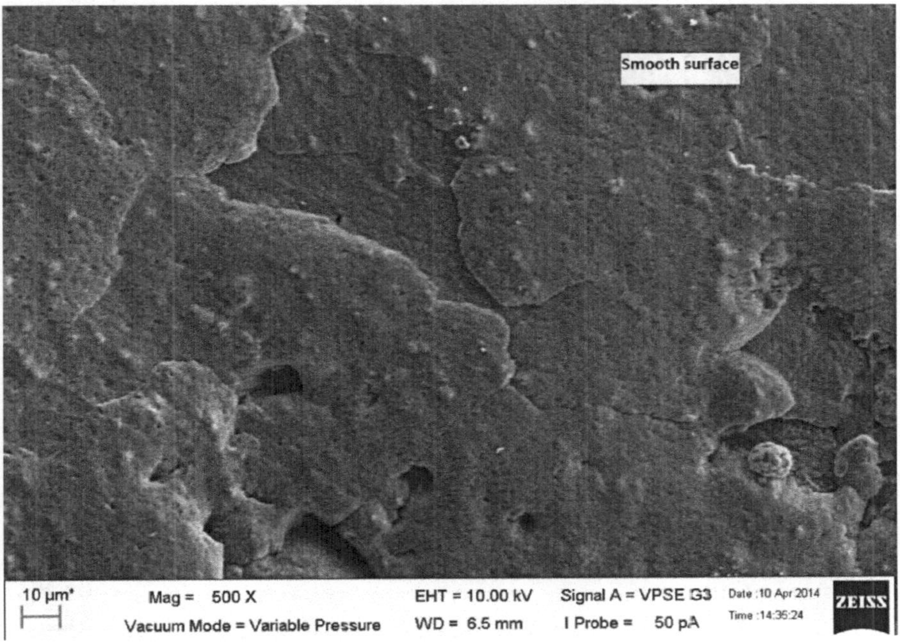

FIGURE 4.32 Surface morphology of eggshells before AMD treatment.

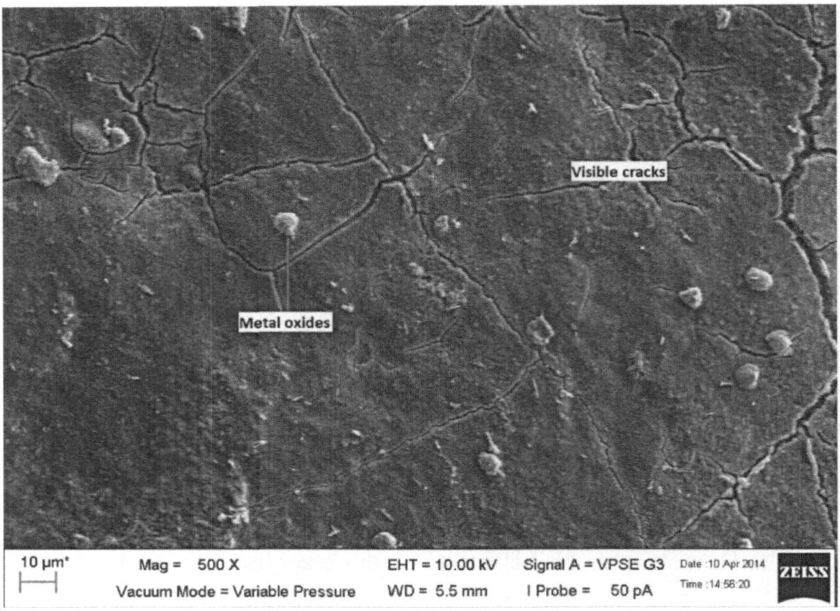

FIGURE 4.33 Surface morphology of eggshells after AMD treatment.

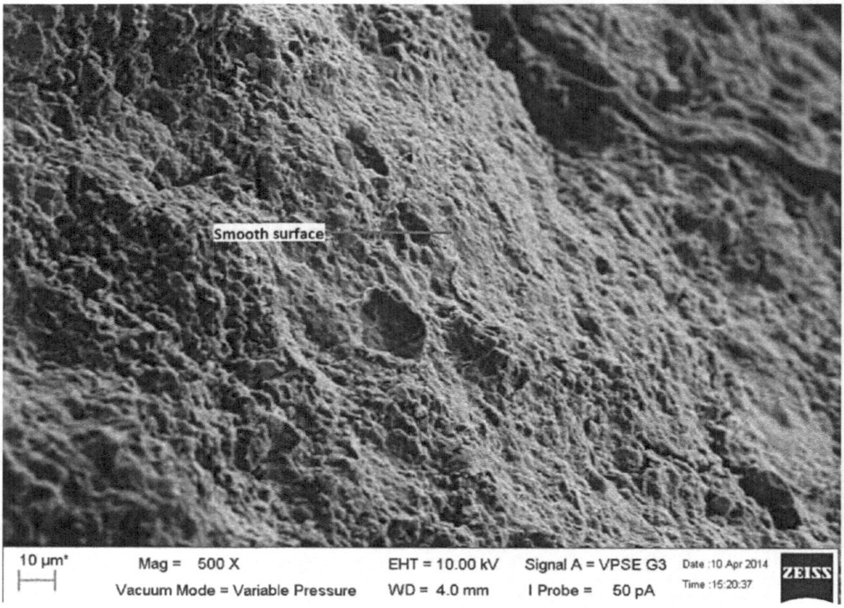

FIGURE 4.34 Surface morphology of limestone before AMD treatment.

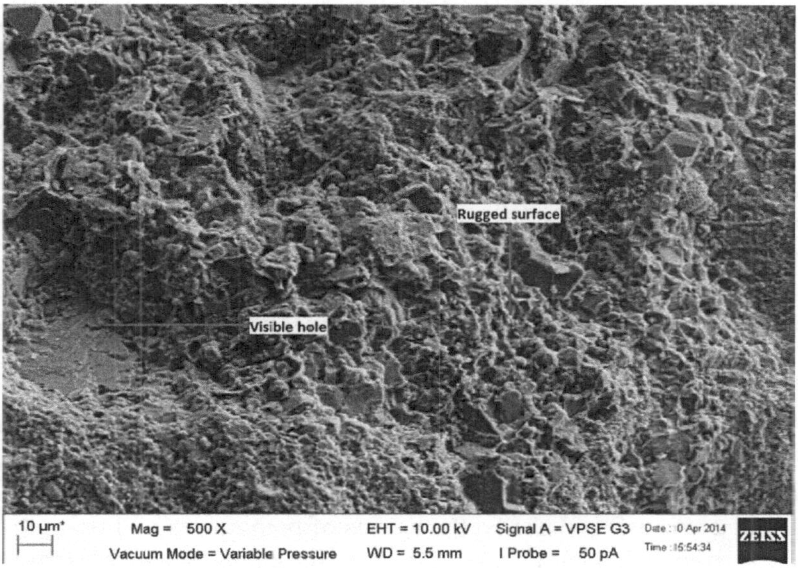

FIGURE 4.35 Surface morphology of limestone after AMD treatment.

In terms of neutralizing mechanisms for AMD treatment in this study, heavy metals generally precipitate as a discrete hydroxide or carbonate in limestone and eggshells systems. For nickel and manganese, however, they may sorb to either a mineral surface, iron precipitate surface, or co-precipitate within the iron precipitates or in the calcite itself. Sorption is an accumulation of metals on any available surface, such as metal precipitate or mineral. In contrast, co-precipitation is an accumulation of metals on an available surface that becomes part of the primary metal mineral lattice and is thus not easily reversible (Lee et al., 2002). The observation of metal oxides on the eggshell surface, as shown in Figure 4.33, has proven that precipitation has occurred. Reduction in the heavy metals concentration is achieved when the rate of heavy metals adsorption exceeds the rate of leaching of heavy metals from the eggshells. This speculation is further supported by the reports Park et al. (2007) published that shows similar result in heavy metal absorption of eggshells.

4.2.3 DURIAN HUSK AND PEANUT HULL FOR GROUNDWATER TREATMENT

4.2.3.1 INTRODUCTION

Durio Zibethinus Murray, well known as durian, is famous among South-east Asia countries, particularly Malaysia, Thailand, Indonesia, and the

Philippines. In Malaysia, a large quantity of durian is generated annually. However, the waste of durian generated, including its entire unwanted skins and seed, had contributed to an increasing volume of waste either at abandoned land nearby durian stalls or landfill sites, notably during the durian season. For instance, durian peel is considered the main contributor to agricultural waste, with approximately 4,80,000 metric tons generated annually. These wastes will usually be disposed to landfills, and some will be burned. Whilst peanut or known as groundnut (*Arachis hypogaea*), has grown in tropics and sub-tropics (Bertioli et al., 2011). In China, peanut shells produce about 5 million tons, and this country is the world's largest producer of peanut shells. The peanut industry is also one of the main generators of agro-industrial waste in terms of the shell or hull (Perea-Moreno et al., 2018).

Apart from that, durian peel and peanut skin could categorize as agricultural solid waste. Generally, the agricultural wastes that generated from durian husks are mostly discarded into landfills or burnt (Lee et al., 2018). By mentioning its husk and shells, the wastes of both durian and peanut could be categorized as biomass wastes. As the disposal usually involves burning and disposed at landfills which are not sustainable in line with their high generation, both of these wastes pose the potential to be utilized for biomass energy production and as a low-cost sorbent alternative for water treatment. One alternative solution is converting those durian peels and peanut skin into an environmentally friendly bio-briquette treated with torrefaction for fuel (Haryati et al., 2018). Furthermore, the material from these agricultural wastes could also be utilized in other applications such as biocomposites, bioethanol production, and physical adsorbents for water treatment due to its richness in cellulosic substances (Abdullah et al., 2016; Obeng et al., 2018).

The durian husk comprises 60.45% cellulose, 15.45% lignin, and 13.09% hemicellulose (Lee et al., 2018). Whilst, peanut shell contains cellulose (40.5%), lignin (26.4%), and hemicelluloses (14.7%) (Bharthare et al., 2014). The availability of cellulose, hemicellulose, and its macro-porosity enables both durian husk and peanut hull to trap water pollutants and resist temperature, pH, mechanical stress and prolonged submersion in water (Shitu et al., 2016).

The physical adsorbent of both durian peel and peanut skin are environmental remediation treatments that are tolerable to remove pollutants, such as dye and heavy metals. Yet, there is no previous study on removing hardness and magnesium and calcium minerals via these adsorbents. On the other hand, AC is an adsorbent material commonly utilized to remove pollutants for water treatment. Even so, AC has limitations as it is costly, hence not affordable to be employed by developing countries, specifically for large-scale utilization (Lee et al., 2018). Thus, durian husk and peanut hull will be utilized in this study for pollutants removal from water effluent. At the same time, it bears another significance by reducing its waste by reusing it as an affordable sorbent.

On the other hand, groundwater can be highly contaminated by mg and ca due to the dissolution of carbonate minerals in the rocks and soil properties of the geological formations, which causes the groundwater hardness to be higher than any other water resource (Kumar & Joshi, 2013). Hard water is a significant concern due to its health implications and disruption to household cleaning tasks (Bertioli et al., 2011). Next, as for the removal of TDS and EC, the synthetic coagulant of aluminum (alum) sulfate is often utilized. However, it lays limitations, including health implications such as inducing Alzheimer and Parkinson (Xu et al., 2017). Therefore, durian husk and peanut hull serve as a suitable alternative to replace alum for water treatment.

4.2.3.2 FINDINGS

From the research findings, the removal efficiency of total hardness (TH) by durian husk and peanut hull, as illustrated in Figure 4.36, indicated a gradual increase from 7 to 18%, a significant rise in removal with an increase of the adsorbent dosage from 0.5 to 2.5 g/l by durian husk. This might be attributed to the increase in the number of possible binding sites and surface area of the sorbent for pollutant removal (Alghambi et al., 2019). It also showed that the mean concentration of hardness (TH) had significantly reduced from the initial concentration of 320 mg/L before the treatment and decreased to 261 mg/L of $CaCO_3$, along with the increased dosage of the modified durian husk. The lower removal percentage that was recorded in this research with the highest peak of 18% might be due to the presence of a higher adsorbent dosage that will cause more aggregation, which will decrease the total surface area of the adsorbent that leading to a decrease in adsorption (Thilagavathi et al., 2012). The other reason could be associated with the adsorbent elements from the durian husk itself leaching out to the water samples with $CaCO_3$, which may consist of multiply constituents and properties such as metals, nutrients, and other trace elements that may compete with each other for the binding sites during adsorption process (Payus et al., 2014).

For the peanut hull, it shows no significant adsorption of hardness ions occurred at dosages of 0.5 to 2.5 g of modified peanut hull was observed. The lowest hardness concentration was reduced from 322 mg/L to 315 mg/L after the treatment with the absorbent addition at 0.5 g of dosage, then the concentrations increased gradually with the increase of the dosage from 1 g up to 2.5 g of dosage, with an inconsistent trend of hardness removal.

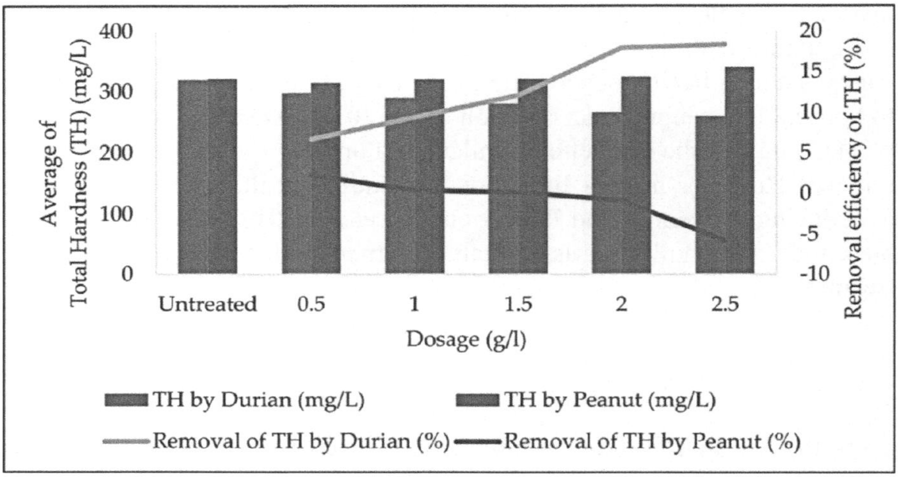

FIGURE 4.36 Effect of adsorbent dosage on TH with its removal efficiency.

This trend might be due to the peanut hull containing very little or almost none of the magnesium and calcium ions as the raw peanuts were being processed with salt and even roasted, so this might increase the hardness by adding these ions from peanut hulls. The other reason might be a higher adsorbent dosage cause aggregation, which decreases the total surface area of the adsorbent, leading to a decrease in adsorption (Al-Qaim, 2011). In addition, it can be seen that the optimum absorption of hardness was recorded at the finest or the smallest grain size of the dosage, which might provide different affinity for adsorption sites. Both adsorbent's average TH and removal efficiency by settling time are shown in Figure 4.37. The settling time was performed to investigate the trend of the efficiency of hardness removal as well as to determine the saturation point towards the TH removal after the rapid and slow mixing in the jar test. The removal efficiency of TH by durian husk was gradually increased from 9 to 25% with an increase in the settling time. The optimum settling time was achieved at 18 hours, with the removal efficiency of hardness recorded at 25.2%, and within 18 hours and 24 hours, the saturation point was obtained. It is also shown that the mean concentration of hardness (TH) had significantly reduced from the initial concentration of 320 mg/L before the treatment and decreased to 240 mg/L of $CaCO_3$ with the increase of the settling time.

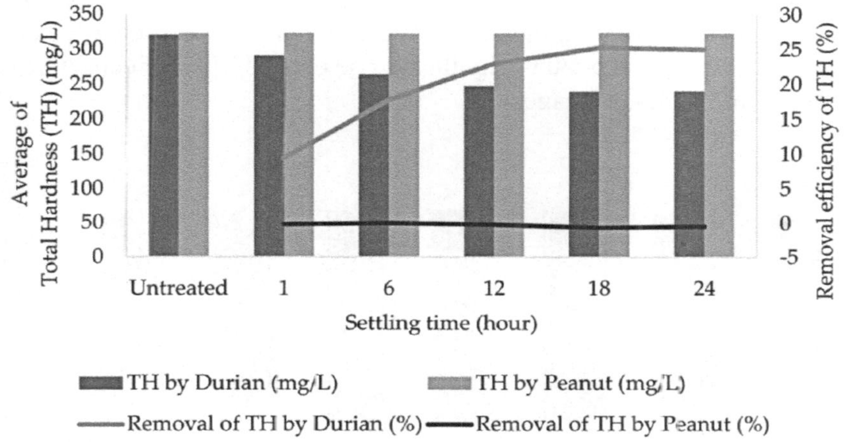

FIGURE 4.37 Effect of settling time on TH with its removal efficiency.

The removal of TH from the groundwater sample by durian husk showed significance with the settling time. The groundwater hardness reduced with settling time could be explained by the physical adsorbent influence from the retention by van der Waals forces that is weak and resulting in changes in the electric charge density of the individual atoms (Crini et al., 2018). Therefore, this makes the hardness particles settle very slowly or maybe not at all, mutually repel each other between the surfaces of the durian husk and hardness particles. However, for the peanut hull, the trend shows no adsorption of hardness ion occurred and is not significant with the settling time in reducing hardness concentrations.

Figures 4.38 and 4.39 show the effect of adsorbent dosage on total dissolved solids (TDS) and electrical conductivity (EC) by modified durian husk and peanut hull. The results in Figure 4.38 illustrate the reduction of mean concentrations of TDS from 426 ppm before treatment by durian husk adsorbent to 424 ppm at 1 g optimum dosage. But then, after the peak, a sudden increase in the TDS concentration was observed during the second interval with more adsorbent dosage and longer contact time. The concentration of TDS shows a significant inconsistent trend for different dosages of durian husk until the rest of the intervals, with lower fluctuations percentage of removal efficiency observed. While for EC in Figure 4.39, the mean concentrations have reduced from 676 μS/cm before the treatment to 665 μS/cm concentration at 1 g optimum dosage achieved. An unstable trend and EC fluctuations were observed throughout the contact time, with a sudden increase even for the first dosage of 0.5 g durian husk addition onto the removal of EC. Another dropped in concentrations at 2 g of dosage can be seen in this study, and another significant rise onto 711 μS/cm concentration at the end of the 2.5 g dosage.

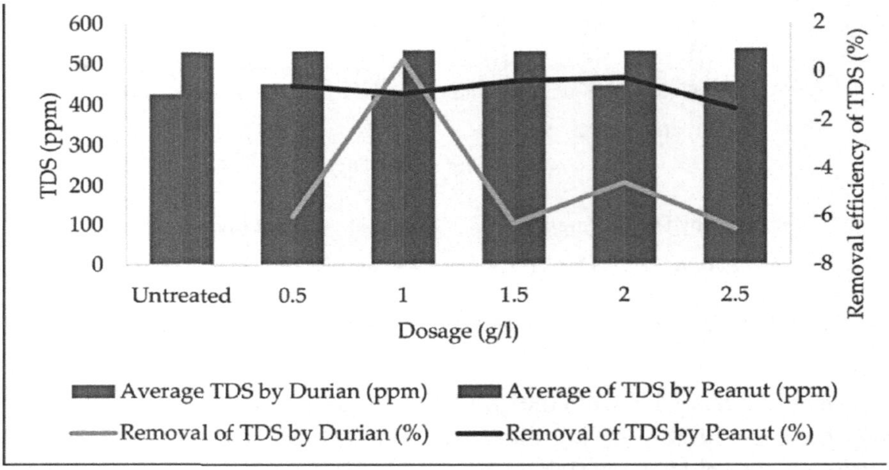

FIGURE 4.38 Effect of adsorbent dosage on TDS and its removal efficiency.

Throughout this study, removal by durian husk was low for TDS and conductivity (EC). However, studies by Abdullah et al. (2016) and Lee et al. (2018) found a high removal ability for durian husks. Their results indicate a significant reduction in the concentration of their hardness synthetic hard water from 709.33 mg/l to 218 mg/l by durian husk as the adsorbent dosage increase from 0.5 g to 2.5 g. The TDS was reduced from 1689.67 ppm to 49 p mm, while the reduction of EC was from 1762.33 to 632.66 μS/cm. In addition, their research primarily used the highest concentration of hardness, up to 700 mg/l of $CaCO_3$, to find out the removal efficiency for water softening in water treatment using the durian husk. Thus, the potential to reduce TH with a high initial concentration of hardness is higher.

Nonetheless, this research serves as the very first study utilizing actual or on-site water samples with a very low initial concentration of hardness,

EC and TDS with 320 mg/L; 676 μS/cm; 426 ppm and below, respectively. It also shows that TDS and EC concentration becomes high with adding the adsorbent dosage. The trend might be attributed to the leaching from the durian husk modified with 1 M NaOH solutions, which contributed to elevated TDS and EC when the dosage was increased.

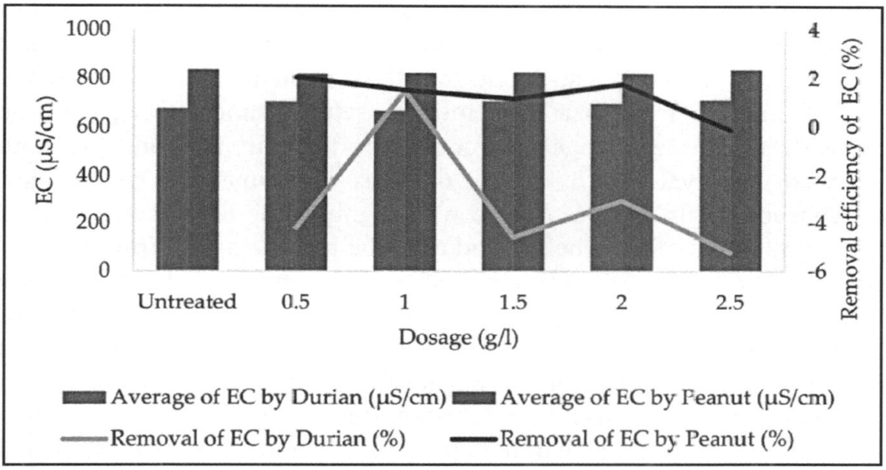

FIGURE 4.39 Effect of adsorbent dosage on EC and its removal efficiency.

For the effect of the peanut hull on the TDS and EC removal, the concentrations range after the treatment was recorded from 528 to an increase of 536 ppm of TDS; and from 819 μS/cm increased to 835 μS/cm. The trend gave slight changes in interactions on TDC and EC concentrations within 0.5 to 2 g of dosages with the reduction from 531 to 529 ppm for TDS and from 704 to 696 μS/cm of EC. However, it increased drastically after the 2 g and 2.5 g dosages of both TDS and EC. In summation, no removal efficiency of TDS and EC were found towards the effect of dosage by peanut hull. The sudden increase of the concentration after addition with more dosage of absorbent reported by Rios et al. (2008) due to the reduction of bioavailability and surface area for the binding sites, if exposed to longer contact time of the adsorbent that may affect the efficiency. Dzombak & Morel (1990) previously discovered that adsorption taken place on the surface usually involves the initial interval phase, which is rapid, followed by slower surface precipitation.

The effect of settling time on the concentrations of TDS and EC are shown in Figures 4.40 and 4.41. Overall the concentrations of TDS and EC before and after the treatment of the groundwater samples by durian husk adsorbent are recorded in the ranges between 407 and 426 ppm of TDS and 640–676 μS/cm of EC, respectively. Whilst peanut hill was recorded at the range between 520 and 533 ppm for TDS and 806 to 822 μS/cm, respectively. It is also shown that the concentrations of TDS and EC have significantly reduced from 426 ppm to 407 ppm of TDS before and after the treatment; and from 676 μS/cm to 645 μS/cm of EC before and after the treatment, with the increase settling time of the modified durian husk from 1 to 24 hours. It also found that there is small removal efficiency of TDS and EC by durian husk towards settling time effects, with the highest removal percentage recorded at 6% for both TDS and EC, both within the interval time 1 hour to 6 hours. The same with peanut hull, the concentrations of TDS and EC went significantly decreased from 528 ppm to 520 ppm of TDS before and after the treatment; and from 819 μS/cm to 803 μS/cm of EC before and after the treatment, with the increase settling time of the modified peanut hull from 1 to 18 hours. The trend for the peanut hull can be seen efficiently removing both TDS and EC within 1 to 18 hours. However, it increased drastically after 18 hours and back at 24 hours intervals. In this research, the settling time shows a positive result in reducing TDS, EC, and hardness due to the ions such as magnesium and calcium from the water samples that eventually bind to the surface of the durian husk (Obeng et al., 2018).

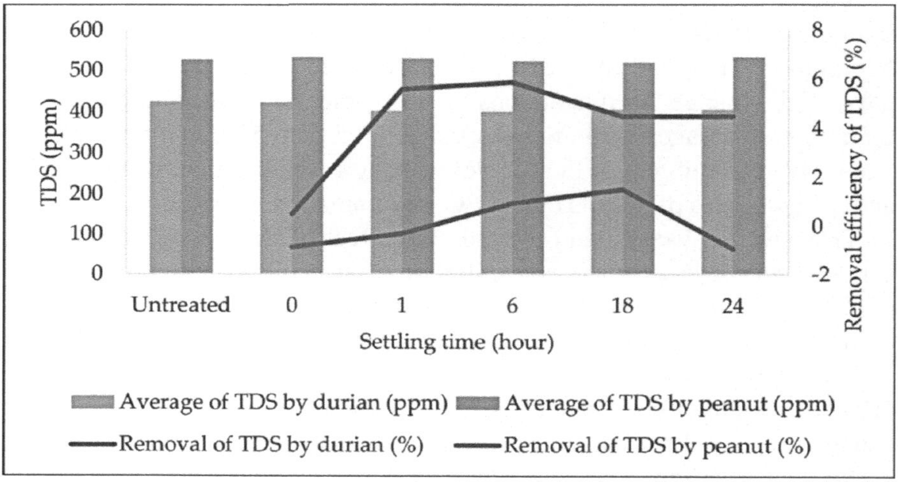

FIGURE 4.40 Effect of settling time on TH with its removal efficiency.

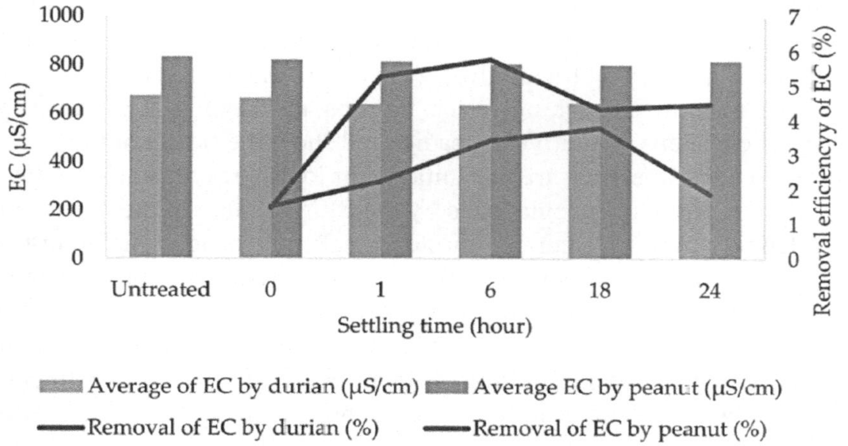

FIGURE 4.41 Effect of settling time on EC with its removal efficiency.

4.3 CONCLUSION

In conclusion, the ability and results of the removal vary per adsorbent in each case, as they have different experimental set-ups. AMD treatment using bone meal shows a significant decrease in pH and concentration of Fe and Zn after the neutralization treatment with bone meal in this experiment, it believes that this method has the potential to be used as a treatment. However, because of the complex mechanism that involves the removal of metals and other factors that affect neutralization, additional studies are needed to optimize the removal of dissolved metals and the neutralization efficiency of bone meal. More research needs to be done for a large-scale AMD treatment neutralization using bone meal to better understand the treatment design system. The batch experiment carried out in this research was to test the potential of bone meal for neutralization and metal removal from the AMD contaminated water samples.

Other than that, the efficiency of eggshells, limestone, and the mixture of these two in neutralizing AMD and heavy metal removal can be identified in this study. The tested neutralizing agents can bring the pH and total acidity of AMD output to within safe limit to be qualified as safe drinking water. Aside from nickel and manganese, the other heavy metals are reduced and meet the criteria set in MNDWQS and WHO. The neutralizing mechanism involved in the AMD treatment of eggshells is speculated to be adsorption and leaching. A similar mechanism is expected in limestone, but it has an additional armoring effect caused by insoluble precipitation by Fe^{3+} and Mn^{2+} on limestone surface that may inhibit the processes. The choice of neutralizing agents for AMD treatment is highly objective-specific, as each agent has its pros and cons. While the performance of eggshells is relatively steadier and show the satisfactory result in shorter contact time, they are not suitable for long-term remediation due to their limited lifespan, as evidenced by the SEM result. On the other hand, the relatively long lifespan of limestone makes them ideal for long-term AMD treatment. The downside is that limestone will temporarily increase the concentration of heavy metals in AMD. However, this disadvantage is negligible given enough contact time (i.e., 3,300 minutes). Therefore, the mixture of eggshells and limestone is recommended for AMD treatment in an ideal situation. Furthermore, while the mixture has been proven effective in neutralizing and heavy metal removal in AMD, it is speculated that it can also provide nutrients to the AMD.

For durian husk adsorbent, the TH, TDS, and EC removal efficiency were found to increase with settling time and dosage, however, went sudden reductions when exposed to higher dosage and longer settling time. Whilst peanut hull shown no adsorption occurs by increasing dosage and settling time. The percentage of removal by durian husk also shows greater efficiency compared to the peanut hull adsorbent. Overall, the percentage of removal efficiency by both adsorbents in this research was found low compared to the previous study. This might be due to the initial concentration of hardness, TDC, and EC used to be too low within 300 mg/l and below. This study is unique as it is the first to use actual on-site samples taken in the field for initial concentrations, as opposed to previous studies that used lab-scale of synthetic hard water with much higher initial concentrations up to 800 mg/L to determine the removal efficiency in water treatment. Higher concentrations will provide a relative abundance of dissolved elements and metals and solubility that will induce and increase the removal efficiency rate of the adsorbent. Thus, exposure rate of the adsorbents' surface area and biding sites is also higher and instigated.

Nevertheless, further research should be examined for the stated mechanism above, as organic solid wastes might have a significant potential to be used as essential adsorbents of physical water pollutant's removal.

KEYWORDS

- **AMD treatment**
- **bone meal**
- **groundwater treatment**
- **limestone**
- **solid wastes**
- **water effluent remediation**

REFERENCES

Abdullah, M., Muhammad, F. M. K., Amirul, A. Z., Nur, H. Md. K., & Nur, A. M. S., (2016). Oil removal using durian peel wastes: Effect of adsorbent condition. *Malaysian Journal of Industrial Technology, 1*(1), 2462–2540.

Akcil, A., & Soner, K., (2006). Acid mine drainage (AMD): Causes, treatment, and case studies. *Journal of Cleaner Production, 14*(12, 13), 1139–1145. doi: 10.1016/j.jclepro.2004.09.006.

Alghamdi, A. A., Abdel-Basit Al-Odayni, Waseem, S. S., Al-Kahtani, A., Fahad, A. A., & Taieb, A., (2019). Efficient adsorption of lead (II) from aqueous phase solutions using polypyrrole-based activated carbon. *Materials, 12*(12), 2020. doi: 10.3390/ma12122020.

Al-Qaim, F. F., (2011). Adsorption of malachite green (MG) on low cost–adsorbent from aqueous solution. *Journal of Babylon University/Pure and Applied Sciences, 1*(19), 48–56.

Bertioli, D. J., Guillermo, S., Fabio, O. F., José, F. V., Leal-Bertioli, S. C., & Marcio, C. M., (2011). An overview of peanut and its wild relatives. *Plant Genetic Resources, 9*(1), 134–149. doi: 10.1017/s1479262110000444.

Bharthare, P., Preeti, S., Pushpendra, S., & Archana, T., (2014). Peanut shell as renewable energy source and their utility in production of ethanol. *Int. J. Adv. Res., 2*, 1–12.

Buksh, N., Yun, C., Ping, X., Hussain, G., & Shai, Y., (2018). Chicken eggshell as a potential eco-friendly, low-cost sorbent: A mini review. *J. Environ. Earth Sci., 8*, 28–39.

Carmen-Mihaela, N., Gérald, J. Z., & Bruno, B., (2007). Passive treatment of acid mine drainage in bioreactors using sulfate-reducing bacteria: Critical review and research needs. *Journal of Environmental Quality, 36*(1), 1–16. doi: 10.2134/jeq2006.0066.

Chan, S. S., (2000). *Kajian Logam Berat Dalam Sedimen Sungai Mamut Dan Sungai Liwagu, Ranau.* BSc diss., Universiti Malaysia Sabah.

Conca, J. L., & Judith, W., (2006). An apatite II permeable reactive barrier to remediate groundwater containing Zn, Pb and Cd. *Applied Geochemistry, 21*(12), 2188–2200. doi: 10.1016/j.apgeochem.2006.08.018.

Crini, G., Eric, L., Lee, D. W., & Morin-Crini, N., (2018). Conventional and non-conventional adsorbents for wastewater treatment. *Environmental Chemistry Letters, 17*(1), 195–213. doi: 10.1007/s10311-018-0786-8.

Daubert, L. N., & Rachel, A. B., (2007). Passive remediation of acid mine drainage using crab shell chitin. *Environmental Engineering Science, 24*(10), 1475–1480. doi: 10.1089/ees.2006.0199.

Dousma, J., & Philip, L. De. B., (1976). Hydrolysis-precipitation studies of iron solutions. I. Model for hydrolysis and precipitation from Fe (III) nitrate solutions. *Journal of Colloid and Interface Science, 56*(3), 527–539. doi: 10.1016/0021-9797(76)90119-3.

Dzombak, D. A., & Morel, F. M. M., (1990). *Surface Complexation Modeling: Hydrous Ferric Oxide*. New York: John Wiley and Sons.

El Haddad, M., Rachid, M., Nabil, S., & Saïd, L., (2016). Evaluation of performance of animal bone meal as a new low-cost adsorbent for the removal of a cationic dye rhodamine B from aqueous solutions. *Journal of Saudi Chemical Society, 20*, S53–S59. doi: 10.1016/j.jscs.2012.08.005.

Evangelou, V. P., (1995). *Pyrite Oxidation and Its Control*. CRC Press.

Haryati, S., Rahmatullah, & Rizka, W. P., (2018). Torrefaction of durian peel and bagasse for bio-briquette as an alternative solid fuel. *IOP Conference Series: Materials Science and Engineering, 334*(1), 1–6. doi: 10.1088/1757-899x/334/1/012008.

Hettiarachchi, G. M., & Perziynki, G. M., (1999). *Effect of Phosphorus and Other Soil Amendments on Soil Lead, Cadmium, and Zinc Bioavailability*. Paper presented at the 5th International Conference on Biogeochemistry of Trace Elements, Vienna, Austria.

Hodson, M. E., Valsami-Jones, E., Cotter-Howells, J. D., William, E. D., Anthony, J. K., Iaian, T., & Alan, W., (2001). Effect of bone meal (calcium phosphate) amendments on metal release from contaminated soils — A leaching column study. *Environmental Pollution, 112*(2), 233–243. doi: 10.1016/s0269-7491(00)00116-0.

Jopony, M., & Felix, T., (2009). Acid mine drainage at mamut copper mine, Sabah, Malaysia. *Borneo Science 24*, 83–94.

Kumar, P., & Laxmi, J., (2013). Pollution caused by agricultural waste burning and possible alternate uses of crop stubble: A case study of Punjab. In: Sunil, N., Rao, K. S., Harald, K., Raju, K. V., & Ruediger, S., (eds.), *Knowledge Systems of Societies for Adaptation and Mitigation of Impacts of Climate Change* (pp. 367–385). Berlie, Hidelberg: Springer. doi: 10.1007/978-3-642-36143-2_22.

Lambert, M., Gary, P., Larry, E., & Jerry, S., (1997). Remediation of lead-, zinc- and cadmium-contaminated soils. In: Ronald, E. H., & Roy, M. H., (eds.), *Contaminated Land and its Reclamation* (pp. 91–102). Cambridge, UK: The Royal Society of Chemistry.

Lee, G., Jerry, M. B., & Gunter, F., (2002). Removal of trace metals by coprecipitation with Fe, Al and Mn from natural waters contaminated with acid mine drainage in the Ducktown mining District, Tennessee. *Applied Geochemistry, 17*(5), 569–581. doi: 10.1016/S0883-2927(01)00125-.

Lee, M. C., Seong, C. K., Ming, Y. C., Ming, M. P., Pui, M. C., & Kim, Y. T., (2018). Preparation and characterization of durian husk fiber-filled polylactic acid biocomposites. *MATEC Web of Conferences, 152*, 1–7. doi: 10.1051/matecconf/201815202007.

Loeppert, R. H., & Lloyed, R. H., (1984). Reactions of Fe^{2+} and Fe^{3+} with calcite. *Clays and Clay Minerals, 32*(3), 213–222. doi: 10.1346/CCMN.1984.0320309.

McQuarrie, D. A., & Peter, A. R., (1984). *General Chemistry*. W.H. Freeman and Company: New York.

Mohd, A., Khairun, A., Tan, C. Y., & Birrel, J., (1995). Technical note, design of the lohan tailing dam, mamut copper mining Sdn. Bhd. *Malaysia. Mineral Engineering, 8*(6), 705–712. doi:org/10.1016/0892-6875(95)00031-K.

Muhammad, S. N., Faradiella, M. K., Mohd, S. Md. Z., Normala, H., & Ferdaus, M. Y., (2015). Passive treatment of acid mine drainage using mixed substrates: Batch experiments. *Procedia Environmental Sciences, 30*, 157–161. doi: 10.1016/j.proenv.2015.10.028.

Obeng, A. K., Duangporn, P., & Siripong, P., (2018). Fermentable sugar production from the peels of two durian (*Durio zibethinus* Murr.) cultivars by phosphoric acid pretreatment. *Resources, 7*(4), 60. doi: 10.3390/resources7040060.

Ojha, N., Karn, R., Abbas, S., & Bhugra, S., (2021). Bioremediation of industrial wastewater: A review. *IIOP Conference Series: Earth and Environmental Science, 796*(1), 012012. doi: 10.1088/1755-1315/796/1/012012.

Park, H. J., Seong, W. J., Jae, K. Y., Boo, G. K., & Seung, M. L., (2007). Removal of heavy metals using waste eggshell. *Journal of Environmental Sciences, 19*(12), 1436–1441. doi: 10.1016/s1001-0742(07)60234-4.

Payus, C., Olga, D., & Moh, P. Y., (2014). Bone meal as alternative treatment for acidic and metal contaminated acid mine drainage water effluent: Lab scale. *American Journal of Environmental Sciences, 10*(1), 61–73. doi: 10.3844/ajessp.2014.61.73.

Perea-Moreno, Miguel-Angel, Manzano-Agugliaro, F., Hernandez-Escobedo, Q., & Alberto-Jesus Perea-Moreno, (2018). Peanut shell for energy: Properties and its potential to respect the environment. *Sustainability 10*(9), 3254. doi: 10.3390/su10093254.

Rajendran, & Michael, R. R., (1981). Environmental health in Malaysia. *Bulletin of the Atomic Scientists*, *37*(4), 30–35. doi: 10.1080/00963402.1981.11458846

Rey, C., (1998). Calcium phosphates for medical applications. In: Zahid, A., (ed.), *Calcium Phosphates in Biological and Industrial Systems* (pp. 217–245). New York: Springer.

Rios, C. A., Craig, D. W., & Clive, L. R., (2008). Removal of heavy metals from acid mine drainage (AMD) using coal fly ash, natural clinker, and synthetic zeolites. *Journal of Hazardous Materials, 156*(1–3), 23–35. doi: 10.1016/j.jhazmat.2007.11.123.

Robinson-Lora, M. A., & Rachel, A. B., (2009). The use of crab-shell chitin for biological denitrification: Batch and column tests. *Bioresource Technology, 100*(2), 534–541. doi: 10.1016/j.biortech.2008.06.052.

Shitu, A., Muhammad, A. I., Yoshida, H., & Siamjam, S. I., (2016). Production of *Phenolic Compounds from Durian Peel Waste Using Sub-Critical Water* (pp. 29–65). Paper presented in International Conference on Agricultural and Food Engineering.

Sneddon, I. R., Orueetxebarria, M., Mark, E. H., Paul, F. S., & Valsami-Jones, E., (2008). Field trial using bone meal amendments to remediate mine waste derived soil contaminated with zinc, lead, and cadmium. *Applied Geochemistry, 23*(8), 2414–2424. doi: 10.1016/j.apgeochem.2008.02.028.

Stella, H. Y. L., (2008). *Acid Mine Drainage at Mamut Copper Mine, Ranau, Sabah*. Msc diss., Universiti Malaysia Sabah.

Sturchio, N. C., Ronald, P. C., Likwan, C., Paul, F. L., Michael, J. B., Yonglin, Q., & Hoydoo, Y., et al., (1997). Lead adsorption at the calcite-water interface: Synchrotron x-ray standing wave and x-ray reflectivity studies. *Geochimica Et Cosmochimica Acta, 61*(2), 251–263. doi: 10.1016/s0016-7037(96)00326-2.

Surasen, C., (2002). *Removal of Cadmium in Synthetic Wastewater by Egg Shell Filter*. Ms. Eng. diss., Kasetsart University.

Thilagavathi, R., Sabarathinam, C., Prasanna, M. V., Chandrasekar, T., & Chelladurai, S., (2012). A study on groundwater geochemistry and water quality in layered aquifers system of Pondicherry region, Southeast India. *Applied Water Science*, *2*(4), 253–269. doi: 10.1007/s13201-012-0045-2.

Valsami-Jones, E., Kristín, V. R., Andrew, P., Dirk, B., Anthony, J. K., & Cressey, G., (1998). The dissolution of apatite in the presence of aqueous metal cations at Ph 2-7. *Chemical Geology, 151*(1–4), 215–233. doi: 10.1016/s0009-2541(98)00081-3.

Won-Kyo, J., Fereidoon, S., & Se-Kwon, K., (2007). 17 calcium from fish bone and other marine resources. In: Fereidoon, S., & Colin, B., (eds.), *Marine Nutraceuticals and Functional Foods* (pp. 419–420). USA: CRC Press.

Xu, N., Michael, F. H. Jr., Gordon, E. B. Jr., & George, A. P., (1996). Co (II) sorption at the calcite-water interface: I. X-ray photoelectron spectroscopic study. *Geochimica et Cosmochimica Acta, 60*(15), 2801–2815. doi: 10.1016/0016-7037(96)00133-0.

Xu, W., Quanlin, Z., Rufan, W., Zhenming, J., Zhenzhong, Z., Xuewen, G., & Zhengfang, Y., (2017). Optimization of organic pollutants removal from soil eluent by activated carbon derived from peanut shells using response surface methodology. *Vacuum, 141*, 307–315. doi: 10.1016/j.vacuum.2017.04.031.

Younas, F., Adnan, M., Zia, Ur. R. F., Xiukang, W., Sadia, Y., Mohy-Ud-Din, W., Muhammad, A. H., et al., (2021). Current and emerging adsorbent technologies for wastewater treatment: Trends, limitations, and environmental implications. *Water, 13*(2), 215. doi: 10.3390/w13020215.

Ziemkiewicz, P. F., Jeffrey, G. S., David, L. B., Sterner, P. L., & Lovett, R. J., (1997). Acid mine drainage treatment with armored limestone in open limestone channels. *Journal of Environmental Quality, 26*(4), 1017–1024. doi: 10.2134/jeq1997.00472425002600040013x.

CHAPTER 5

From Waste to Building Blocks: How to Transform a Major Environmental Problem into a Solution?

JOSÉ DE JESÚS VÁZQUEZ CHABOLLA[1] and ERIKA ACOSTA-CRUZ[2]

[1]*Advisor, The Office of the Attorney General for Environmental Protection of the Subnational Government of Quintana Roo, México; Member, Latin-America and the Caribbean Community of Practice on Managing for Development Results (CoPLAC-MfDR)*

[2]*Department of Biotechnology, School of Chemistry, Autonomous University of Coahuila, Saltillo, Coahuila, México*

ABSTRACT

Due to the current pandemic (COVID-19), there is an expectation of a return to normality in an environment of garbage generated day by day. This essay calls into question the concept of 'normality' in the context of waste management, which have become obsolete by not including determining factors, showing that the state of 'pre-pandemic normality,' to which it is aspired to return, presents much more worrying and tragic socio-environmental indicators. Evidence is presented to argue that the principle – sine qua non- of economic growth is a determining and irreplaceable engine of development, and even more so of the capitalist model of development, is metabolically unsustainable to demand for permanently incremental amounts of energy and goods of nature on a planet incapable of sustaining such a claim. Without even counting on the practice of transformation technologies that allow 100% of the recycling of materials without generating disturbances to water, soil, or air. Pandemics are not the only fatal consequences of the destruction

Waste Biorefineries: Future Energy, Green Products, and Waste Treatment.
Abu Zahrim Yaser, Siew-Xian Chin, Leonardo Sepúlveda Torre, & A. K. Haghi, (Eds.)

of nature, nor are they the most serious. They are symptoms of a model in crisis, which sustainable development aims to alleviate without calling for the establishment of other technologies as promised by transforming materials that have been used and now discarded, transforming them back into useful materials called "Vomer" that can be used in urban infrastructure or movable and immovable property with economic reward for the effort applied, achieving sustainability as a priority goal and solving the problem about lack of affordable housing. Likewise, it is proposed with industrial production, which has been modified due to the change in consumption habits, requiring new technologies and collection centers for its treatment and transformation of this waste into materials that are again useful for the population, thus increasing the reuse of materials that have been used, which will prevent the subtraction of more of this raw material. In this sense, the need to build a "new normal" is proposed, different from the unviable "pre-pandemic normality," which implies evaluating the fundamentals of the transformation model that is presented in this publication for analysis and incorporation into life in a sustainable community. It calls for an urgent response, diverse, and inclusive, from within the scientific community, with enough detachment to admit other visions and knowledge that have much to contribute from other ways of coexisting with nature.

5.1 Introduction

To address the problem represented by waste, it is necessary to work integrally and with a vision of circular economy, where the use of raw materials is a priority in all stages of the life cycle of a product.

In this sense, the hierarchy of waste management remains in force, where prevention and minimization put before the final disposition. It is necessary to correct the poor disposal of the country's waste, and to meet the needs faced by management, to develop policies and instruments to expand the coverage of services, as well as to increase the capacity of recovery of the economic system, to establish the separation of waste from the origin, to promote recycling and to establish an effective inspection and surveillance body, as regards the final provision; all of the above, under a sustainable business model and circular economy.

This document aims to be the vision that concretizes the formulation and implementation of public policies, as well as the execution of instruments, programs, and plans, to strengthen waste management in any environment that may arise.

Each entity has its own characteristics around the generation and management of its waste. Therefore, it must be considered that there is no single solution to the challenge posed by sustainable waste management. The problem of waste has no borders. For this reason, the best practices in Integral Waste Management are carried out with a territorial approach to articulate all the policies that concur in the matter, such as policies for the protection of soil, water, and emissions into the atmosphere, as well as those aimed at reducing the generation of waste, to valorize it and promote its integral management. With this territorial approach, associative models have been developed that make Integral Waste Management possible with the creation, adoption, and operation of models for sustainable waste management.

An alternative involves an articulated and interrelated set of normative, operational, financial, planning, administrative, social, educational, monitoring, supervision, and evaluation actions for the management of waste, from its generation to final disposal. For this, it is necessary to establish a regional sustainable model for waste management, sensitizing the inhabitants, institutions, commerce, and industry of the region about the importance of the management of their waste, dignify the urban waste collectors, strengthen the collection segregated by type of waste and material to prevent and minimize the impacts to health and the environment.

To achieve the above, it is necessary to change management to the current one, adopting a scheme that allows institutional strengthening of those in charge of waste to achieve an integral management of municipal solid waste. The proposal to develop decentralized associations aims to make the waste collection and management service transparent and efficient, through the establishment of decentralized and autonomous entities.

To this end, the guidelines for creating these bodies will be developed, which will have a Governing Board and Technical and Advisory Councils. This figure will make its management transparent and integrate the component of citizen participation in decision-making and sustainable waste management.

The Operating Bodies will have the responsibility of providing a professional, efficient, and orderly service. They will integrate urban waste collectors and the informal sector in terms of recycling into their work. They will have the mission of transforming the traditional operation into a circular economy scheme, with respect for the hierarchy of waste management.

The formation of the operating bodies will have the participation of cooperative societies to organize urban waste collectors (commonly known as "pepenadores" in Spanish) and in this way have formal jobs and the benefits that this entails such as social security, housing credits, and retirement funds. Through the formation of the operating bodies, certification mechanisms will be created for service providers in the different stages of the management of the waste: collection, transport, collection, storage, recycling, recovery, and final disposal.

The new normal is a challenge that invites us to review some notions, such as the notion of normality itself, and with certainty, it demands to characterize some notions, such as the pre-pandemic reality. Returning to the "usual" condition does not mean in any way returning to a "natural state." The fact that the pre-pandemic state has been assumed as a "norm or rule" is nothing more than accepting by default that this precondition is normal or even natural, just because it "conforms to certain norms established in advance." In this way, the return to pre-pandemic normality, marked by the norms of development and globalization, does not presuppose in any way to return to a "natural" condition, given that the economic, social, and environmental indicators offered by the sources or activities of multilateral organizations (to resort to conservative sources whose objectivity are not unanimously recognized but which enjoy a generalized acceptance), they move rapidly away from a supposed socio-environmental sustainability. Given that the pandemics themselves are one more symptom of a development model aimed at depleting the planet's capacities to provide conditions for life, it is an act of wisdom to take advantage of the profound reflection that this extraordinarily tragic and collective circumstance gives us the opportunity to promote a new normal that starts from the questioning of development as an aspiration. It is expected that the "return" will not be to an "inter-pandemic normal," whose cycles will become the new "normal" and that the transformation of waste will prevent viruses, bacteria, etc., from continuing to circulate in the populations.

5.2 TRANSFORMATION OF THE LAST WASTE IN THE RECYCLING CHAIN

The development of the world economy in the environmental aspect is closely related to the technologies that are developed for the solution of problems in the mission of fulfilling a better application for our natural resources, as well as achieving the creation of more and better materials product of recycling and efficient energy flows. The economic evaluation that demonstrates an effectiveness of zero waste technologies to determine the effect on the reduction of the economic cost of recycling at all stages but especially in the last stage where there is always a part of material that is despised and does not manage to be incorporated into the defined production chains, this waste as there is no technology that transforms them into new consumer materials, is that it makes the calculation of the damage caused by the entire industrial sector or commercial activity to the environment continue to be carried out, especially on the basis of the comparison of companies that use zero waste technologies and companies with traditional technologies. The trend towards the creation of low- and zero-waste territorial and industrial sectors is vital direction in the development of the post-pandemic economy of any nation that seeks the reasonable use of natural resources and the conservation of ecosystem balance.

It is important to mention the environmental issues generated by fast economic development around the globe, which are countless: the consumption of natural resources, particularly non-renewable ones, large amounts of industrial waste, environmental pollution with toxic elements, among others, are causing the great problems that are now being added looking for answers: reduction of the fertility of the earth; reduction of areas allocated to agricultural land; less productivity; the use of obsolete and unsafe equipment; the deterioration of people's working conditions; lack of rationalization of land management and conservation of water resources in their natural state. Therefore, we can mention that human activities have negative effects on environmental conditions, which is why many people expect on the large-scale introduction of new zero-waste technologies that must solve the problem of air, hydrosphere, and lithosphere pollution, always with a single objective that is to end the amount of garbage that is accumulating on the planet.

The possible positive environmental effect of the implementation of zero-waste technology is immense. It is mentioned in several investigations that up to 20 tons per capita of different raw materials are processed every year in average populations of up to 2,00,000 inhabitants, and only 5 to 10% is recycled to finished products, so we can infer that the rest of the materials are now waste deposited somewhere on the planet without any use, this being the unconsumed part of raw materials, resulting from the consumption of industrial outcome as they wear out or become outdated, so they also fall into the category of consumer waste. Therefore, practically the whole part of materials taken from nature is recovered by the planet, but already with new properties, such as waste, which lead to an abuse of the ecosystem balance and therefore causing damage.

Now, economic and ecosystem issues are closely related and only some of them are being solved, the others have been classified as those that cannot be solved, usually destined for landfills. As this waste is deposited in a single geographical point, it is necessary to redefine it in times of post-pandemic COVID-19-Coronavirus – as a single unit to be resolved environmentally, since they form a single economic potential of raw material. To give the traditional example known to humans, natural resources are for industrial companies their environment in which their natural circle is formed and is the one that will define their production capacity of installed plant and therefore determine the manufacture of their products which will depend on the amount of those natural resources that are obtained within defined parameters of acquisition to create economically marketable products, which will include in their costs the expense in the purchase and installation of treatment plants to eliminate pollution of water, air, and soil, which in sum will depend on their benefits to achieve.

In short, countries have their own list of ecosystem issues that are directly related to its economy. And it is in this way that they must be committed to the transformation of their waste at the local or regional level. First, you should always establish responsibility for the consequences of the burden of managing your waste, which is the responsibility of each company individually. In this way, it is that the active growth of the economy in each demarcation should not negatively affect the state of our ecosystem. This is how individual companies and collectively regulated by their authorities must control the economic and environmental situation by achieving balance and solving existing global problems. In short, the way in which the State-Business interaction to solve economic and environmental problems is the introduction of zero-waste inventions (in all their varied technologies) and achieve their multiplier effect. This opportunity shows the relationship between the innovative development of the environment-oriented economy and the introduction of zero-waste technologies.

Analysis of scientific publications shows that zero-waste technologies can be developed in four essential categories (Cehlár et al., 2013, 2019; Rybár, Beer, & Cehlár, 2016; Zhironkin et al., 2018):

- The development of technological systems that involve the reuse and reduction of non-drained water consumption based on existing water cleaning methods, although as a rule, secondary contamination is formed in the form of solids or saturated solutions.
- Development of technologies for the processing-transformation of production and consumption waste, which should not be expressed

as an ecosystem burden, but as a source of additional income. It should be pointed-out that during the operation of modern gas and water cleaning systems, solid waste is formed (which is a concentrated mixture of contaminants).

- Organization of fundamentally new processes to produce traditional types of goods, which allow to eliminate or reduce the processing stages or technological stages in which the main amount of waste is generated.
- Development and creation of territorial and industrial complexes with a closed structure of raw material and waste flows with a minimum of emissions.

The main goals for the implementation of zero-waste technologies originated from the following conditions:

- Most environmental pollution is the result of inadequate development of industrial demands.
- Waste of unused production is natural resources damages; the production and use of recycled materials (waste) with growth in demand.
- Natural materials could be considered as vital source for increasing labor productivity.
- The requirement for the justification of industrial technologies is the development of technical solutions for “closed” technologies (the circulation of materials).
- The global resource protection to solve the basic problems in the field of metabolism between man and nature must be carried out at the state level.

The discharge of toxic components from wastewater and waste gases had to be accomplished mostly for the conversion of these components into a harmless form and was rarely combined with their reuse (Tyulenev, Zhironkin, & Litvin, 2015). In many cases, companies tried to lower the concentration of toxic waste during its disposal in the biosphere (Tyulenev, Zhironkin, Kolotov, & Garina, 2016). Measures to lower industrial waste and heat waste in the manufacturing of goods, as well as the recycling of this waste were mainly implemented to save materials and energy and measures to protect the environment were not considered.

The constant increase in the use of natural resources, the increase in environmental pollution requires the implementation of a zero-waste technology strategy. The foundation of this technology means the fact that unutilized production waste is not both fully used natural resources and a source of ecosystem pollution. Lowering the amount of waste used in relation to the number of manufactured products will make it possible to produce more goods from the same number of raw materials and will be a great accomplishment of ecosystem protection.

The biosphere contributes the natural resources from which goods are produced in the industrial sector, and waste is generated. In many cases, after proper treatment, they can be used as secondary raw materials or as secondary energy sources (Kovalev, Gerike, Khoreshok, & Gerike, 2014). If it is not possible or economically feasible for technical or technological reasons, they must be introduced into the biosphere in such a way that they do not harm the natural environment.

The development of the activities of humanity since the beginning of the industrial revolution concludes in our times with the recognition of the pollution caused by man who in turn has come to understand after facing the pandemic by COVID-19 (coronavirus) that it is necessary to introduce manufacturing techniques that produce minimum emissions and by which the planet has the capacity of self-cleaning with what it will sufficiently prevent the phenomenon of irreversible ecosystem changes. Experts have proposed the definition of circular economy using zero waste technology, which should be theming source of materials for later use (Gilyazidinova, Rudkovskaya, & Santalova, 2018; Gianoncelli et al., 2013; Kononchuk, Alekseev, Zubkova, & Udovitsky, 2017).

Zero waste technology is the practical use of new methods alongside with knowledge to ensure the use of generated waste that has ended its useful life at first to produce new materials that replace the extraction of natural resources, use energy from green sources and therefore, the protection of the environment within human needs is a reality.

Zero waste technology means an optimal manufacturing model, which in most cases has not been fully realized, but with the development of technical advancement it is already possible. To have the first technology that makes it possible to use the waste that is useful in the productive chains of recycling and/or reuse bringing's closer and closer to the ideal proposed in this work. This technology shown in the following section is called "Solid Waste Transformation." More specifically, the zero-waste technological system should be understood because of the activity of which there are no discharges to the ecosystem. Zero waste production is a set of organizational and materials, equipment, technological processes alongside with technical measures, to ensure the feasible and integrated use of raw materials and lowering the negative effects of waste on the ecosystem.

Zero waste manufacturing can be accomplished by the full extent of possible disposal of waste produced in direct technological processes. Low-waste technology is a zero-waste intermediate stage and differs from it when it provides the finished product with non-fully recyclable waste.

The final accomplishment of zero waste production is due to the existence of several stages capable of processing waste of all kinds. The system becomes "zero waste" when in the umpteenth stage no amount of waste is released with the known negative impact on the environment.

The main recommendations on the development of new zero-waste technologies should include:

- Its development must include:
  - Maximizes the reuse of waste and its waste heat.
  - Minimize the use of semi-finished products, raw materials and working energy in the extraction and manufacture of large amount of industrial waste and waste heat, such as electricity or cooling water.
  - Only use technical devices with long service life and low weight, manufactured in accordance with the requirements of zero-waste technology.
  - Use devices of the optimum principle of action, for example, with a high degree of separation or with a high rate of heat and mass transfer, with minimal pressure loss and low heat loss.
- Main industrial processes:
  - Use only processes that consume the lowest energy expenditure; and
  - Applies highly efficient catalytic processes.
- Technological design:
  - Applies the principle of counterflow or circulation.
  - Avoids the principle of forward flow and mixing.
- Process parameters:
  - Selects the optimal reaction temperature.
  - Small driving forces.
  - Excludes extreme technological parameters, e.g., temperature and pressure.

- Goods produced:
  - Put in the design of the product a small specific mass, provide a long service life, as well as a minimum of waste and waste heat during use.
  - Ensures the suitability of the worn product as a secondary raw material (secondary energy source).
- Waste, waste heat:
  - Receive only waste in recyclable form.

For the reasonable management of an integrated system of collection, transport, neutralization, and disposal of waste and pollution locally or in an entire region, it is necessary to have current information on the location of waste along with its quantity, composition, and properties, possibilities of recycling or deletion. Information systems make it possible to define and establish relationships "waste with raw materials," "supplier to consumer." Management centers for the mutual exchange of waste for the purpose of its succeeding disposal.

At present, localization connections and combinations of different technological methods with municipal consumption areas are increasingly important for the use of waste and waste heat. Therefore, in many cases it is possible to use water first for domestic purposes, and then after cleaning, which requires fairly low costs, to use for manufacturing purposes.

For the organization of industrial production without waste and low waste, the cooperation of companies from different industries is of special importance. The most supportive opportunities for co-production are formed in the conditions of the territorial and productive complex, where a set of interrelated and interdependent facilities is planned, in proportional development of various sectors of the national economy. These objects are created for the joint solution of one or more particular economic issues They are known by the size of production and a clear specialization all over the country and its economic sectors. They are concentrated in a limited territory, necessarily compact, which has the set and the number of resources necessary to solve the corresponding tasks of development of zero waste production, including now the incorporation of a territorial environmental complex with solid waste transformation plants.

In addition, the territorial and productive environmental management complex will effectively use local and external resources (from the point of view of the local-national economy), provide environmental protection and will have a unified infrastructure for production and environmental savings.

The economic advantages with an adequate and optimal development of the territorial and productive industrial complex will allow in turn an advantageous and expeditious transport of waste over relatively short distances, which facilitates the solution of many problems related to the territorial distribution of companies.

5.3 SOLID WASTE PROCESSING PLANT

Process through plants all the solid waste that is generated in a community, which transforms the waste into a homogeneous mass, which ends in a molding process in which it takes the form that is desired, with the aim of definitively transforming the accumulation of waste called garbage and its consequences that are increasing the current Climate Change. The result of the plant is a harmless and non-dangerous product for humans called "Vómer."

The solid waste processing plant, which respects all the principles that determine zero waste technologies, has a minimum installed plant calculation industrial transformation capacity of 30 tons of waste per working time and a maximum capacity that depends on the generation of total waste in the place where it is installed, due to its conception from its planning for its installation that in its concept is linear by multiples of 30 to 50 tons, reaching the amount of production/transformation that is required, in which they can assemble the necessary lines to transform the amount of total waste generated by the target population. For the installation of the plants, 210 and 206 m^2 of land are necessary. This technology is based on a specific process to identify garbage as its raw material (garbage is the solid waste generated in communities, industry, commerce, and housing) and the adhesion of supplies (additive V208 MC, V209 CS, V210 BS and V211 PZ; which are designated with keys because they are basic to the private rights of the invention). Raw materials and supplies converge in four phases that are intimately related and that; to carry out in a good term, they use various unit operations (Figure 5.1). These phases are:

- Phase A or reception;
- Phase B or neutralization;
- Phase C or decontamination; and
- Phase D the ecotransformation.

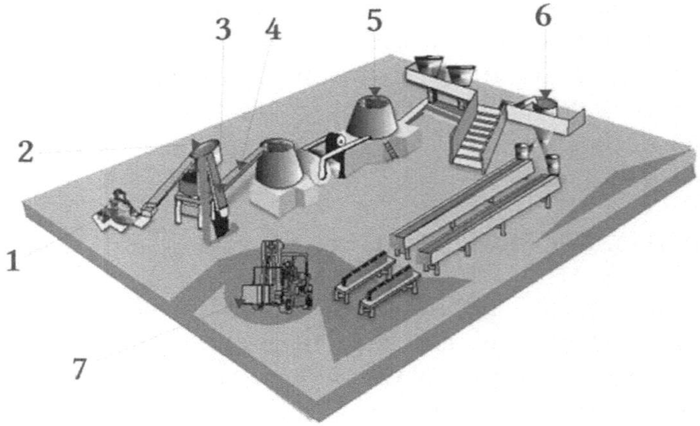

FIGURE 5.1 Project outline. (1) residues reception; (2) decontamination and classification; (3) grinding; (4) drying and storage; (5) mixing of raw material and components; (6) product elaboration; and (7) storage and commercialization.

5.3.1 COLLECTION

Starting with the waste collection vehicles for the different needs in the industries, shops, administrative centers, or homes to the facilities of the plant, receiving the waste in its natural form or in appropriate containers which are deposited in a receiving tank. In this reception tank, neither the different containers are broken to release the waste with the mechanical help of the crusher.

5.3.2 NEUTRALIZATION

The neutralization phase begins with the step of placing the waste on a conveyor belt that ends in a stainless-steel hopper. This hopper allows the waste to reach the vertical chain mill, where it is cut. Subsequently, the waste is transported, by means of another conveyor belt, to a trammel where the products V210 BS and V211 PZ are added through dosing systems. The addition of these products neutralizes the physicochemical and microbiological processes, which would otherwise be carried out and which are the cause of the biodegradation processes that lead to the generation of unpleasant odors and liquids with high concentration of organic matter very polluting and toxic to the environment.

5.3.3 DECONTAMINATION

In the decontamination phase, the waste is transferred through sorting conveyor belts (in which the metal parts are removed because of magnetization) to a hopper that ends in a hammer mill that commissions the waste to be crushed. After being crushed, the solid waste passes into the sieving section. In this section we find two sieves (one that selects particles of one inch and another of half inch in diameter) in a vibratory system to accelerate the process. Solid waste that meets the required particle size is transported to the drying furnace, where 88% of the moisture is removed from the solid waste at a temperature of 200° to 250C. This equipment has a system that condenses the water vapor and takes it to a storage tank to use when conditioning the dough and giving it the appropriate level of humidity to be able to pass to the formation press.

5.3.4 ECOTRANSFORMATION

In this phase, the residues are placed in the mixers, where the additives V208 MC and V209 CS are added through a dispensing system, which attends each additive according to the established formulation. This solution provides resistance to the mass. The raw material of this phase is the product of phase C, which is a compact mass that results from the mixture of crushed solid waste (with a humidity of 27%) and the solution with additives. After this process, the compact dough by means of a gate drive, passes to an injection hopper that ends in a training chamber where the material is molded into a bar (Figure 5.2).

FIGURE 5.2 Ecotransformation. The product of Phase 3, a compact mass that results from the mixture of crushed solid waste (with a humidity of 27%) and the solution with additives, passes to an injection hopper that ends in a training chamber where the material is molded into a bar.

After passing through the training chamber or press, the waste enters the furnace section, where it is subjected to temperatures of 120°C and subsequently the compressed dough bars of 25 centimeters per side come out. The bar made comes out of the press as a continuous strip to a roller conveyor, where the cutting is made by means of water pressure filament and so the bar can form blocks (called Vómer) of the required measurement. Once they find all the unit operations of the plant active (a process that lasts about an hour, since the first residue enters the system and until the product begins to leave the press), the production of this is seven meters of compressed mass bar per hour. After the quality supervision of the finished product, the material passes to the warehouse, where it remains for three days before being marketed (Figure 5.3).

FIGURE 5.3 Product ready for distribution.

This shows total compliance as a complete zero waste technology with the final product of the objective achieved using waste as a raw material of this innovative industry of recycling all materials without distinction or classification of waste to conform useful products in the construction industry that have not generated pollution to the air, soil, or water. The process includes the treatment and incorporation of leachate formed by the washing that occurs with the action of the water element on the waste, so it can be affirmed that this process fulfills the objective of transforming 100% of the waste generated by a locality in any of its activities, whether industrial, commercial, administrative, or housing. With this scenario, we could affirm that it is possible that any activity can develop without harming the environment that surrounds them, if they have this process indicated for the transformation of the waste that they produce, complying with the agreements established by the planet in the face of Climate Change, fully complying with the principle of Zero Waste Technologies and Solutions in Developing Countries.

5.4 FIRST PLANT CONSTRUCTION

The construction of the first plant was carried out in the town of Conkal, Yucatan, Mexico, which was certified in its procedure to demonstrate that they do not produce emanations neither to the air, nor to the soil nor to the water, as the demonstration that the products produced are harmless for the human being and for the environment. A project authorized and financed by CONACYT Mexico was carried out, Key Registration Project 129364: "Development and implementation of clean technology and demonstration of the safety of the products generated through the start-up of a pilot plant and experimental prototype for the treatment and transformation of municipal solid waste," confidential by the company Vantech Company, S.A. de C.V., in the Announcement 2009–2007 of the Mixed Fund for the Promotion of Scientific and Technological Research CONACYT—Government of the State of Yucatan, concluding with the final certification of the Technological Institute of Mexico on June 23, 2014, proceeding to the construction of the plant with a capacity of 100 tons per day of production on February 11, 2016.

The product obtained from the solid waste processing plant, called a Vómer, has characteristics like some materials used for the construction of urban furniture. These vómers can be converted into block form, roof conditioning and construction vault, and can have different uses for sticks, bards, truck bus stops, cattle feeders, drinkers, roofs, sidewalk garrison, structures for water collection and storage, office division, among others. The vomer can also present variants in its composition, depending on which one can be classified into common vómer (Vc), product of all the urban solid waste that it receives and plans to use it in the construction of rural houses, animal facilities, wineries, among others; elite vómer (Go), a mixture of Vc, to which 20% of patent products (OP) are added, is presented to improve the finish and obtain less weight than Vc, and will be used for the same uses as this; executive vómer (Vex), mix of Vc to which it adds 40% of (OP) and for its characteristics of finish and weight will be used for the conservation of energy in industrial plant (Figure 5.4).

FIGURE 5.4 House constructed with vómers.

The vómer can be exposed to the open air without liquids or gases coming out of it and, unlike untreated solid waste, does not promote the development of cockroaches, rats, and birds of prey that transmit diseases to nearby populations.

Over time, the idea that it is possible to develop economic activities within the concept of sustainability has gained more traction. These ideas fit the need for "development that meets the needs of the present without jeopardizing the ability of future generations to meet their own needs." It will be indispensable in this task, to have an adequate interaction between the conservation of the environment, the application of clean technologies such as this one that this work presents for the full fulfillment of the existing rule and the generation of income to be distributed jointly and severally. For this, we continue to build the path that allows us to achieve a life without waste, not to hide garbage. It is necessary that human beings have a vision in which the sustainable use of natural resources becomes a fundamental part of the economic and social development of people, incorporating innovative processes for the transformation of the waste generated.

A process of transformation of waste that seeks to contribute to economic development since it is possible to convert the garbage that was extracted from natural resources and with this process it is possible to produce materials necessary for the future development of the populations and that if they are in harmony with the Environment and others will face Climate Change because the products overcome the resistance and temporality in the use of the existing, with this adoption of technological tools to manage the waste of the populations and all their economic and social activities. It will be due to the inclusion of the current technological process of waste transformation that Solid Waste Transformation Plants can be built, ending the innumerable problems generated by the presence of waste. This desire, which is increasingly recognized by governments and social sectors, must lead countries to develop policies and laws that promote economic development framed in processes of environmental conservation and generation of social welfare. Making possible a sustainable, sustainable, and conservation development of natural resources, was seen as a cross-cutting issue of development, and that not only touches the environment by itself, but is strictly related to issues such as agriculture, health, housing, and education; so that waste is not an isolated subject of forward-looking political decisions in the face of Climate Change.

5.5 CONCLUSION

We can conclude that the Implement the waste management centers through transformation technologies must include for transversal principles such as:

- Effective communication.
- Promote effective proactive transparency of all waste generated.
- Design and develop a communication and dissemination strategy that supports the implementation of the dissemination of this proposal and is transparently informed about the actions to be implemented

by each center that is established. In this framework, it would seek to sensitize the population about the importance of the integral management of solid waste and responsible consumption, as well as the potential benefits of the vomer product in the construction of housing or urban infrastructure.

- Effective education for sustainability.
- Integrate the circular economy in educational processes. Considering the pyramid of the waste management hierarchy, prevention, and minimization are the first phases. In this sense, citizen awareness is vital for the development of prevention and minimization activities.

Environmental education proposes:

- Strengthen formal and non-formal environmental education processes on responsible consumption and integrated management of solid waste.
- Promote the updating of teachers, educators, and environmental promoters in both subjects.
- Promote environmental management processes in educational institutions (green schools) and public and private institutions (green offices).
- Guide the informal sector on the integral management of solid waste.
- Promote the environmental training of public servants linked to the integral management of solid waste and responsible consumption, its transformation for the production of vomers and its use in the construction of social housing.
- Raise awareness among stakeholders about their co-responsibility in the production and consumption of goods and services.

To achieve waste management, the participation and co-responsibility of all actors is required. The incorporation of merchants and all those activities that generate waste in the collection, transport, and valorization is an efficient way to generate jobs with small investment. The large industry could focus its investments on transformation technology to produce vomers, for self-consumption or incorporation into the social housing construction network, as well as integrate entrepreneurs, merchants, and industrialists who make up its value chain with a circular approach.

Companies dedicated to the extraction of raw materials and the manufacture of parts or products will also be involved in obtaining the Circular Economy seal, under which companies that reflect the successful circular economy model that in turn integrates their value chain will be recognized. In this sense, alliances must be created between manufacturers, marketers, and recyclers to track their waste in the life cycle of the products. The extension of responsibility to the producer or importer of the good is an integral part of this effort. As there is no need for a separation of waste now, it generates greater comfort in the management from its collection to achieve its management. In the same way, the operating bodies will correspond to this action by society, by carrying out an effective collection of all their waste.

5.5.1 CONCURRENCE

Articulation of the government and the active participation of the private sector and civil society. Responsibility for waste management is shared by all stakeholders. In this sense, the articulation between government, civil society, and the private sector is required to give attention to the management of the different sources of waste that are of national and global interest.

5.5.2 CITIZEN PARTICIPATION

Strengthen the participation of citizens in waste management without the need now to be separated, which will facilitate cooperation in the management of their waste. The society will be aware of the management of waste in the localities. In the same way, it will be represented in decision-making through the mechanisms that are designed for its inclusion. To this end, the appropriate participation bodies for implementation will be formed.

The training of environmental promoters for the integral management of solid waste and responsible consumption will be favored.

5.5.3 MONITORING REPORT AND VERIFICATION (MRV)

Strengthen the monitoring of the performance of sustainable waste management. Any initiative, program, project, and activity must have a scheme that allows you to monitor achievements and results. Three main components will be established for measurement, reporting, and verification for waste management:

i. Information system for the analysis of waste generation, service coverage, material bank, energy use, energy generation and transformation of 100% of the waste collected without pollution generation.

ii. Estimation of emissions avoided by the reduction of waste in open dumps and landfills.
iii. Estimation of economic, environmental, and human health benefits obtained.

There is no single position on the meaning and implications of transformation technology and industrial waste collection centers, the context in which they are framed has given rise to conflicting positions that justify and question the way in which an economy based on the intensive use of biomass could be operating, held by developing countries with the technology and scientific knowledge of developed countries.

The position of the OECD and the European Union indicate that within the bioeconomy, the development of innovations that meet environmental sustainability criteria is sought to move from an economy based on fossil resources to production methods based on the use of biomass and then control with the same mission its waste generated so that it is transformed from scientific knowledge and technological, with the expectation that this will solve multiple current and future problems such as the degradation of ecosystems, poverty, and lack of public health.

In opposition, there is the uncertainty and the skepticism of those who will be the real beneficiaries of the transformation technology, when considering the weakness of the institutional systems of the countries that hold the natural wealth, compared to the main actors of the world markets.

Latin America faces this institutional weakness to innovate in terms of transformation technology even though it has a great wealth of natural resources to conserve and sufficient human capital. However, countries such as Argentina and Brazil are currently leading initiatives of a national policy that promotes the bioeconomy in specific areas and with it, they are generating synergies to promote technological development in the management of their generated waste.

In the case of Mexico, the experiences that academia and industry have developed innovations that meet the characteristics of the bioeconomy can be encouraging, in the sense that an initiative is growing to align with the new demands of the market, which are oriented towards the paths of sustainability and "greening." However, the constant is that they are ideas and developments of great disinvestment, but that they are instrumented at extremely limited levels which prevents the wastes from taking on an importance to generate a presence in the markets or an environment of scientific research and development.

In the rest of the countries of the region, there is still a strong lag in the development of the technology of the transformation of residues, given the lack of trained human resources, infrastructure, investments, and a scientific policy that encourages.

It is necessary to document experiences and analyze the institutional context to identify the potentialities and possibilities of generating scientific, technological, social, political, and economic conditions that allow an adequate and convenient transition of the countries of Latin America towards the technology of the transformation into industrial centers of handling and production of useful materials from garbage for the construction of affordable social housing as required in post-pandemic times.

KEYWORDS

- **blocks**
- **construction**
- **housing**
- **monitoring report and verification**
- **pollution**
- **recycling chain**
- **solid waste processing plant**
- **transformation**
- **waste**

REFERENCES

Cehlár, M., Janočko, J., Šimková, Z., Pavlik, T., Tyulenev, M., Zhironkin, S., & Gasanov, M., (2019). *Resources, 8*(1), 21.

Cehlár, M., Rybár, R., Pinka, J., Haxhiu, L., & Beer, M., (2013). *Arch. Min. Sc., 58*(2), 557–568.

Gianoncelli, A., Zacco, A., Struis, R., Borgese, L., Depero, L. E., & Bontempi, E., (2013). *Env. Chem. Sust. World, 4*, 103.

Gilyazidinova, N., Rudkovskaya, N., & Santalova, T., (2018). *E3S Web Conf., 41*, 02017.

Kononchuk, O., Alekseev, A., Zubkova, O., & Udovitsky, V., (2017). *E3S Web Conf., 21*, 02003.

Kovalev, V., Gerike, B., Khoreshok, A., & Gerike, P., (2014). *Taishan Academic Forum – Project on Mine Disaster Prevention and Control, 184.*
Rybár, R., Beer, M., & Cehlár, M., (2016). *Measurement, 88*, 153–164.
Tyulenev, M., Zhironkin, S., & Litvin, O., (2015). *Pollution Research, 34*, 825.
Tyulenev, M., Zhironkin, S., Kolotov, K., & Garina, E., (2016). *Poll. Res., 35*, 221.
Zhironkin, S., Khloptsov, D., Skrylnikova, N., Petinenko, I., & Zhironkina, O., (2018). *E3S Web Conf., 41*, 04010.

CHAPTER 6

Biomass to Biobutanol: Current Trends and Challenges

MUHD. ARSHAD AMIN,[1] HAFIZA SHUKOR,[2] NOOR FAZLIANI SHOPARWE,[3] MUAZ MOHD. ZAINI MAKHTAR,[4] and RAFIDAH JALIL[5]

[1]*Faculty of Chemical Engineering Technology, University Malaysia Perlis, Arau, Perlis, Malaysia*

[2]*Center of Excellence for Biomass Utilization, Faculty of Chemical Engineering Technology, University Malaysia Perlis, Arau, Perlis, Malaysia*

[3]*Faculty of Bioengineering and Technology, University Malaysia Kelantan, Jeli Campus, Jeli, Kelantan, Malaysia*

[4]*Bioprocess Technology Division, School of Industrial Technology, University Sains Malaysia, Gelugor, Pulau Pinang, Malaysia*

[5]*Forest Research Institute Malaysia (FRIM), Kepong, Selangor, Malaysia*

ABSTRACT

The term "biofuel" refers to a liquid or gas fuel obtained mostly from biomass for use in transportation. Biobutanol is a potential replacement biofuel for fossil-based liquid fuels as they become depleted. Biobutanol is a transportation fuel that may be simply combined with either gas or petrol at any ratio. Clostridia are the most prevalent fermentative organisms used in biobutanol production. It might well be recognized for its ability to use the acetone-butanol-ethanol (ABE) fermentation route to convert various types of renewable biomass to biobutanol. Various aspects of biobutanol fermentation, including butanol toxicity and product titer, have also been addressed. The

Waste Biorefineries: Future Energy, Green Products, and Waste Treatment.
Abu Zahrim Yaser, Siew-Xian Chin, Leonardo Sepúlveda Torre, & A. K. Haghi, (Eds.)

recent advancement in lignocellulosic biomass treatment technology, which is significantly greener and safer for the environment, has been clearly articulated. This chapter also covers several metabolic engineering and simple engineering laboratory approaches such as Adaptive Laboratory Evolution (ALE) for strain improvement to overcome butanol toxicity. These biobutanol difficulties could be solved to improve microorganism resistance to high solvent concentrations and hence increase biobutanol output. The final section of this chapter will discuss the future of biobutanol production as a new sustainable and renewable future transportation fuel. Overall, this chapter will provide a better grasp of current trends and issues in biomass-based biobutanol production.

6.1 INTRODUCTION

Limited supply, fast depletion and increasing price of fossil fuels resources, and environmental concerns on greenhouse gas (GHG) emissions due to over consumption of fossil fuels have increased the demand for sustainable and green energy produced from biomass (Raganati et al., 2014; Cao et al., 2016; Yangling et al., 2022). An evolution from fossil fuels to green energy could reduce the dependency on fossil fuel resources and overcome global warming issues related to GHG emissions (Luo et al., 2021).

Biobutanol is an advanced biofuels and important source of biochemicals that has great attention due to its excellent properties such as high energy density, more hydrophobic property, compatible with existing pipeline infrastructures, low volatility and corrosiveness (Cao et al., 2016; Guo et al., 2022; Yangling et al., 2022). The Environmental Energy Company (US) has declared that biobutanol can be used as a total replacement for gasoline without any modification on the existing motor engine (Yangling et al., 2022). Presently, butanol is produced mainly via synthesis of petrochemical known as oxo process) where propylene is hydroformylated to butyraldehyde, and then hydrogenated to produce n-butanol (Xue et al., 2017).

Biobutanol has been produced through acetone-butanol-ethanol (ABE) fermentation process from sugars or starch as substrates by using bacteria from *Clostridium* spp. (Dong et al., 2011; Rafidah et al., 2020; Guo et al., 2022). ABE fermentation is a securing bioprocess technology to produce biofuels and biochemicals for sustainable development (Guandong et al., 2022). It has been characterized as heterotrophic since its produce's multiple types of by-products (Sukumaran et al., 2011; Ibrahim et al., 2015). During ABE fermentation process, sugars or starch substrates has been used by bacterial strains to raise up its rapid growth and produce acetic acid, butyric acid and other organic acids (Figure 6.1). The biobutanol production involves two main stages which were acidogenesis and solventogenesis (Ibrahim et al., 2015; George et al., 2016). During acid production known as acidogenesis (first phase of ABE fermentation), organic acid such as acetic acid and butyric acid promptly accumulate. This phase occurs at pH values more than 5. Accumulation of organic acid the culture pH tends to decrease due to the metabolic switch of *C. acetobutylicum* from acidogenesis to solventogenesis (Ndaba et al., 2015; Guo et al., 2022).

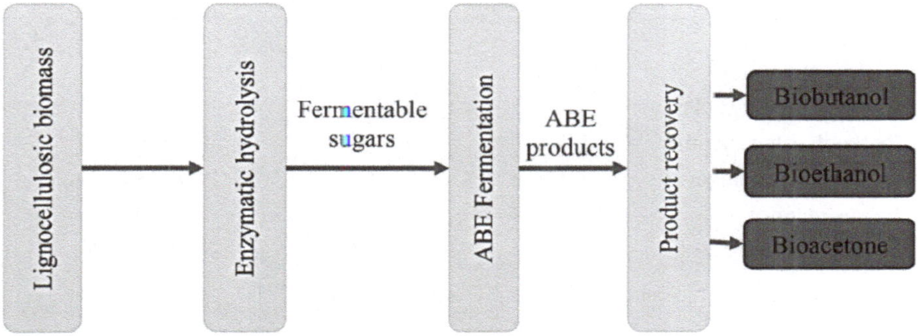

FIGURE 6.1 ABE fermentation process.

During solventogenesis (second phase of ABE fermentation), organic acids utilized to produce solvent and this solvent production increases when butyric acid accumulates and pH decreases. Organic acids are converting into main products of ABE fermentation which were acetone, butanol, ethanol, acetic acid and butyric acid as liquid products and hydrogen and carbon dioxide (CO_2) as gaseous products (Ndaba et al., 2015, 2017; Guo et al., 2022). Recovery of biobutanol can be done using several methods includes distillation, gas stripping, liquid-liquid extraction, adsorption, supercritical fluid extraction and pervaporation (Nanda et al., 2017).

Like bioethanol, global demand for biobutanol also increased and the market size estimated to achieve USD 4.0 billion by 2025 (Yanglin et al., 2022). Luo et al. (2021) also reported that the global demand for 1-butanol has achieved USD 6.4 billion in 2019 and targeted to increase to USD 7.6 billion in 2024.

6.2 CHARACTERISTICS AND APPLICATIONS OF BIOBUTANOL

As compared to bioethanol, biobutanol has several advantages to be used as next generation biofuel, such as higher energy content, can be blending with gasoline and being used directly in motorcar engines without modification, lower volatility, less corrosive, higher viscosity than other alcohols and lower water absorption (Dong et al., 2011; Ibrahim et al., 2015; Ndaba et al., 2015). The properties of biobutanol in comparison with bioethanol and gasoline are shown in Table 6.1. Biobutanol also has higher boiling point causes it takes longer time to be burned in motor engine (Ndaba et al., 2015).

TABLE 6.1 Comparison of Biobutanol Properties with Bioethanol and Gasoline

Properties	Bioethanol	Biobutanol	Gasoline
Molecular weight	46	74	111
Boiling point (°C)	78	118	25–215
Energy density (MJ/kg)	27	33	32
Research octane number	129	96	91–99
Motor octane number	102	78	81–89
Cetane number	5–8	12	0–10
Density (g/mL) at 20°C	0.79	0.81	0.72–0.78

Source: Adapted from Ndaba et al. (2015); Obergruber et al. (2021).

Biobutanol comprises of 25% more energy than bioethanol. Biobutanol is not as hygroscopic as bioethanol. Due to that it is less corrosive and versatile since it can be supplied using current pipelines while bioethanol needs railway transport. Other than that, it is safer due to lower vapor pressure and its volatility is approximately 7.5 times lesser than bioethanol and 32 times lesser than gasoline (Pugazhendhi et al., 2019; Obergruber et al., 2021).

Butanol is an important solvent, chemical, chemical intermediate and extractant for cosmetics and pharmaceutical industries. It can be used directly as biofuels as well as for the production of butyl acrylate, methacrylate esters, diluent for brake fluid formulations and as solvent for the production of hormones, vitamins, and antibiotics (Dong et al., 2011; Ndaba et al., 2015). Other than that, biobutanol also suitable to be used as fuel additives since it has similar properties to gasoline (Luo et al., 2021).

6.3 BIOBUTANOL PRODUCTION VIA ACETONE-BUTANOL-ETHANOL (ABE) FERMENTATION

Biobutanol is frequently produced from food crop biomass. However, non-food crop biomass has higher demand to ensure it does not compete with food crop and land use (Ndaba et al., 2015). Clostridia are among the most important industrial production strains for biobutanol. It is an anaerobes and spore-forming bacteria with complex life cycle (Guo et al., 2022). All clostridial bacteria strains can be used for ABE fermentation to produce acetone, butanol, and ethanol including *Clostridium acetobutylicum*, *C. beijerinckii*, *C. saccharobutylicum*, *C. saccharoperbutylacetonicum* and *C. aurantibutyricum* (Dong et al., 2011; Raganati, 2014).

6.3.1 TYPE OF FEEDSTOCKS

Biobutanol can be produced from various types of feedstocks using various strains and different technology approach. Selection of feedstocks for the production of biobutanol is very important to produce high yield and high quality of products. Currently, four generation of feedstock has been identified to produce biobutanol. Each type of feedstocks has its own benefits and limitations (Guo et al., 2022).

6.3.1.1 FIRST-GENERATION FEEDSTOCK

First generation biobutanol requires simple process mostly by fermentation of hexose (C6) sugars. Source of first-generation feedstocks came from starch and sucrose feedstocks such as sugarcane, corn, maize, rice, wheat, and cassava (Ndaba et al., 2015; Cao et al., 2016; Guo et al., 2022). Summary of biobutanol production that can be produce from first-generation feedstocks were shown in Table 6.2.

TABLE 6.2 Biobutanol Production from First Generation Feedstocks

Feedstock	Bacteria strain	Biobutanol yield (g/L)	References
Cassava starch	*Clostridium beijerincki*	6.66	Li et al., (2013)
Oil palm sap	*Clostridium acetobutylicum DSM 1731*	14.4	Komonkiat and Cheirsilp (2013)
Sago starch	*Clostridium saccharoperbutylacetonicum N1-4 (ATCC 13564)*	9.83	Al-Shorgani et al., (2012)
Combination of glucose, starch and food waste	*Clostridium spp. HN4*	5.23	Qin et al., (2018)
Sweet corn	*Clostridium saccharoperbutylacetonicum N1-4*	8.76	Darmayanti et al., (2020)

The benefit of using this type of feedstocks is its producing high yield of biobutanol and its fermentable substrates is sufficient to produce biobutanol with simple pretreatment processes (Guo et al., 2022). Darmayanti et al. (2020) has studied the potential of starch sources (potato, rice, sweet corn and sweet potato) to produce biobutanol resulting in biobutanol yield ranging from 7.58 to 8.76 g/L. The limitations and problems of using this feedstock are competes with food supply, give negative impact on food security, increased food prices, this type of feedstocks not sustainable and not cost effective (Ndaba et al., 2015; Cao et al., 2016; Guo et al., 2022). According to Luo et al. (2021), the cost of feedstock for biobutanol production from edible corn occupies 60 to 70% from the total production cost.

6.3.1.2 SECOND GENERATION FEEDSTOCK

Lignocellulosic biomass is the source for second generation feedstocks. It is very cost effective since it can be obtained abundantly, has huge carbon resources and no competition with food crop (Guo et al., 2022). An abundant and cheaper lignocellulosic biomass like agricultural and forestry residues are the potential feedstock for biobutanol production and efficient bioconversion is important to ensure economic success of the commercial production (Ndaba et al., 2015). Biobutanol production from various type of second-generation feedstocks were shown in Table 6.3.

TABLE 6.3 Biobutanol Production from Second Generation Feedstocks

Feedstock	Bacteria strain	Biobutanol yield (g/L)	References
Sweet potatoes residues	*Clostridium spp.*	7.96	Yangling et al., (2022)
Oil palm trunk fibre	*C.beijerincki* TISTR 1461	10.0	Komonkiat and Cheirsilp, (2013)
Rice straw	*C.sporogenes* BE01	5.52	Gottomukalla et al., (2013)
Palm kernel cake	*C.acetobutylicum YM1*	3.55	Al-Tabib et al., (2017)
Acacia mangium	*C.acetobutylicum YM1*	4.20	Rafidah et al., (2020)
Rice waste	*Clostridium spp. BOH3*	17.7	Guandong et al., (2022)
Oil Palm Frond Juice	*C. acetobutylicum* ATCC 824	9.24	Syazana et al., (2016)
Rice straw	*C.acetobutylicum NCIM233*	13.5	Ranjan et al., (2013)
Wheat straw	*C.acetobutylicum ATCC824*	12.64	Qi et al., (2019)
Switchgrass	*C.saccharobutylicum DSM13864*	13	Gao et al., (2014)

However, biobutanol production from this type of feedstock has several limitations such as it requires complex and efficient pre-treatment process due to recalcitrance to degrade and difficult to achieves sufficient fermentable sugars due to unique chemical compositions (Cao et al., 2016; Guo et al., 2022). Besides that, cultivation land and storing of feedstock after harvesting also a challenge when using this type of feedstocks (Ndaba et al., 2015).

6.3.1.3 THIRD GENERATION FEEDSTOCK

Algal biomass including both macroalgae and microalgae is third generation feedstock for biobutanol production (Table 6.4) an advanced liquid biofuel having potential to replace petroleum-based fuels (Konjan et al., 2021). Its advantages as a feedstock are no competition with cultivation land and food supply as well as its produce fewer or no lignin and less or no inhibitors during ABE fermentation (Guo et al., 2022). Algal biomass is used as feedstock for biodiesel production since it has high oil content up to 50%. Waste of algal biomass generated after oil extraction can be used as feedstock for biobutanol production (Ndaba et al., 2022).

TABLE 6.4 Biobutanol Production from Third Generation Feedstocks

Feedstock	Bacteria strain	Biobutanol yield (g/L)	References
Dry algae	*Clostridium saccharoperbutylacetonicum N1-4*	3.74	Castro et al., (2015)
Microalgae biodiesel residues	*Clostridium acetobutylicum*	3.86	Cheng et al., (2015)
Chlorella sp.	*C. acetobutylicum*	6.23	Onay, (2018)
Jamaica Bay Macro Algae	*Clostridium beijerinckii* and *C. saccharoperbutylacetonicum*	4.00	Potts et al., (2012)

The limitations to use this type of feedstock are the cost of downstream processes is high due to low production of algal biomass and indirectly reduce the productivity of biobutanol. So that, it is very difficult to get sufficient fermentable sugars for biobutanol production (Guo et al., 2022).

6.3.1.4 FOURTH GENERATION FEEDSTOCK

New generation feedstock for biobutanol production is from syngas that has been produced via gasification process. The benefits of using syngas are it has high ability to capture carbon dioxide (CO_2) with no complex pretreatment required. It can directly utilize *Clostridium* spp. with high production rate of biobutanol (Guo et al., 2022). However, this technology is still new and there is unknown parameter that limit the production of biobutanol as well as poor mass transfer from gas to liquid (Guo et al., 2022).

6.4 CURRENT CHALLENGES ASSOCIATED WITH BIOBUTANOL PRODUCTION

Numerous challenges have been identified associated with biobutanol production from biomass such as low yield, contamination, and biobutanol toxicity (Peabody & Kao, 2016). Instead of biobutanol toxicity, other factors inhibiting ABE fermentation include type of substrate, nutrient concentration, phage contamination, oxygen diffusion, macromolecular accumulation, low water activity, presence of dead cells and acidogenic fermentation (Jones et al., 2000; Nanda et al., 2017; Los, 2020). Due to that, extensive research has been done to overcome all of these challenges.

6.4.1 BIOBUTANOL TOXICITY

Biobutanol toxicity is a limiting factor of ABE fermentation to obtain final yield of biobutanol which is more severe as compared to acetone and ethanol (Nanda et al., 2017). The toxicity of biobutanol to microbial cellular systems is very complex. Form of toxicity includes inhibition and disruption of cellular membrane structure. Biobutanol produced has been detected to inhibit the transport of nutrients such as glucose into the *Clostridium* spp. cell and interrupts membrane stability as well as inhibiting ATPase function (Peabody & Kao, 2016).

Furthermore, biobutanol toxicity is also related to its hydrophobic properties and other long-chain aliphatic alcohols that disrupt the phospholipid component of bacterial cell membranes. High level of butanol would encourage an adverse modification of the cell membrane structure of *C. acetobutylicum* resulting in membrane fluidity. Exposure of 1% of biobutanol rise the membrane fluidity up to 30% that effect the final product yield (Nanda et al., 2017). Other effects of biobutanol toxicity on the bacterial cells are disruption of cell membranes and several physiochemical features including glucose uptake, solute transport, membrane permeability, intracellular ATP level, chemiosmosis, and activity of intrinsic membrane proteins (Kumar & Gayen, 2011).

During pre-treatment of lignocellulosic biomass, inhibitors compound has been generated. The type of inhibitors and its amount depending on the feedstock and pre-treatment processes. These inhibitors can be classified into weak organic acid, furan derivatives and phenolics that inhibit the biobutanol productivity (Luo et al., 2021). Its toxicity ranked formic acid > phenols > furfurals with same dosage during biobutanol production (Guo et al., 2022). *C. acetobutylicum* is very sensitive to formic acid and 0.1 M amount of this acid could induce an acid crash that completely inhibits cell growth and biobutanol production (Guo et al., 2022). Gu et al. (2015) also reported that lignin-derived phenolic including phenolic acids and phenolic aldehydes are amongst the fermentable inhibitors that is very toxic to *Clostridia* spp. According to Luo et al. (2021), the glucose uptake rate was decreased by 21% when the fermentation was added with a combination of phenolic acids and phenolic aldehydes and the product yield has been decreased by 50%.

6.4.2 ACIDOGENIC FERMENTATION

The pH values of medium are very crucial especially for ABE fermentation. During acidogenesis phase, the formation of acetic acid and butyric acid lowering the pH values. The solventogenesis phase started immediate after the pH reaches the critical point where the acids start to produce solvents. However, formation of acid started at pH values less than 4.5 that will decrease the effectiveness of solvent production. So that, the buffering capacity of the fermentation medium should be increased to enhance the *Clostridium* spp. growth, carbohydrate conversion rate and biobutanol production yield (Nanda et al., 2017).

According to Maddox et al. (2000), when the pH values of ABE fermentation medium very close to neutral (pH – 7), acidogenic fermentations start to occur. This is due to the fast acids production followed by inhibition of solventogenesis when the concentration of total acid reaches 240–250 mmol/l. At this pH, solventogenesis can be attained by decreasing the glucose uptake or acid production rate by using higher glucose or lowered yeast extract concentrations in the growth medium.

6.4.3 PHAGE CONTAMINATION

Phage infections or contamination by bacteria has become a major problem in ABE fermentations and it is a potential cause of failure of bacteria-driven production processes (Jones et al., 2000; Los, 2020). According to Los (2020), phase contamination is divided into two types as shown in Figure 6.2.

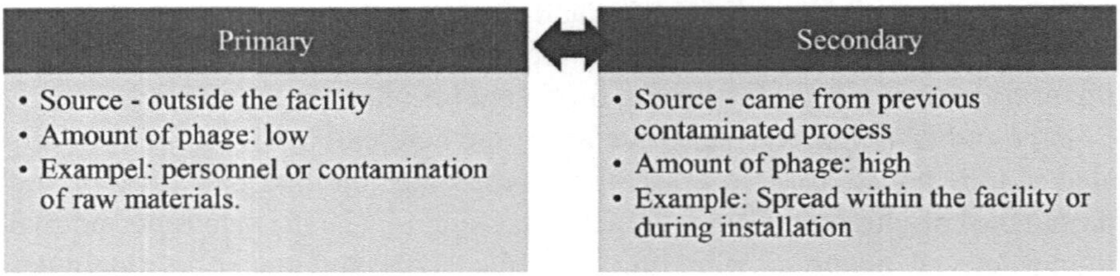

FIGURE 6.2 Phage contamination.

Source: Adapted from Los (2020).

Phage contamination caused fermentation to stop immediately, low quality of products and subsequently economic loss. Phage infection have occurred in many industrial processes relating to fermentation of milk and cheese by lactic acid bacteria, ABE by *Clostridium* spp. and other products (Zou et al., 2022). The indications of phage contamination as reported by Nanda et al. (2017) were shown in Figure 6.3.

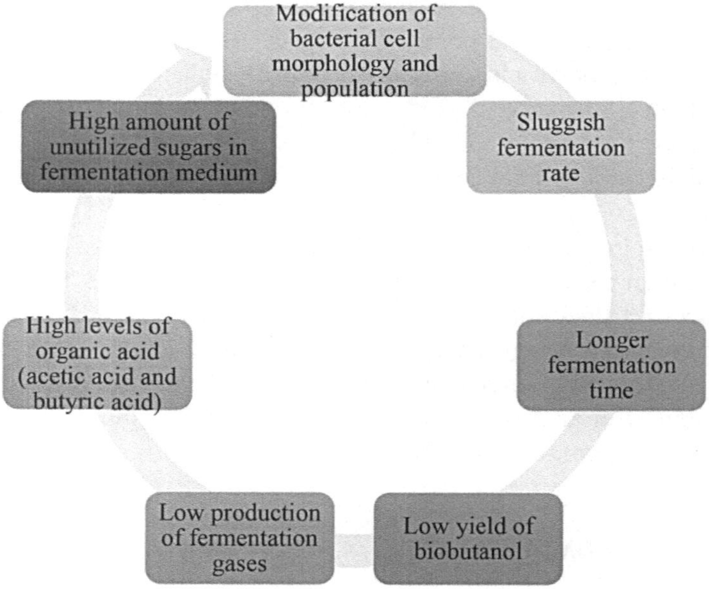

FIGURE 6.3 Phage contamination symptoms.

Source: Adapted from Nanda et al. (2017).

Infections usually happened during seed culture development are expected to cause extensive contamination in all fermenters. However, infections occur at the beginning of the fermentation processes shows more difficulties as compared to later stages which has less widespread effects. Typically, low level of phage infections difficult to detect. It can be made by visual observation of the fermentation medium when the substrate color changed from lighter to dark. While at high level of phage infections, the fermentation processes tend to stop completely with majority of the cells may lyse within hours (Jones et al., 2000; Nanda et al., 2017).

Phage infection occurs due to several number of factors includes phage involved, the number of phage relative to the number of cells, composition of fermentation substrate, physical and chemical conditions in fermenters and the stage of the fermentation where the infection occurs. The most successful approach to overcome the effect of phage infections during ABE fermentation was to sustain the energetic of the strain that were resistant to numerous ranges of different phage. Selection and isolation of phage-resistant mutants that engaged with solvent-producing strains proficiencies referred as phage immunization (Jones et al., 2000). Also, one of the common sources of phage infection are raw materials which are used in bulk that originate from far distance and variable destinations, thus covering more environments and potential phage sources (Los, 2020).

Strain rotation and manipulation is also one of the strategies to prevent or minimize phage contamination. Commonly, seed culture rotation has been applied in diary industry as a standard method to avoid irregularities of production. Additional features to prevent primary phage contamination are process design and selection of equipment as well as cleanliness of production facilities. To some extent, it also can help to prevent secondary phage contamination if the contaminant contaminated material inside the production line until it can be appropriately deactivated (Los, 2020). Moreover, prevention strategies such as good laboratory and factory practice to maximize the hygiene, sterilization, disinfection, and decontamination are needed to overcome phage contamination (Los, 2014).

6.5 STRATEGIES FOR INCREASING BIOBUTANOL PRODUCTION FROM BIOMASS

Although various type of feedstocks starting from first till fourth generation has been identified to produce biobutanol with different *Clostridium spp.*, there are numerous drawbacks related to ABE fermentation that needs to be improve. Some of the limitation that has been determined were low yield of biobutanol caused by inhibition during fermentation and high cost of biobutanol recovery. Not only inhibitors, selection of feedstock also important to reduce the cost of raw materials as well as selection of an efficient pretreatment and hydrolysis method to produce higher fermentable sugars (Ndaba et al., 2015; Cao et al., 2016). Additionally, cost of biobutanol production using *Clostridium acetobutylicum* is not economically viable. The efficacy of this strain and downstream process has high impact on economics of biobutanol production. Due to that a few strategies such as strain improvement to develop advanced genome study and deep knowledge of genes functioning in metabolic and regulatory pathways must be taken to increase biobutanol production from biomass (Xue et al., 2017).

6.5.1 INHIBITORS DETOXIFICATION

Generally, levels of inhibitors are depending on the type of feedstock and conditions of pretreatment and hydrolysis processes. According to Guo et al. (2022), several steps has been developed to overcome the inhibition problems by using less recalcitrant feedstocks, mild pretreatment conditions, enzymatic hydrolysis which is effectively decrease the formation of inhibitor, simultaneous pretreatment, hydrolysis, fermentation, and product recovery and integrating several detoxification methods. Some of the detoxification methods are electrodialysis, over-liming, activated carbon (AC)/charcoal, filtration/centrifugation, resin treatments and *in situ* extraction with Oley alcohol (Luo et al., 2021; Guo et al., 2022). Other detoxification strategy to increase biobutanol yield were minimal sugar loss during detoxification, low energy consumption, reduce the present of metal salts in fermentation medium and minimize the water discharges (Yangling et al., 2022). Additionally, *in situ* product recovery methods efficiently decrease the production cost by continuous elimination and purification of inhibitory products during fermentation that promote the enhancement of biobutanol production (Xue et al., 2017).

Yangling et al. (2022) have studied several detoxification methods on the hydrolysate which were over-liming using $Ca(OH)_2$, activated charcoal adsorption and XAD-4 resin adsorption. The yield of biobutanol after detoxification using these three methods were 3.53 g/L, 3.45 g/L and 8.02 g/L, respectively. Cao et al. (2013) has treated dilute acid and hot water pretreated corn stover with *Coniochaeta ligniaria* NRLL30616. It was found that most of the acetate and more than 50% of furfural, HMF, phenolic compounds were effectively eliminated.

However, the development of simultaneous fermentation and connected with butanol recovery processes can dismiss butanol toxicity and inhibitors (Nanda et al., 2017).

6.5.2 PROCESS INTEGRATION AND OPTIMIZATION

Several strategies can be applied to optimize the biobutanol production through process integration process. Traditionally, biobutanol has been produced via separate hydrolysis and fermentation (SHF) process. This process method has produced higher inhibitors and lower yield of biobutanol. Recently, simultaneous saccharification and fermentation (SSF) has been implemented to decrease the inhibitors. This process method using fewer processing steps that reduce the operating cost, it reduces the contamination risks and produce less inhibitors that could inhibit the biobutanol production during the fermentable sugars' conversion process (Guo et al., 2022). According to Cao et al. (2016), biobutanol production involves pretreatment, saccharification, fermentation, and recovery process. Further decrease the cost for each individual process and efficient combination of these processes would direct the biobutanol production towards commercial scale. Comparison of biobutanol production using different process methods were shown in Table 6.5.

TABLE 6.5 Comparison of biobutanol production using different method

Feedstock	Fermentation process	Biobutanol yield (g/L)	References
Corn stover	SSF	10.8	Guo et al., (2022)
Rice straw	SSFR	24.80	Guo et al., (2022)
Xylan	CBP	13.28	Guo et al., (2022)
Rice straw	SHF	2.92	Nakayama et al., (2011)
Crystalline cellulose	CBP	7.90	Qureshi et al., (2008)
Wheat straw	SSF	5.05	Wang et al., (2013)
Monosaccharides	CBP	12.21	Xin et al., (2017)

Note: SSF-simultaneous saccharification and fermentation; SSFR-simultaneous saccharification, fermentation and recovery; CBP-consolidated bioprocessing.

Besides that, it is important to utilize a variety of carbohydrates to increase biobutanol product yield including pentose (C5) and hexose (C6) that will be the source of glucose and xylose, respectively. Utilizing both sugars optimally could reduce the cost of substrates. The conversion of xylose and glucose mixture simultaneously has produced maximum biobutanol yield (12.05 g/L) which is almost the same with biobutanol produced from maize or molasses (Dong et al., 2011). Furthermore, utilization of inexpensive feedstocks such as agricultural residues with powerful butanol-producing strains to utilize various sugars such as glucose, xylose, fructose, mannose, and arabinose would increase the biobutanol product yield (Xue et al., 2017).

According to Gu et al. (2014) the use of inexpensive feedstock is essential to cut the biobutanol production cost, due to that the selected *Clostridium* spp. strains should be vigorous that can be modified for wide range of carbohydrates sources. Consumption of related sugars uptake and transport system could enable these strains to utilize various type of substrates during solventogenesis phase to produce biobutanol and its by-products. The process of sugar transport could be done by applied many mechanisms including Na+ and H+-symport, ABC system and PEP-dependent phosphotransferase system (PTS) (Mitchell et al., 2016; Xue et al., 2017).

Supplementary of nutrients in the fermentation medium also could help to enhance the biobutanol product yield. According to Qin et al. (2018), supplementary of 3 g/L $CaCO_3$ and 0.5% Tween 80 (non-ionic surfactant) in the ABE fermentation of mixed glucose, starch, and food waste, has stimulated the amylase enzymes activities and increased biobutanol yield of 17.64 g/L. The absent of elements such as nitrogen, magnesium, potassium, calcium, and sulfur as well as vitamins like A, B, and C have helped the biobutanol-producing strain for ABE fermentation growth well and obtain better viability (McGill et al., 2013; Liu et al., 2019; Darmayanti et al., 2020).

6.5.3 STRAIN IMPROVEMENT

One of the biggest challenges for commercialization of biobutanol is low biobutanol yield due to the formation of by-products (acetone and ethanol) and low tolerance capability of microorganism. Modification of the strain through mutagenesis, metabolic, and evolutionary engineering has been done to improve the biobutanol yield (Cao et al., 2016). Besides that, robust and powerful strain with high tolerance to inhibitors should be constructed (Luo et al., 2021). According to Dong et al. (2011), *Clostridium acetobutylicum* strain could produce a higher yield of butanol in solvent which is about 70% and this strain has been studied by Institute of Microbiology, Chinese Academy of Science. The studied focused on strain improvement by metabolic engineering to increase solvent production, improve biobutanol yield and enhance the microbial tolerance against solvents. Jiang et al. (2009) has increased the biobutanol ratio by weakening the acetone production pathway where acetoacetate decarboxylase (AADC) which is the main enzymes to produce acetone has been interrupted by using TargeTron technology. As a result, the biobutanol ratio has increased of about 14.4% and acetone produced has reduced to only 0.21 g/L.

Bui et al. (2015) has improved tolerance to biobutanol by applied current knowledge through strain modification via exploiting recognized mechanisms either individually or together. Higher composition of cis-vaccenic acid C18:1 and efflux pump *SrpABC* from *Pseudomonas putida* in the *Escherichia coli* membrane and also iron uptake gene *feoA* has increased 1-butanol tolerance. In addition, the metabolic engineering of *E. coli* for biobutanol production has been studied where the mutant of *E. coli* has produced 30 g/L of biobutanol which is beyond the level of production that has been obtained by butanol producers (Shen et al., 2011; Al-Shorgani et al., 2012).

Current approach towards genetic and metabolic engineering of butanol-producing strains still in progress to overcome some restrictions such as low biobutanol yield, titer, and productivity. This approach could help the *Clostridium* spp. to improve its efficiency during solventogenesis phase and make it viable for ABE fermentation on a commercial scale. The targeted metabolic studies of *Clostridium* spp. were shown in Figure 6.4 (Nanda et al., 2017).

Biobutanol-producing strains with no spore formation is most required for ABE fermentation because *Clostridium* spp. produces solvents only during a short time of sporulation phase with cigar-shaped cell structures (Jones & Woods, 1986; Nanda et al., 2017).

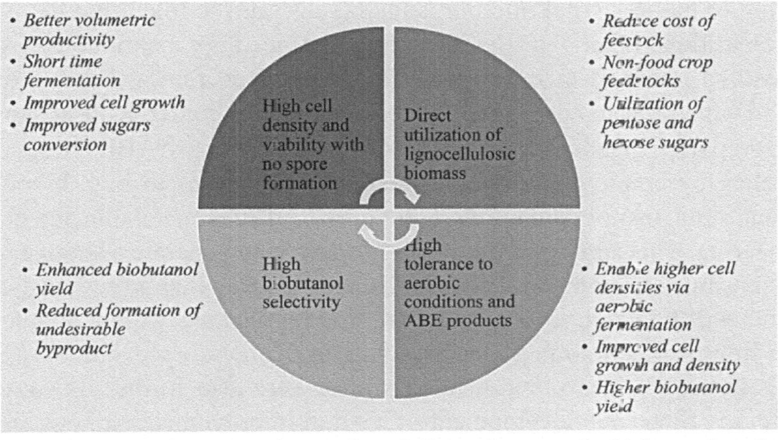

FIGURE 6.4 Metabolic engineering studies of *Clostridium* spp. for biobutanol production.

Source: Adapted from Nanda et al. (2017).

6.6 RECENT NOVELTY ON BIOBUTANOL STRAIN IMPROVEMENT STRATEGY TO ENHANCE BUTANOL TOLERANCE

In conventional biobutanol production that commonly undergoes clostridial fermentation processes, bacteria are poisoned by butanol and this action on the bacterial cell is referred to as chaotropic. This effect is associated with membrane disruption and cell fluidity. In addition, it was reported that 13 to 15 g/L of butanol fully inhibited Clostridial cell growth and pH gradient maintenance. This inhibitory mechanism explains why solvent-producing Clostridium cannot produce solvents at concentrations beyond 15 g/L butanol. Table 6.6 shows the common butanol tolerance concentration for every different species of clostridium.

TABLE 6.6 Common Butanol Tolerance Concentration for Every Different Species of Clostridium

Strain	Butanol Titer	References
Native *C. acetobutylicum*	10–14 g/L	Bao et al. (2020)
C. acetobutylicum YM1	15 g/L	Al-Shorgani (2016)
Clostridium saccharobutylicum DSM 13864	11.60 g/L	Dong et al. (2016)
Clostridium sp. strain NJP7	12.21 g/L	Xu et al. (2017)
C. acetobutylicum ATCC 824	11.2 g/L	Bao et al. (2020)
Clostridium sp. strain GL11	14.7 g/L	Xin et al. (2016)

Therefore, as a way to mitigate this issue, strain improvement strategy is gaining appealing interest in many researchers as it provides a long-term solution to the industrial microorganism (substrate) in tolerating the inhibitory condition rather than necessitating the need for continuous product removal as intended in method like in-situ product removal that is vastly applied which is also not cost effective. In the same context, Kolesniska et al. (2019) reported that this strategy is also known to improve ABE fermentation efficiency by employing strategies such as adaptations to high biobutanol concentration, osmotolerance augmentation, and even metabolic and genetic engineering. This strategy's ultimate objective is to generate a strain capable of increasing both biobutanol yield and tolerance level. In this section, detailed insights will be given to several major strategy that are widely adopted for strain improvement strategy such as adaptive laboratory evolution (ALE), mutagenesis, and gene modification which were also further illustrated in Figure 6.5. Since the development of a strain improvement strategy specifically for biobutanol is still in progress, other successful applications utilizing these strategies will be demonstrated to further illustrate their feasibility.

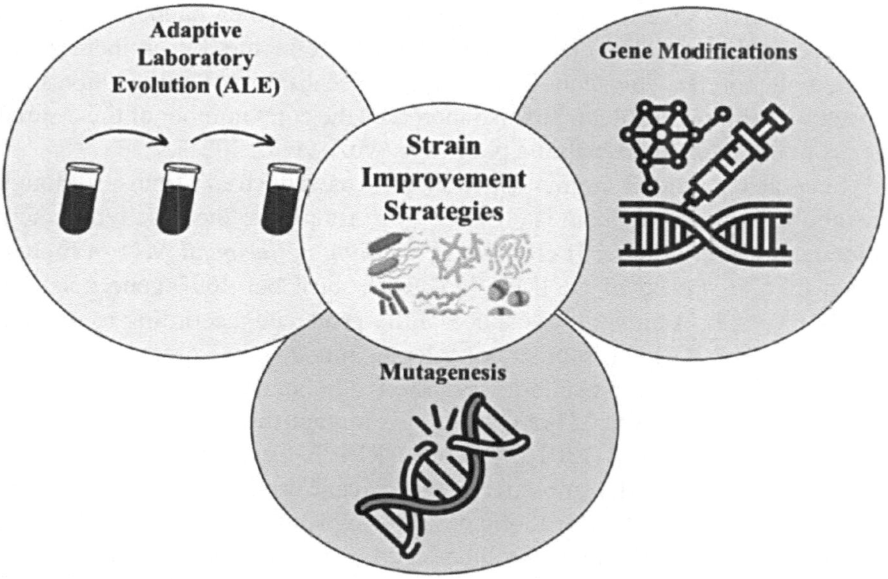

FIGURE 6.5 Strain improvement strategies that are widely incorporated in most studies.

6.6.1 GENE MODIFICATIONS

Gene modification is a genetic engineering technique that alters both genotype and phenotype by inactivating or overexpressing genes in order to produce strains with greater yield and tolerance than their wild-type strains (Wang et al., 2016). The effectiveness of genetically modified strains is not only long-lasting, but it also overcomes the limitations of natural microorganisms, such as low productivity and metabolite inhibition. Südfeld et al. (2017), for instance, demonstrated the efficacy of gene modification specifically through gene deletion in which lipid accumulation in *N. oceanica* increased by 40% in comparison to the wild type of strain. This was further supported by Xu et al. (2015) who conducted overexpression of genes on malic enzyme and produced 2.5-fold increase in total lipid concentration in *P. tricornutum*. In addition, the inactivating genes strategy serves as the premise for blocking the by-product genes that result in significant energy consumption during subsequent purification operations. For instance, Bassi et al. (2020) in their study found that the generation of 1,3-propanediol was greatly enhanced as a result of the elimination of by-products made possible by gene knockout, a type of genetic engineering modification that tries to render the targeted genes inactive. Overexpression of genes, on the other hand, focuses on optimizing the overall metabolic pathway by producing a prominent gene-related phenotype. This technique is crucial because it not only resolves the growth-inhibiting problem, but also increases the copy number of the desired genes in the intended metabolic pathways (Veza et al., 2021).

Several Clostridial strains function well as industrial strains, although metabolic engineering them is difficult. For instance, to increase ferulic acid tolerance, Liu et al. (2015) created *Clostridium beijerinckii* M11, a mutant strain of *C. beijerinckii* NCIMB 8052 with the Cbei 4693 gene deletion. About 6.23 g/L of butanol was successfully produced, according to reports. *C. beijerinckii* BT14, a biobutanol-tolerant mutant, was created by Kong et al. (2016) using atmospheric and room temperature plasmas (ARTP). Butanol production was 25% higher in this mutant than in the parent strain. In addition, Wang et al. (2016) created CRISPR-Cas9 for genome editing in *C. beijerinckii*. Furthermore, to boost solvent production, Herman et al. established a reasonable metabolic engineering strategy for *C. saccharoperbutylacetonicum* N1-4 that combined gene overexpression with targeted knockouts of critical genes. Biobutanol and total ABE titers both increased by 18% and 30%, respectively. Using metabolic engineering to promote ABE fermentation, Zhang et al. (2021) boosted sucrose metabolism in *C. saccharoperbutylacetonicum* N1-4 by removing a transcriptional repressor gene followed by endogenous sucrose overexpression (Veza et al., 2021).

6.6.2 MUTAGENESIS

Mutagenesis is a genetic engineering technique that differs from gene modification in that it focuses on altering the deoxyribonucleic acids (DNA) of an organism, resulting in a gene mutation. A gene may be transformed unintentionally ('spontaneous mutation') or intentionally ('induced mutation') via mutation (Xu et al., 2017). Although mutations are typically deleterious and destroyed by natural selection, some mutations are advantageous to microorganisms (Gu et al., 2017). In biobutanol production, the basis of mutagenesis employed as strain improvement strategy is that the typically employed wild type strain *Clostridium acetobutylicum* suffers major bottleneck issue that is primarily on limitation of tolerance level towards solvents toxicity that leads to low biobutanol titer, yield, and productivity (Li et al., 2016). Therefore, a mutant strain that is developed through mutagenesis could accelerate the increase of metabolic level for industrial purposes. In terms of general achievements of mutagenesis, Arora et al. (2021) exhibited random mutagenesis in a microalgal strain, resulting in microalgal mutants with high biomass productivity, high lipid content, and abiotic stress tolerance. Mutagenesis utilizing a recombinant technique on *Streptomyces clavuligerus* NRRL3585 has increased clavulanic acid production and demonstrated the viability of the mutant strain, which is capable of producing 30 times more clavulanic acid than the wild type of strain (Sexena et al., 2015). In addition, the type of mutagens has an essential effect in boosting the efficiency of mutagenesis. For instance, the use of chemical mutagens (N-methyl-N′-nitro-N-nitrosoguanidine) and physical mutagens (ultraviolet irradiation at 254 nm) in Heerd et al. (2014)'s study on *Aspergillus sojae* resulted in a 1.5-fold increase in pectinolytic enzyme production in the mutated strain, *A. sojae* ATCC 20235, compared to the wild strain.

Apart from that, as for one of the major achievements of this strategy for biobutanol production is that a maximum, 20.3 g/L biobutanol has been successfully produced by hyper biobutanol tolerant of *Clostridium acetobutylicum* JB200 which obtain from *C. acetobutylicum* ATCC 55025 after mutagenesis and adaptation in a fibrous bed bioreactor (Xu et al., 2017). Gu et al. (2017) studied ARTP mutations in combination with genome shuffling of TSH06 and discovered that the symbiotic system on the evolved TSH06, TS4-30 enabled butanol titer enhancement and micro-oxygen tolerance. TS4-30 had the highest butanol titer of 15.63 g/L, which was 34% higher than TSH06 (12.19 g/L). Li et al. (2016) used a different method of mutagenesis and genomic shuffling. The shuffled strain GS4-3 of *C. acetobutylicum* showed improved fermentation performance after four rounds of genome shuffling, generating 20.1 g/L of butanol. Despite recent advances in genetic engineering to produce a recombinant strain that can tolerate biobutanol toxicity, the improvement of strain with superior properties also can be achieved by using ALE.

6.6.3 ADAPTIVE LABORATORY EVOLUTION (ALE)

Adaptive laboratory evolution (ALE) is a novel approach for generating evolved microbial strains with desired features on the laboratory bench by implementing the Darwinian Theory's natural selection rules. William Dallinger performed ALE around a 100 years ago, in the middle of the last century. However, there has been an ever-increasing number of similar trials in the last 25 years, with *Escherichia coli* and *Saccharomyces cerevisiae* being the most common organisms under examination (Dragosits & Mattanovich, 2013). ALE is a scientific approach used to artificially induce strain evolution based on a predetermined laboratory setting. It has been frequently used to gain insight into the basic mechanism of molecular evolution via adapting the microbial population under specific growth conditions during a long-term selection of stress. This method offers some major benefits as compared with classical genetic engineering but with some inherent limitations like efficiency issues. ALE has been applied to studies on various organisms, include microalgae, mammalian cells and viruses, to standard model organisms such as *E. coli* and yeast (Lee & Kim, 2020). Many successful stories on the application of ALE on the improvement of microorganisms had been reported previously.

One successful recent ALE adaptation in strains improvement can be seen by improved acid tolerance and succinic acid production in *Actinobacillus succinogenes.* In this study, *A. succinogenes mutant BC-4* was developed through adaptive evolution to boost cell growth and succinic acid production at weak acid under anaerobic conditions. At pH 5.8, the evolved strain showed remarkable acid stress resistance, producing 20.77 g/L succinic acid with a yield of 0.63 g/g (Zhang et al., 2020). *Synechococcus elongatus* PCC 11801 has successfully underwent ALE to improve its solvent tolerance (Srivastava et al., 2021). The most recent achievement for strain improvement was achieved for *E. coli* which successfully mitigate acid stress using the ALE strategy (Du et al., 2020) and ALE also was applied to characterize and improve native methanol metabolism (methylotrophy) in *S. cerevisiae* (Monica et al., 2020). On another hand, Wang et al. (2016) has successfully performed ALE for *Chlorella sp.* To remove phenol in wastewater. The resulting strain was obtained after 31 cycles for about 95 days, and the strain could grow under 500 mg/L and 700 mg/L phenol without significant inhibition. The glucose-tolerant strain of *C. cohnii* (Li et al., 2017), biodegradation of furfural by *Pseudomonas pseudo alcaligenes* CECT 5344 (Macias & Blasco, 2019) and solvent stress cross-tolerance conferred to *Corynebacterium glutamic* also has been successfully performed by ALE method (Oide et al., 2015). Apart from that, ALE approach was incorporated at study on increasing *C. saccharoperbutylacetonicum*'s tolerance to acetic acid and 5-hydroxymethyl furfural (Alves et al., 2021). Table 6.7 shows summary of the achievements for strain improvement with respective type of strain that incorporated ALE method.

TABLE 6.7 Summary of the Achievements for Strain Improvement with Respective Type of Strain That Incorporated Adaptive Laboratory Evolution (ALE) Method

Type of Strain	Achievements	References
E. coli	Successfully mitigate acid stress	Du et al. (2020)
S. cerevisiae	Characterize and improve native methanol metabolism	Monica et al. (2020)
Chlorella sp.	Remove phenol in wastewater	Wang et al. (2016)
Pseudomonas pseudo alcalitgenes	Biodegradation of furfural	Macias & Blasco (2019)
C. cohnii	Produce glucose-tolerant strain	Li et al (2017)
Corynebacterium glutamic	Produce solvent stress cross-tolerance that conferred to the strain	Oide et al. (2015)

However, limited studies have been conducted on ALE approach for the improvement of biobutanol tolerance and in fact, no studies have been directed to relate microbial properties and fermentation kinetics of biobutanol tolerance strain of *Clostridium* sp. for biobutanol production. At present, study by Tian et al. (2019) shows that growth adaptation in the chemostat using biobutanol tolerant of *C. thermocellum* has improved the strain tolerance up to 15 g/L from 5 g/L. This method shows a great improvement of biobutanol tolerance via growth adaptation. Recent study on using ALE that also had successfully produce an improved *Clostridium cellulovorans* (Wen et al., 2019) also give a pathway and guidance on improving *C. acetobutylicum* whereby the n-butanol production was increased 138-fold, from less than 0.025 g/L to 3.47 g/L. Table 6.8 shows the summary of improvement on the butanol titer after incorporating strain improvement strategies.

6.7 CURRENT AND FUTURE PERSPECTIVE OF BIOBUTANOL DEVELOPMENT FOR SUSTAINABLE LARGE-SCALE PRODUCTION

Biobutanol is traditionally produced from lignocellulosic biomass such as corn, cassava, wheat, and potato from an industrial standpoint. Researchers have discovered and enhanced strains of solventogenic Clostridium species in an effort to produce biobutanol from inexpensive and widely available raw materials. Using *Clostridium* species like as *Clostridium acetobutylicum* and *Clostridium beijerinckii* directly during ABE fermentation has significant disadvantages, such as the potential development of an inhibitory chemical that would hinder the overall product recovery (Kolesinska et al., 2019). Therefore, this demonstrates the necessity for viable strain improvement for

TABLE 6.8 Summary of Improvement on the Butanol Titer after Incorporating Strain Improvement Strategies

Methods	Type of Strain	Butanol Titer	Improvement	References
Adaptive laboratory evolution (ALE)	*C. thermocellum*	5 g/L	Tolerance level of butanol increased to 15 g/L.	Tian et al. (2019)
	Clostridium cellulovorans	0.025 g/L	Yield of butanol increased to 3.47 g/L.	Wen et al. (2019)
Mutagenesis	*Clostridium acetobutylicum* JB200.	16 g/L	Maximum 20.3 g/L of butanol produced.	Xu et al. (2017).
	GS4-3 of *C. acetobutylicum*	–	20.1 g/L of butanol generated.	Li et al. (2016)
	Genome shuffling of TSH06.	12.19 g/L	Highest butanol titer of 15.63 g/L was produced.	Gu et al. (2017)
Gene modifications	*C. beijerinckii* BT14	–	Butanol production was 25% higher.	Kong et al. (2016)
	C. saccharoperbutylacetonicum N1-4	–	Biobutanol and total ABE titers both increased by 18% and 30%, respectively.	Herman et al. (2017)
	C. beijerinckii NCIMB 8052	–	6.23 g/L of butanol was successfully produced.	Liu et al. (2015)

sustainable production. Biobutanol production on a large scale is still in the process of being developed, and most of the recent success that has been widely reported by many other researchers has focused on scaling up strategies (Kushwaha et al., 2019).

Scaling up provides a framework for identifying and developing a process that will successfully create a desired product when manufactured on a commercial scale. To transition from small scale, which is often undertaken in the form of a pilot plant, to large scale, it is essential to comprehend the effects of size variations on a variety of physical and chemical processes (Ibrahim et al., 2018). ButaNEXT, for instance, is one of the successful projects that illustrates the techno-economic feasibility of converting lignocellulosic biomass (wheat) into biobutanol through the integration of several technological advancements. Development of a Clostridial strain able to ferment cellulosic sugars effectively with a 20% higher butanol tolerance than the baseline strain, validation of the technical performance from handling to product recovery using 100 L of bioreactor, and, most importantly, design specifications for a pilot prototype with a lower energy consumption and the ability to be scaled up to industrial size were among the significant accomplishments revealed by the study (Campo et al., 2018). Figure 6.6 shows the significance of scaling up strategy.

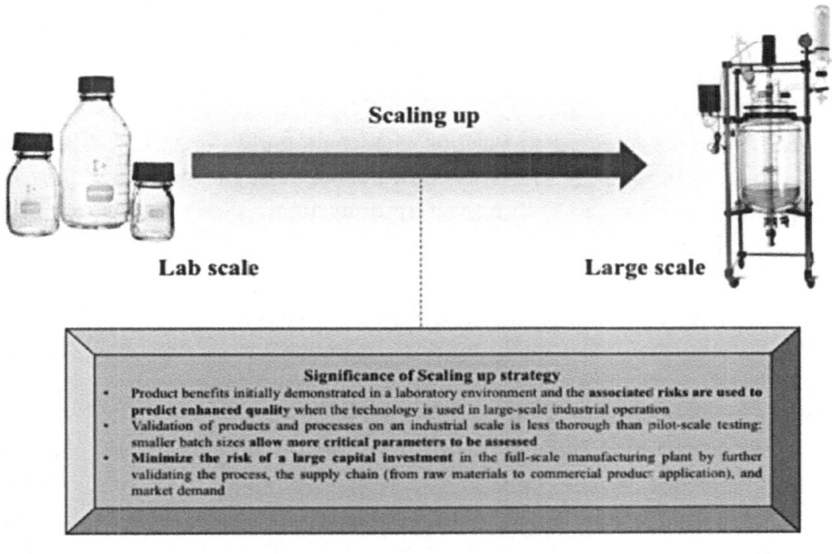

FIGURE 6.6 Significance of scaling up strategy.

In addition, other biobutanol development that are growing at the moment are geared toward integrated product recovery for downstream processing. In order to prevent butanol inhibition, in-situ product removal is vastly applied on industrial level which also includes various other techniques such as liquid-liquid extraction, pervaporation, adsorption, and gas stripping (Basu et al., 2021). Each technique has its own functionality that aids in enhancing the productivity of biobutanol on large scale production. For instance, De Guido et al. (2019) found that gas stripping reduces fermentation duration from 53 hours to 48 hours. Xu et al. (2019), who proved 60% (w/v) butanol was recovered from the fermentation medium after separation using this method, also support the benign nature of gas stripping. Aside from that, as for butanol recovery in pervaporation that incorporates membrane, up to 105.4 g/L was produced through silicon-silicon membrane, and the majority of researchers, such as Sarabia et al. (2020); De Guido et al. (2019); and Basu et al. (2021), concur that pervaporation significantly increases the yield of biobutanol and is cost effective.

Apart from that, for adsorption method product recovery, the common adsorbents used in the current development of butanol recovery are AC, zeolite, and polymeric resins. In the same context, Basu et al. (2021) emphasized the importance of having a suitable adsorbent with criteria such as high adsorption capacity, high selectivity, stability (mechanical, thermal, and chemical), and no toxicity to the fermenting microorganisms in the case of in-situ recovery. In most circumstances, however, the butanol recovery obtained by adsorption is not considerably greater than other product recovery methods. Raganati et al. (2020) found that butanol recovery was less than 75–85% when subjected to various charcoals and resins (IRC-50, XAD-2, XAD-4, XAD8, and XAD-16) type of adsorbents. In contrast, liquid-liquid extraction is performed within the fermenter using ionic liquids, which does not harm the bacteria. In the same context, the significant component of this method is ionic liquid solubility and solvent solubility, where ionic liquid solubility and temperature-dependent polarity of water play a predominating role in butanol recovery, enabling 74% of butanol to be extracted (De Guido et al., 2019). Table 6.9 shows the product recovery approach for biobutanol production that includes biobutanol recovery.

In terms of future perspective of biobutanol for sustainable fuel production, as mentioned in the preceding section, biobutanol is a promising renewable biofuel that has the potential to serve as a replacement for non-renewable fossil fuel for future generations (Kolesinska et al., 2019). The increasing need for fuel has increased researchers' interest in discovering alternate,

TABLE 6.9 Product Recovery Approach for Biobutanol Production that Includes Biobutanol Recovery

Methods	Type of Strain	Biobutanol Recovery	References
Pervaporation	*Clostridium acetobutylicum* XY 16	105.4 g/L	Basu et al. (2021)
Adsorption	*Clostridium beijerinckii* CC 101	54.6 g/L	Molasses et al. (2021)
Gas stripping	*Clostridium acetobutylicum* JB 200	400 g/L	Wen et al. (2018)
Liquid-liquid extraction	–	70%	De Guido et al. (2019)

eco-friendly, and renewable energy sources (Kushwaha et al., 2019). Utilization of renewable feedstock supplies the globe with economically viable and environmentally acceptable fuel. One of the future prospects of biobutanol for sustainable fuel production is the advancement of scaling up strategy, as mentioned in the previous section. This strategy can be further integrated through techniques such as Life cycle assessment (LCA), which employs an improved metabolic engineering approach for large-scale biobutanol synthesis with *Cyanobacteria synechocystis* PCC6803. The significance of this work lies in the fact that the results demonstrate that biofuels produced with genetically modified microbes are a possible option for future sustainable transportation, particularly where renewable electricity or hydrogen-powered cars are not practical (Villacreses et al., 2021).

Not only that, the outlook for the future of integrated product recovery is shifting toward different hybrid solutions that are utilized to enhance butanol production and purification. According to Table 6.3, the high biobutanol recovery obtained via gas stripping is due to the adoption of hybrid strategies involving a two-stage gas stripping process. The first stage focuses on removing the ABE solvent from the fermentation broth to alleviate their inhibition on cells, while the second stage concentrates butanol from the aqueous phase formed with the condensate collected from the first-stage gas stripping process, resulting in total biobutanol recovery of 400 g/L that is equivalent to 20-fold of conventional ABE fermentation (Wen et al., 2018). In the same context as hybrid techniques, the pervaporation method that employs PMS membrane and a two-stage approach can produce up to 167 g/L of butanol (Talluri et al., 2019). Combining two procedures, such as gas stripping and pervaporation, as undertaken by Wukovits et al. (2018), yielded 59 g/L of butanol, which widens the research gap and necessitates more research into other feasible integration methods that could offer significant butanol recovery.

6.8 CONCLUSIONS

In conclusion, based on the review, biobutanol has demonstrated remarkable commercialization potential and serves as an excellent alternative to fossil fuels, as its properties are nearly identical to those of gasoline. However, biobutanol toxicity, which causes inhibition and disrupts product recovery, is the most significant drawback that many researchers are currently attempting to address. Numerous efforts have been made to address this issue, with an emphasis on strain improvement strategies. In comparison to genetic modification, mutagenesis, and adaptive laboratory evolution (ALE), ALE is the most practical and cost-effective strategy because it does not require an understanding of the genetic study of the strain. ALE is used to gain insight into the fundamental mechanism of molecular evolution by adapting a microbial population to specific growth conditions during a long-term selection stress. This method offers significant advantages over traditional genetic engineering. Many prior accounts of the successful use of ALE to the enhancement of microbes had been published. However, to date, very limited study related to strain improvement for *Clotridium* sp. in biobutanol production is reported using ALE. Future studies may therefore apply ALE as a viable strain enhancement approach to enhance biobutanol tolerance for high biobutanol production. Apart from that, numerous researchers are currently engaged in extensive efforts to generate biobutanol on a sustainable, large scale. Recent accomplishments from the ButaNEXT project that are able to identify all the necessary elements, including design specification, strain improvement, and performance validation in a 100 L bioreactor, for the transition from small-scale to large-scale production demonstrate the need for additional research on scaling up strategy. Furthermore, it can also be concurred that out of all the integrated product recovery approach, gas stripping has shown promising result with high butanol yield produced.

ACKNOWLEDGMENT

The author would like to acknowledge the support from the Fundamental Research Grant Scheme (FRGS) under a grant number of FRGS/1/2021/TK0/UNIMAP/02/31 from the Ministry of Higher Education, Malaysia.

KEYWORDS

- **ABE fermentation**
- **adaptive laboratory evolution**
- **biobutanol**
- **biomass**
- **clostridium**
- **mutagenesis**
- **sustainable large-scale production**

REFERENCES

Al-Shorgani, N. K. N., (2015). *Biobutanol Production by a Novel Local Strain of C. Acetobutylicum YM1*. Doctoral Thesis Universiti Kebangsaan Malaysia, Universiti Kebangsaan Malaysia (UKM).

Al-Shorgani, N. K. N., Kalil, M. S., & Yusoff, W. M. W., (2016). Fermentation of sago starch to biobutanol in a batch culture using *Clostridium saccharoperbutylacetonicum* N1-4 (ATCC 13564). *Annals of Microbiology, 62*(1), 1059–1070.

Al-Tabib, A. I., Al-Shorgani, N. K. N., Hassimi, A. H., Aidil, A. H., & Mohd, S. K., (2017). Production of acetone, butanol, and ethanol (ABE) by *Clostridium acetobutylicum* YM1 from pre-treated palm kernel cake in batch culture fermentation. *Bioresources, 12*(2), 3371–3386.

Alves, R. F., Zetty-Arenas, A. M., Demirci, H., Dias, O., Rocha, I., Basso, T. O., & Freitas, S., (2021). Enhancing acetic acid and 5-hydroxymethyl furfural tolerance of *C. saccharoperbutylacetonicum* through adaptive laboratory evolution. *Process Biochemistry 4*(12), 179–189.

Arora, N., & Philippidis, G. P., (2021). Microalgae strain improvement strategies: Random mutagenesis and adaptive laboratory evolution. *Trends in Plant Science, 26*(11), 1199–1200.

Bao, T., Feng, J., Jiang, W., Fu, H., Wang, J., & Yang, S. T., (2020). Recent advances in n-butanol and butyrate production using engineered *Clostridium tyrobutyricum*. *World Journal of Microbiology and Biotechnology, 36*(9), 32–39.

Bassi, M. A., Lopez, M. A., Confalone, L., Gaudio, R. M., Lombardo, L., & Lauritano, D., (2020). Enhanced reader.pdf. *Nature, 3*(8). 539–547.

Basu, C., (2021). *Biofuels and Biodiesel, 4*(1), 2–78.

Bui, L. M., Lee, J. Y., & Geraldi, A., (2015). Improved n-butanol tolerance in *Escherichia coli* by controlling membrane related functions. *Journal of Biotechnology, 20*(4), 33–44.

Campo, I. D., Alegria, I., Munárriz, M., Davies, T., Smith, H., Pallares, Á., & Mazzagatti, V., (2018). Scaling-up lignocellulosic butanol production (Butanext). *European Biomass Conference and Exhibition Proceedings, 4*(1), 4–9.

Cao, G., Sheng, Y., Zhang, L., Song, J., Chong, H., & Zhang, J., (2016). Biobutanol production from lignocellulosic biomass: Prospective and challenges. *Journal of Bioremediation and Biodegradation, 7*(4), 345–363.

Castro, Y. A., Ellis, J. T., Miller, C. D., & Sims, R. C., (2016). Optimization of wastewater microalgae saccharification using dilute acid hydrolysis for acetone, butanol, and ethanol fermentation. *Applied Energy, 14*(1), 14–19.

Cheng, H. H., Whang, L. M., Chan, K. C., Chung, M. C., Wua, S. H., Liu, C. P., Tien, S. Y., et al., (2015). Biological butanol production from microalgae-based biodiesel residues by *Clostridium acetobutylicum*. *Bioresource Technology, 18*(4), 379–385.

Darmayanti, R. F., Susanti, A., Setiawan, F. A., Rizkiana, M. F., Muharja, M., Aji, B. B., Prasiefa, M. G., et al., (2020). Exploring starch sources for the refreshment process of acetone-butanol-ethanol fermentation with *Clostridium saccharoperbutylacetonicum* N1-4. *International Journal of Technology, 12*(2), 309–319.

De Guido, G., Lodi, G., & Pellegrini, L. A., (2019). Study of the integrated product recovery by gas stripping for the ABE fermentation with a fed-batch fermenter. *Chemical Engineering Transactions*, *74*(11), 793–800.

Dong, H., Tao, W., Dai, Z., Yang, L., Gong, F., Zhang, Y., & Li, Y., (2011). Biobutanol. *Advances in Biochemical Engineering/Biotechnology, 4*(5), 1–17.

Dong, J. J., Ding, J. C., Zhang, Y., Ma, L., Xu, G. C., Han, R. Z., & Ni, Y., (2016). Simultaneous saccharification and fermentation of dilute alkaline-pretreated corn stover for enhanced butanol production by *Clostridium saccharobutylicum* DSM 13864. *FEMS Microbiology Letters, 363*(4), 55–64.

Dragosits, M., & Mattanovich, D., (2013). Adaptive laboratory evolution – principles and applications for biotechnology. *Microbial Cell Factories, 12*(1), 64–68.

Du, B., Olson, C. A., Sastry, A. V., Fang, X., Phaneuf, P. V., Chen, K., Wu, M., et al., (2020). Adaptive laboratory evolution of *Escherichia coli* under acid stress. *Microbiology, 166*(2), 141–148.

Gao, K., Boiano, S., Marzocchella, A., & Rehmann, L., (2014). Cellulosic butanol production from alkali-pretreated switchgrass (*Panicum virgatum*) and phragmites (*Phragmites australis*). *Bioresource Technolology, 174*(1), 176–181.

Gottumukkala, L. D., Parameswaran, B., Valappil, S. K., Mathiyazhakan, K., Pandey, A., & Sukumuran, R. K., (2013). Biobutanol production from rice straw by a non-acetone producing *Clostridium sporogenes* BE01. *Bioresource Technology, 145*(2), 182–187.

Gu, C., Wang, G., Mai, S., Wu, P., Wu, J., Wang, G., Liu, H., & Zhang, J., (2017). ARTP mutation and genome shuffling of ABE fermentation symbiotic system for improvement of butanol production. *Applied Microbiology and Biotechnology, 101*(5), 2189–2199.

Gu, H., Zhang, J., & Bao, J., (2015). High tolerance and physiological mechanism of *Zymomonas mobilis* to phenolic inhibitors in ethanol fermentation of corncob residue. *Biotechnology and Bioengineering, 112*(2), 1770–1782.

Gu, Y., Jiang, Y., Yang, S., & Jiang, W., (2014). Utilization of economical substrate-derived carbohydrates by solventogenic clostridia: Pathway dissection, regulation, and engineering. *Current Opinion on Biotechnology, 29*(4), 124–131.

Guandong, S., Claire, C., & Jianzhong, H., (2021). Enhanced biobutanol production from starch waste via orange doping. *Renewable Energy, 193*(2), 576–583.

Guo, Y., Liu, Y., Guan, M., Tang, F., Wang, Z., Lin, L., & Pang, F., (2022). Production of butanol from lignocellulosic biomass: Recent advances, challenges, and prospects. *RSC Advances, 12*(1), 18848–18863.

Heerd, D., Tari, C., & Fernández-Lahore, M., (2014). Microbial strain improvement for enhanced polygalacturonase production by *Aspergillus sojae. Applied Microbiology and Biotechnology, 98*(17), 7471–7481.

Herman, N. A., Li, J., Bedi, R., Turchi, B., Liu, X., Miller, M. J., & Zhang, W., (2017). Development of a high-efficiency transformation method and implementation of rational metabolic engineering for the industrial butanol hyperproducer *Clostridium saccharoperbutylacetonicum* strain N1-4. *Applied and Environmental Microbiology, 83*(1), 32–44.

Ibrahim, M. F., Kim, S. W., & Abd-Aziz, S., (2018). Advanced bioprocessing strategies for biobutanol production from biomass. *Renewable and Sustainable Energy Reviews, 9*(1), 23–44.

Jiang, Y., Xu, C., Dong, F., et al., (2009). Disruption of the acetoacetate decarboxylase gene in solvent-producing *Clostridium acetobutylicum* increases the butanol ratio. *Metab. Eng., 11*(2), 284–291.

Jones, D. T., & Woods, D. R., (1986). Acetone-butanol fermentation revisited. *Microbiology Reviews, 50*(1), 484–524.

Jones, D. T., Shirley, M., Wu, X., & Keis, S., (2000). Bacteriophage infections in the industrial acetone butanol (AB) fermentation process. *Journal of Molecular Microbiology and Biotechnology, 2*(1), 21–26.

Kolesinska, B., Fraczyk, J., Binczarski, M., Modelska, M., Berlowska, J., Dziugan, P., & Kregiel, D., (2019). Butanol synthesis routes for biofuel production: Trends and perspectives. *Materials, 12*(2), 33–45.

Komonkiat, L., & Cheirsilp, B., (2013). Felled oil palm trunk as a renewable source for biobutanol production by *Clostridium* spp. *Bioresource Technology, 145*(2), 200–207.

Kong, X., He, A., Zhao, J., Wu, H., Ma, J., Wei, C., & Jiang, M., (2016). Efficient acetone-butanol-ethanol (ABE) production by a butanol-tolerant mutant of *Clostridium beijerinckii* in a fermentation-pervaporation coupled process. *Biochemical Engineering Journal, 105*(2), 90–96.

Kongjan, P., Usmanbaha, N., Khaonuan, S., Jariyaboon, R., O-Thong, S., & Reungsang, A., (2021). Chapter 18 – butanol production from algal biomass by acetone-butanol-ethanol fermentation process. *Clean Energy and Resources Recovery, 4*(3), 421–446. Elsevier.

Kumar, M., & Gayen, K., (2011). Developments in biobutanol production: New insights. *Applied Energy, 88*(2), 1999–2012.

Kushwaha, D., Srivastava, N., Mishra, I., Upadhyay, S. N., & Mishra, P. K., (2019). Recent trends in biobutanol production. *Reviews in Chemical Engineering, 35*(4), 475–504.

Lee, S., & Kim, P., (2020). Current status and applications of adaptive laboratory evolution in industrial microorganisms *J. Microbiol. Biotechnol., 30*(6), 793–8035.

Li, L., Ai, H., Zhang, S., Li, S., Liang, Z., Wua, Z., Yang, S., & Wang, J., (2013). Enhanced butanol production by coculture of *Clostridium beijerinckii* and *Clostridium tyrobutyricum*. *Bioresource Technology, 143*(2), 397–404.

Li, X., Pei, G., Liu, L., Chen, L., & Zhang, W., (2017). Metabolomic analysis and lipid accumulation in a glucose tolerant *Crypthecodinium cohnii* strain obtained by adaptive laboratory evolution. *Bioresource Technology, 235*(2), 87–95.

Liu, K., Atiyeh, H. K., Pardo-Planas, O., Ezeji, T. C., Ujor, V., Overton, J. C., & Tanner, R. S., (2015). Butanol production from hydrothermolysis-pretreated switchgrass: Quantification of inhibitors and detoxification of hydrolyzate. *Bioresource Technology, 189*(1), 292–301.

Liu, K., Zheng, J., Wang, Z., & Chen, F., (2019). Effects of household cooking processes on mineral, vitamin B, and phytic acid contents and mineral bioaccessibility in rice. *Food Chemistry, 28*(2), 59–64.

Los, M., (2020). Strategies of phage contamination prevention in industry. *Open Journal of Bacteriology, 4*(1), 20–23.

Luo, H., Liu, Z., Xie, F., Bilal, M., & Peng, F., (2021). Lignocellulosic biomass to biobutanol: Toxic effects and response mechanism of the combined stress of lignin-derived phenolic acids and phenolic aldehydes to *Clostridium acetobutylicum*. *Industrial Crops and Products, 17*(1), 11–37.

Macias, D., & Blasco, R., (2019). *A Case of Adaptive Laboratory Evolution (ALE): Biodegradation of Furfural by Pseudomonas Pseudoalcaligenes CECT 5344, 3*(11), 5–8.

Maddox, I. S., Steiner, E., Hirsch, S., Wessner, S., Gutierrez, N. A., Gapes, J. R., & Schuster, K. C., (2000). The cause of "acid-crash" and "acidogenic fermentations" during the batch acetone-butanol-ethanol (ABE-) fermentation process. *Journal of Molecular Microbiology and Biotechnology, 2*(1), 95–100.

McGill, C. R., Kurilich, A. C., & Davignon, J., (2013). The role of potatoes and potato components in cardiometabolic health: A review. *Annals of Medicine, 45*(7), 467–473.

Molasses, S., (2021). Capability of immobilized *Clostridium beijerinckii* TISTR 1461. *Processes*, *9*(1), 56–77.

Monica, E., Gonzalez-Garcia, R. A., Kaspar, V., Manuel, R. P., Colin, S., Isak, S. P., Esteban, M., et al., (2020). Adaptive laboratory evolution of native methanol assimilation in *Saccharomyces cerevisiae. Nature Communications, 16*(1), 1–12.

Nakamaya, S., Kiyoshi, K., Kadokura, T., & Nakazato, A., (2011). Butanol production from crystalline cellulose by cocultured *Clostridium thermocellum* and *Clostridium saccharoperbutylacetonicum* N1-4. *Applied Environmental Microbiology, 7*(1), 6470–6475.

Nanda, S., Golemi-Kotra, D., McDermott, J. C., Dalai, A. K., Gökalp, I., & Kozinski, J. A., (2017). Fermentative production of butanol: Perspectives on synthetic biology. *New Biotechnology, 3*(7), 210–221.

Obergruber, M., Vladimir, H., Petr, P., Viera, K., Martin, K., Jiří, B., & Jakub, M., (2021). Physicochemical properties of biobutanol as an advanced biofuel. *Materials*, *14*(1), 914–920.

Oide, S., Gunji, W., Moteki, Y., Yamamoto, S., Suda, M., Jojima, T., & Yukawa, H., (2015). *Thermal and Solvent Stress Cross-Tolerance Conferred to Corynebacterium Glutamicum by Adaptive Laboratory Evolution, 81*(7), 2284–2298.

Onay, M., (2018). Investigation of biobutanol efficiency of *Chlorella* sp. cultivated in municipal wastewater. *Journal of Geosience and Environment Protection, 6*(2), 40–50.

Peabody, G. L., & Kao, K. C., (2016). Recent progress in biobutanol tolerance in microbial systems with an emphasis on *Clostridium. FEMS Microbiology Letters*, *63*(5), 1–6.

Potts, T., Jianjun, D., Michelle, P., Peter, M., Robert, B., & Jamie, H., (2012). The production of butanol from Jamaica bay macro algae. *Environmental Progress & Sustainable Energy, 31*(3), 33–40.

Pugazhendhi, A., Mathimani, T., Varjani, S., Rene, E. R., Kumar, G., Kim, S. H., Ponnusamy, V. K., & Yoon, J. J., (2019). Biobutanol as a promising liquid fuel for the future – recent updates and perspectives. *Fuel, 2*(3), 637–646.

Qi, G. X., Dongmei, H., Jianhui, W., Yu, S., & Xu, G., (2019). Enhanced butanol production from ammonium sulfite pre-treated wheat straw by separate hydrolysis and fermentation and simultaneous saccharification and fermentation. *Sustainable Energy Technologies and Assessments, 3*(6), 10–54.

Qin, Z., Duns, G. J., Pan, T., & Xin, F., (2018). Consolidated processing of biobutanol production from food wastes by solventogenic *Clostridium* spp. strain HN_4. *Bioresource Technology, 2*(6), 148–154.

Qureshi, N., Ezeji, T. C., Ebener, J., Dien, B. S., & Cotta, M. A., (2008). Butanol production by *Clostridium beijerinckii* Part 1: Use of acid and enzyme hydrolyzed corn fiber. *Bioresource Technology, 9*(2), 5915–5922.

Rafidah, J., Sahaid, M. K., Norliza, A. R., Al-Tabib, A. I., Aidil, A. H., & Mohd, F. A., (2020). Effect of initial fermentation medium on bioacetone, biobutanol and bioethanol (BioABE) production from fermentable sugars of *Acacia mangium* using *Clostridium acetobutylicum* YM1. *Bioresources, 15*(3), 6912–6927.

Raganati, F., Procentese, A., Olivieri, G., Russo, M. E., Salatino, P., & Marzocchella, A., (2020). Bio-butanol recovery by adsorption/desorption processes. *Separation and Purification Technology*, *2*(12), 11–23.

Raganati, F., Procentese, A., Olivieri, G., Salatino, P., & Marzocchella, (2014). Biobutanol production from hexose and pentose sugars. *Chemical Engineering Transactions, 3*(8), 193–198.

Ranjan, A., Khanna, S., & Moholkar, V. S., (2013). Feasibility of rice straw as alternate substrate for biobutanol production. *Applied Energy, 10*(3), 32–38.

Sarabia, C. A., Revuelta, D. G., Fallanza, M., et al., (2020). Polymer inclusion membranes containing ionic liquids for the recovery of n-butanol from ABE solutions by pervaporation. *Sep. Purif. Technol., 2*(3), 34–45.

Saxena, S., (2015). Applied microbiology. *Strategies of Strain Improvement of Industrial Microbes*, *3*(4), 1–190.

Shen, C. R., Lan, E. I., Dekishima, Y., Baez, A., Cho, K. M., & Liao, J. C., (2011). Driving forces enable high-titer anaerobic 1-butanol synthesis in *Escherichia coli. Applied Environment and Microbiology, 77*(9), 2905–2915.

Srivastava, V., Amanna, R., Rowden, S. J. L., Sengupta, S., Madhu, S., Howe, C. J., & Wangikar, P. P., (2021). Adaptive laboratory evolution of the fast-growing *cyanobacterium Synechococcus elongatus* PCC 11801 for improved solvent tolerance. *Journal of Bioscience and Bioengineering, 131*(5), 491–500.

Südfeld, C., Hubáˇcek, M., Figueiredo, D., Naduthodi, M. I., Van, D. O. J., Wijffels, R. H., Barbosa, M. J., & D'Adamo, S., (2021). High-throughput insertional mutagenesis reveals novel targets for enhancing lipid accumulation in *Nannochloropsis oceanica. Metab. Eng., 66*(2), 239–258.

Syazana, N., Nasrah, M., Khushairi, A., Zahari, M., & Masagut, N., (2016). *Biobutanol Production by Clostridium Acetobutylicum ATCC 824 Using Oil Palm Fround (OPF) Juice, 2*(2), 4–10.

Talluri, P., Patakova, P., Tomas, M., & Vopicka, O., (2019). *Transient and Steady Pervaporation of 1-Butanol – 1-Propyne (PTMSP) Membrane, 1*(9), 4–7.

Tian, L., Cervenka, N. D., Low, A. M., Olson, D. G., & Lynd, L. R., (2019). A mutation in the AdhE alcohol dehydrogenase of *Clostridium thermocellum* increases tolerance to several primary alcohols, including isobutanol, n-butanol and ethanol. *Scientific Reports, 9*(1), 1–7.

Veza, I., Muhamad, S. M. F., & Latiff, Z. A., (2021). Recent advances in butanol production by acetone-butanol-ethanol (ABE) fermentation. *Biomass and Bioenergy, 4*(2), 45–49.

Villacreses-Freire, D., Ketzer, F., & Rösch, C., (2021). Advanced metabolic engineering approaches and renewable energy to improve environmental benefits of algal biofuels:

LCA of large-scale biobutanol production with *Cyanobacteria Synechocystis* PCC6803. *Bioenergy Research, 6*(2), 66–75.

Wang, L., Xue, C., Wang, L., Zhao, Q., Wei, W., & Sun, Y., (2016). Strain improvement of *Chlorella sp* for phenol biodegradation by adaptive laboratory evolution. *Bioresource Technology, 2*(5), 264–268.

Wang, Z. Y., Cao, G. L., Jiang, C., Song, J. Z., & Zheng, J., (2013). Butanol production from wheat straw by combining crude enzymatic hydrolysis and anaerobic fermentation using *Clostridium acetobutylicum* ATCC824. *Energy Fuels, 27*(1), 5900–5906.

Wen, H., Chen, H., Cai, D., Gong, P., Zhang, T., Wu, Z., & Tan, T., (2018). Integrated in situ gas stripping-salting-out process for high-titer acetone-butanol-ethanol production from sweet sorghum bagasse. *Biotechnology for Biofuels*, *11*(1), 1–12.

Wen, Z., Ledesma-Amaro, R., Lin, J., Jiang, Y., & Yang, S., (2019). Improved n -butanol production from *Clostridium cellulovorans* by integrated metabolic and evolutionary engineering. *Applied and Environmental Microbiology, 85*(7), 44–54.

Wukovits, W., Kirchbacher, F., Miltner, M., & Friedl, A., (2018). Assessment of hybrid processes for bio-butanol purification applying process simulation. *Chemical Engineering Transactions*, *70*(1), 319–324.

Xin, F., Chen, T., Jiang, Y., Dong, W., Zhang, Q., Zhang, M., Wu, H., et al., (2017). Strategies for improved isopropanol-butanol production by a *Clostridium* strain from glucose and hemicellulose through consolidated bioprocessing. *Biotechnology for Biofuels*, *10*(1), 11–23.

Xin, F., Wang, C., Dong, W., Zhang, W., Wu, H., Ma, J., & Jiang, M., (2016). Comprehensive investigations of biobutanol production by a non-acetone and 1,3-propanediol generating *Clostridium* strain from glycerol and polysaccharides. *Biotechnology for Biofuels, 9*(1), 24–28.

Xu, M., Zhao, J., Yu, L., & Yang, S. T., (2017). Comparative genomic analysis of *Clostridium acetobutylicum* for understanding the mutations contributing to enhanced butanol tolerance and production. *Journal of Biotechnology, 26*(1), 36–44.

Xue, C., Zhao, J., Chen, L., Yang, S. T., & Bai, F., (2017). Recent advances and state-of-the-art strategies in strain and process engineering for biobutanol production by *Clostridium acetobutylicum. Biotechnology Advances, 5*(2), 310–322.

Xue, J., Niu, Y. F., Huang, T., Yang, W. D., Liu, J. S., & Li, H. Y., (2015). Genetic improvement of the microalga *Phaeodactylum tricornutum* for boosting neutral lipid accumulation. *Metab. Eng., 27*(4), 1–9.

Yangling, J., Liang, Z., Zhuolin, Y., Yang, F., & Hai, Z., (2022). Waste to energy: Biobutanol production from cellulosic residue of sweet potato by *Clostridia acetobutylicum. Environmental Engineering Research, 27*(5), 66–70.

Zhang, B., Wu, J., & Meng, F., (2021). Adaptive laboratory evolution of microalgae: A review of the regulation of growth, stress resistance, metabolic processes, and biodegradation of pollutants. *Frontiers in Microbiology, 12*(1), 10–15.

Zhang, W., Tao, Y., Wu, M., Xin, F., Dong, W., Zhou, J., Gu, J., et al., (2020). Adaptive evolution improves acid tolerance and succinic acid production in *Actinobacillus succinogenes*. *Process Biochemistry, 98*(1), 76–82.

Zou, X., Xiao, X., & Mo, Z., (2022). Systematic strategies for developing phage resistant *Escherichia coli* strains. *Nature Communications, 13*(1), 44–65.

CHAPTER 7

The Potential of Developing Aquaculture Feed From Organic By-Products

ROSSITA SHAPAWI

Borneo Marine Research Institute, University Malaysia Sabah, Jalan UMS, Kota Kinabalu, Sabah, Malaysia

ABSTRACT

The growth of aquaculture is in tandem with the growth of the sustainable aquafeed industry, with about 70% of the farmed species dependent on formulated feeds. In recent years, the aquaculture feed industry has intensified the use of alternative ingredients characterized by competitive price, good nutritional profile and feasibility of mass production, in an effort to reduce dependency on fish meal and fish oil. Several alternative ingredients from different origins have been investigated to ensure the dependable availability of feed ingredients to meet the demand of fast-growing aquaculture industry. In general, aquatic species have better tolerance to many animal-based ingredients compared to those derived from plants plant-based because of their limited physiological ability to digest complex carbohydrates. Among the alternative ingredients, animal and plant by-products have high potential to be further explored not only due to their low cost but also the benefits of reducing the organic waste and pollution in the environment. This chapter will discuss the potential of several by-products that have been utilized in the industry and research findings on incorporation of protein, lipid, carbohydrate, and micronutrients (vitamins and minerals) from biological waste, advancing the concept of circular economy in aquaculture production. Poultry by-product, feather meal, meat, and bone meal, blood meal, and fishery

Waste Biorefineries: Future Energy, Green Products, and Waste Treatment.
Abu Zahrim Yaser, Siew-Xian Chin, Leonardo Sepúlveda Torre, & A. K. Haghi, (Eds.)

waste are examples of by-product ingredients derived of animal origin. Plant by-product meal such as rice bran, vegetable, and fruit waste, microalgae culture by-products and brewery waste deserve more attention, given the very promising findings on certain high-value aquaculture species. Meanwhile, kitchen food waste requires more investigation on how to improve the utilization of this abundant organic matter. Use of these organic by-products has several advantages as well as challenges that need to be addressed. Some by-product ingredients have shown the potential to improve growth, immunity, and disease resistance. Deficiency of essential nutrients, presence of anti-nutritional factors, low digestibility, and palatability issues are the main constraints in replacing fish-based dietary ingredients by those from organic waste. Nevertheless, several approaches have been adopted to overcome these issues. The success in using by-product-based ingredients in aquaculture feed formulation will help the industry to grow sustainably and consistent with the principles of circular bioeconomy.

7.1 INTRODUCTION

7.1.1 AQUACULTURE INDUSTRY STATUS

Globally, the demand for protein from aquatic species has continued to increase during the past decades. The fish supply has been relying on capture fisheries for so long despite the reported static landings over the years. Fortunately, aquaculture has emerged as the sustainable alternative to capture fisheries in supplying protein from aquatic species. Being the world's fastest-growing food production sector, aquaculture provides high-quality food, employment opportunities, boost economic growth and improve food security. Today, half of the seafood supply is from the aquaculture sector with more than 600 species cultivated all over the world. Global fish production is expected to hike from 179 million tons in 2018 to 204 million tons in 2030, with aquaculture production projected to reach 109 million tons in 2030 (FAO, 2020).

Approximately 70% of the total farmed species are those requiring feed input, especially in the intensive fish farming system. The traditional approach of using trash fish in marine fish farming has been steadily replaced by formulated feeds over the past 10 years which can supply balance nutrients and several other advantages such as longer shelf-life, better feed conversion ratio, and more hygienic. A complete feed contains nutrients such as protein, lipid, carbohydrate, vitamins, and minerals, which inclusion levels will be influenced by the cultured species and their life stage and culture system used. Protein is the most expensive macronutrient in the fish diet, incorporated at levels between 30% and 60% in feeds depending on the species and life stages. Traditionally, the fish meal has been used as a source of protein in fish diets mainly due to its balanced nutritional profile, high digestibility and palatability. Meanwhile, fish oil is used to supply dietary lipid and essential fatty acids. Both ingredients are produced from wild-captured pelagic fish with several advantages such as excellent nutritional profile, good digestibility and palatability for the farmed fish. Unfortunately, these commodities becoming more expensive and in limited supply worldwide, with production is forecasted to be shrunk further in the near future. As a result, the inclusion rates of fish meal and fish oil in formulated aquaculture feeds have shown a clear downward trend for the past five years. One approach to reducing cost and dependency on fish meal and fish oil is to partially or totally substitute them with less expensive ingredient sources.

As the aquaculture industry grows, the demand for feeds is also increasing every year and sustainable ingredients must be sourced. Feed production is a very significant contributor to the environmental footprint of aquaculture. Thus, efforts to reduce the dependency on fish-based ingredients are one of the major highlights in the industry. It is important to note that both fish meal and fish oil are not indispensable in aquaculture feed production as the essential nutrients available in fish meal and fish oil, such as amino acids and fatty acids can be obtained from other sources. However, a thorough evaluation of alternative ingredients is necessary before they can be used safely. There are two major groups of alternative ingredients being extensively evaluated so far: terrestrial animal- and plant-based ingredients. Total replacement of fish meal and fish oil with these alternative ingredients might be possible in certain species or in some cases requires the addition of some necessary feed supplements such as essential nutrients (essential fatty acids and amino acids, vitamins, enzymes, etc.) or by subjecting the ingredients to certain treatments which can improve feed utilization. In general, several alternative ingredients have been successfully used to partially replace fish meal and fish oil with promising output and are being used in commercial feed production. The potential demand for these alternative ingredients depends primarily upon their price, nutritional value and other factors related to their overall performance.

7.1.2 SELECTION OF AQUAFEED INGREDIENTS

There are several criteria that need to be evaluated in the selection of dietary feed ingredients which include ingredient characterization, ingredient digestibility, palatability, nutrient utilization and functionality (Glencross et al., 2007). Nutritional profiling, costing, and supply are the major basis to be considered during the characterization of an ingredient. Information on the ingredient composition such as protein, lipid, carbohydrate, vitamins, minerals, amino acids and fatty acids are often evaluated to determine if they are at a desirable level. Similarly, the anti-nutritional substance that may be present in the ingredients is also warranted to be evaluated to make sure the diet that will be produced is safe for the farmed fish. Both costing and raw material suppliers are very important factors that can directly influence the success of the animal feed business. Recently, rising feed ingredient costs in animal feed production pose a significant challenge to the industry. Therefore, there must be careful planning on how to balance the cost and quality of the selected ingredients. It is wise to prioritize locally available ingredients to reduce the footprint of aquaculture.

Assessment of ingredient's digestibility is conducted to determine the available nutrient for the target species after the feed is ingested, and also the proportion of waste that is excreted in the environment. Determination of digestibility can be done by using direct or indirect methods. In the direct method where total feces collection is applied, the amount of taken feed and excreted feces are measured and its ratio is considered to be the digestibility of the feed or nutrients. Meanwhile, in the indirect method, which is widely used in aquatic animals, digestibility is measured gravimetrically by measuring feed intake and the subsequent fecal production by measuring the concentration of an inert marker in the feed and in the feces.

Palatability in aquaculture nutrition can be described as fish's response to the feed offered to them in terms of joy in accepting them. Ingredients characterized by high nutrient content alone could not be considered good aquaculture feed ingredients if they are having issues with palatability. Fortunately, the challenges with palatability can be overcome by the addition of feed attractant and/or stimulant in the diet formulation. The palatability test can be conducted in a behavioral study through the measurement of feed intake.

Once the characterization, digestibility, and palatability of an ingredient are defined, it is important to determine the biological response of the target species to the formulated feed containing the test ingredients. The dose-response approach is often used to determine the optimum level of fish meal replacement with alternative ingredients at different inclusion levels. Growth (e.g., weight gain and specific growth rate), survival rate, and feed utilization efficiency (e.g., feed conversion ratio, protein efficiency ratio) of the farmed species fed different diets are measured at the end of the feeding trial. Recently, more in-depth investigations are conducted such as understanding the effect of alternative ingredients on blood hematology, body chemical composition, immunity, and disease resistance.

7.1.3 ORGANIC BY-PRODUCTS

Organic by-products are biodegradable materials originating from either plants or animals, mostly produced in the agriculture, food and beverage industry and households. They can exist either in solid or liquid forms. Solid organic waste is generally regarded as organic-biodegradable waste, normally with high moisture content. Dispose of organic waste is often a huge problem in many countries and contributes to the wastage of nutrients such as minerals and proteins and it also causes environmental pollution. Not only that, the management of waste materials will have an impact on the economic and social sectors. They can generate unpleasant odors, promotes insects and rodents, and spread diseases.

Organic by-products could provide a new, low cost and readily accessible nutrient source for the aquaculture feed industry. Recently the use of organic by-products as an alternative aquaculture feed ingredient has gained more attention. There is a wide array of available by-products and waste with high potential to be used as feed ingredients. This chapter will be discussing the potential and challenges of different types of ingredients based on organic by-products originating from different sources.

7.2 INGREDIENTS BASED ON ANIMAL BY-PRODUCTS

Ingredients based on animal by-products can be derived from unwanted parts of the animal body such as skin, bones, feathers, blood, offal, and horn. In general, animal-based ingredients has better digestibility than those of plant origin due to the fact that plant contains high fiber.

7.2.1 POULTRY BY-PRODUCT

Poultry by-product meals (PBM) are one of the most promising animal protein sources that have been included in fish diets. PBM are materials obtained from the wastes of poultry production and processing plants. It is usually exclusive of feathers and intestines but may contain the feet and head in association with rendered meat from the carcass. These meals vary in nutritional value and composition depending on the processing applied and the materials that are included in the meal (Nengas et al., 1999). PBM is produced in many parts of the world, including the Southeast Asia region, which accounts for approximately one-quarter of the global poultry trade (FAO, 2004). It has a high potential to be incorporated into the diet of aquaculture species due to its high protein content, a good profile of amino acids, and lower price compared to fish meal. Table 7.1 shows the nutritional profile of different grades of PBM (pet food grade and feed grade) as compared to fish meal. In general, PBM is characterized by high protein and moderate lipid contents. In terms of the amino acid composition of PBM, tryptophan is the most limiting amino acid. However, in most farmed fish, methionine, and lysine are often the limiting essential amino acids. In PBM, methionine content is normally less than lysine which encourages the industry to supplement PBM-based diets with this methionine, especially when a significant amount of PBM is used to replace FM. PBM is typically priced at 50–70% of fish meal, thus making this commodity very attractive to be used as a major protein source in aquaculture feed. In addition, studies on the apparent digestibility of PBM revealed that this product is well-digested by several fish species (Bureau et al., 1999; Shapawi et al., 2007; Yang et al., 2006). Initially, PBM was only able to partially replace fish meal at a level not exceeding 50%. In recent years improved performance of PBM utilization was observed, with total replacement of fish meal is possible in several fish species (Nengas et al., 1999; Takagi et al., 2000; Gaylord & Rawles, 2005). The improved performance of PBM was mainly due to the improvement of processing technology which can preserve the nutritional value and improve the digestibility of PBM (Bureau et al., 1999). Nevertheless, the use of PBM as a single protein source in the diets is challenged by reduced feed digestibility and limiting essential amino acids, especially in carnivorous fish species (Shapawi et al., 2007), but the use of feed supplements may overcome this issue. Utilizing PBM as a feed protein source is definitely make a contribution to the protection of the natural environment as vast quantities of poultry by-product waste were discarded into the environment. Table 7.2 shows the performance of PBM when used to replace FM in different fish species.

TABLE 7.1 Nutrient Composition of Feed Grade and Pet-Food Grade Poultry By-Product Meal (Percentage Dry Matter)

Proximate Composition	FBM¶	PBM§
Dry matter	95.6	96.8
Ash	12.7	15.52
Crude lipid	10.9	10.11
Crude fiber	0.5	‡ND
NFE*	6.7	6.8
Amino Acid Composition		
Aspartic acid	4.58	4.87
Glutamic acid	9.47	10.11
Serine	3.94	2.72
Glycine	7.57	7.51
Histidine	1.48	1.57
Arginine	6.00	5.64
Threonine	3.15	2.70
Alanine	4.33	4.51
Proline	5.51	4.66
Tyrosine	2.12	1.80
Valine	4.14	3.30
Methionine	1.43	1.76
Cystine	1.70	0.87
Isoleucine	3.10	2.80
Leucine	5.55	4.99
Phenylalanine	3.30	2.84
Tryptophan	0.62	0.63
Lysine	3.51	4.12

**NFE*: Nitrogen-free extract = 100 – (%ash + %protein + %lipid + %fiber).

¶Feed grade poultry by-products (Dindings Ltd., Malaysia).

§Pet food-grade poultry by-products (National Renderers Association, Inc., USA).

‡ND: Not detected.

Source: Adapted from Shapawi et al. (2007).

7.2.1.1 FEATHER MEAL

Feather meal is one of the by-products of the rendering industry. In general, feather contains high keratin, a fibrous protein characterized by very low digestibility. Therefore, to increase the digestibility, the feather is often

TABLE 7.2 The Performance of Poultry By-Product Meal as an Alternative Protein in Different Aquatic Species

Species	Findings	References
Rainbow trout (*Oncorhynchus mykiss*)	Partial replacement (50%) of fish meal produced fish with weight gains and FCR not significantly different to the control fish.	Steffens (1994)
Nile tilapia (*Oreochromis niloticus*)	The growth of fish-fed poultry by-product meal (100% replacement) was not significantly different from FM based diet. However, FCR was significantly lower in poultry by-product meal diet.	El-Sayed (1998)
Gilthead seabream (*Sparus aurata*)	Good quality poultry by-product meal can be used at a level up to 50% without affecting the growth performance.	Nengas et al. (1999)
Red seabream (*Pagrus major*)	FM in red seabream can be totally replaced by poultry by-product meal without affecting growth performance.	Takagi et al. (2000)
Gibel carp (*Carassius auratus*)	Poultry by-product meal could replace up to 500 g/kg of fish meal protein without affecting the growth rate. However, the apparent digestibility of protein decreases at this level.	Yang et al. (2004a)
Hybrid striped bass (*Morone chrysops* X *M. saxatilis*)	Amino acid supplementation of poultry by-product meal can be used to totally replace fishmeal in diets for hybrid striped bass.	Gaylord & Rawles (2005)
Cuneate drum (*Nibea miichthioides*)	When fish meal was replaced with 30% and 50% poultry by-product meal, growth, feed intake and FCR were not affected.	Wang et al. (2006)
Hybrid striped bass (*Morone chrysops* X *M. saxatilis*)	Reduced performance of fish at 70% inclusion of poultry by-product meal when compared to the generic diets.	Rawles et al. (2006)
Black sea turbot (*Psetta maeotica*)	Replacement of FM at a level above 25% caused reduction in growth, nutrient utilization and nitrogen retention.	Yigit et al. (2006)
Humpback grouper (*Cromileptes altivelis*)	Good quality poultry by-product meal can successfully replace more than half the protein from marine fish meal in the diets for humpback grouper.	Shapawi et al. (2007)
Tiger grouper (*Epinephelus fuscoguttatus*)	Apparent digestibility coefficients were influenced by the inclusion of poultry by-product meal in the feeds. The present study shows that poultry by-product meal is an excellent alternative protein source for farming the tiger grouper juveniles with fish meal protein replacement level of 50% resulted in the best overall performances.	Gunben et al. (2014)
Asian seabass (*Lates calcarifer*)	Total substitution of fish meal protein by methionine supple-mented poultry by-product meal negatively influenced the growth performance, liver health, histological traits of different organs, immune, and antioxidant response, and expression of stress-related genes in juvenile barramundi.	Chaklader et al. (2020)

subjected to hydrolysis with strong acid and grinding process under elevated heat and pressure. Apart from this, fermentation and enzymatic treatments of feathers are also practiced improving digestibility (Arunlertaree & Rakyuttithamkul, 2006; Poolsawat et al., 2021). The advantage of using feather meal lies in its low cost (feather meal is approximately two-thirds the cost of other animal protein) and commercial availability. Table 7.3 shows the chemical composition of feather meal. Crude protein, crude lipid, and crude ash of feather meal ranged from 80–84%, 2–6%, and 2.8–14%, respectively. Meanwhile, calcium and phosphorus were reported to be around 0.33–1.4 and 0.5–1.3, respectively. Feather meal is characterized by high sulfur-containing amino acids, especially cystine. Methionine is deficient in methionine, tryptophan, histidine, and lysine.

TABLE 7.3 Chemical Composition of Feather Meal

*Proximate and Mineral Composition (Percentage Dry Matter)	
Crude protein	80–84
Crude lipid	2–6
Crude ash	2.8–14
Calcium	0.33–1.4
Phosphorus	0.5–1.3
**Amino Acid Composition	
Arginine	5.6
Histidine	0.9
Isoleucine	3.9
Leucine	6.9
Lysine	2.3
Methionine	0.6
Phenylalanine	3.9
Threonine	3.8
Tryptophan	0.6
Valine	5.9
Cystine	4.3
Tyrosine	2.5

*Hassan et al. (1997); Ewing (1997); Yu (2007); **NRC (1994).

Studies on the replacement of fish meal with feather meal have been conducted on several species including Chinook salmon Oncorhynchus tshawytscha (Fowler, 1990), Japanese flounder Paralychthys olivaceus

(Kikuchi et al., 1994), tilapia (Bishop et al., 1995), Indian major carp Labeo rohita (Hasan et al., 1997), hybrid catfish (Arunlertaree & Rakyuttithamkul, 2006), and African catfish Clarias gariepinus (Chor et al., 2013). In most studies, feather meal can be used as a substitute for fish meal at a level of less than 50% due to the issues with low digestibility and lacking essential amino acids.

7.2.1.2 BLOOD MEAL

Blood meal is an animal by-product readily available in abattoirs, mostly from cattle, hogs, and chickens. It is derived from fresh clean blood undergoing a similar process of rendering. It is subjected to a drying process that can be done via spray-drying, oven-drying, drum-drying, flash-drying, or solar-drying methods. Nutritionally, blood meal consists of almost all protein (90–95%) and a little bit of fat and ash. Blood meal can be a good source of minerals. Cattle blood meal was reported to contain iron, calcium, magnesium, and phosphorus at 95.46 ± 19.84 mg/100 g, 10.24 ± 4.41, 16.50 ± 3.65, and 130.00 ± 45.21, mg/100 g, respectively (Kikafunda & Sserumaga, 2005). Digestibility of blood meal was reported to be around 80–99%, depending on the processing methods (Hamed et al., 2017). The use of blood meal as a substitute for fish meal in aquaculture feeds is not as intensive as other animal by-products. Blood consumption is prohibited in Islam, thus Muslim consumers have doubt to consume fish that were fed with blood meal-based feeds.

Studies in the past revealed that blood meal can substitute fish meal at up to 10% with no effect on performance gilthead sea bream (Sparus aurata), but may lead to sensory differences compared with fish fed the control diet, while the inclusion of 5% blood meal had no effect on growth or sensory characteristics (Martínez-Llorens et al., 2008). Meanwhile, in pirarucu (Arapaima gigas), blood meal was successfully included at a level of not more than 6% (Ribeiro et al., 2011). Feeds for rainbow trout were successfully formulated for rainbow trout, Oncorhynchus mykiss with 22.7% blood powder and 17.0% fish meal which contained 65.3% less fish meal and 38% less total phosphorus than present in the control diet (Luzier et al., 1995). In a study using juvenile catfish, Clarias gariepinus, fish meal can be totally replaced by blood meal with no adverse effect on growth, survival, and feed conversion (Agbebi et al., 2009). Approximately 35% of fish meal protein could be replaced by both fermented and unfermented blood meal for juvenile silver pompano, Trachinotus blochii without compromising growth performance and feed efficiency, potentially leading to significant cost savings (Hamed et al., 2017).

Studies on the mixture of blood meal with other alternative proteins were also evaluated with promising findings. For example, blood meal cooked with broken rice meal successfully replaces fish meal at 25% replacement level in tambaqui, Colossoma macropomum (Martins & Guzman, 1994), and up to 80% of fish meal protein can be replaced by processed meat meal and blood meal with no adverse effects on growth, survival, and feed conversion ratio of grouper, Epinephelus coioides juveniles (Millamena, 2002). These findings show that the performance of blood meal is strongly influenced by the target species and processing methods used in meal production. Low palatability and deficiency in certain essential amino acids such as methionine have limited the high inclusion of blood meal in fish feed formulations.

7.2.1.3 MEAT AND BONE MEAL

Meat and bone meal is derived from the rendered product from mammal tissues, including bone, but exclusive of blood, hair, hoof, horn, hide trimmings, manure, and stomach contents. The product must contain a minimum of 4% phosphorus with a calcium level not exceeding 2.2 times the phosphorus content to be defined as meat and bone meal (Ockerman & Basu, 2014). Meat and bone meal contains high protein (48–66%), moderate lipid (8–16%), and high ash (19–31%). It is a good source of leucine and lysine but is deficient in methionine (Donkoh et al., 1994; Li et al., 2008; Rossi & Davis, 2014). Due to the presence of bones, meat and bone meal contains appreciable amounts of total nitrogen (~8%), phosphorus (~5%), and calcium (~10%) (Jeng et al., 2006). In general, high variation in the chemical composition of meat and bone meal was observed, mostly due to the different raw materials quality, processing technology used, and the ratio of meat to bone content.

Substitution of 50% fishmeal did not compromise the growth, feed, and nutrient efficiency of gilthead seabream (Sparus aurata) juveniles. At the same time, essential amino acid retention was not affected by meat and bone meal inclusion and this replacement of fishmeal protein with meat and bone meal can improve economic efficiency (Moutinho et al., 2017). Meat and bone meal could replace up to 50% of fish meal protein in diets for river prawns, Macrobrachium nipponense (Yang et al., 2004b). Using Florida

Pompano, Trachinotus carolinus as a target species, fish meal was successfully replaced by meat and bone meal in a soybean meal-based diet with the addition of taurine and methionine (Rossi & Davis, 2014). In contrast, the substitution of fish meal with meat and bone meal had resulted in reduced growth performance and feed efficiency in juvenile redrum, Sciaenops ocellatus (Kureshy et al., 2000). Overall, the utilization of meat and bone meal in aquaculture feeds is limited by the high ash content and deficiency in essential amino acids.

7.2.2 FISHERY BY-PRODUCT

By-products from the fishery industry account for about 25% of the total production (Rustad, 2003). These by-products can be in the form of the head (contribute the highest proportion), viscera, skin, bones, meat attached to the bones, and fins, which are made up of about 15% of the total fish body weight. In the country where the fish is normally consumed as filets, the by-products are more abundant compared to those consumed as whole fish. Fish by-products are often ended in the environment and cause pollution (Figure 7.1). This will not only help in reducing the amount of waste but also gain an economic advantage. The nutritional content of fishery by-products is influenced by the quality of the raw fish or crustaceans and the proportion of different parts of the body. In general, the nutritional value of the by-product is slightly lower than the fish meal. Protein, lipid, ash, and chitin normally ranged from 44–52%, 8–9.5%, 23–28%, and 12–15%, respectively (Begum et al., 2006). In general, skin presented the most significant protein source, trimmings, and bones were high in calcium, and the head, intestines, and bones were a good source of lipids (Kandyliary et al., 2020). Meanwhile, fishery by-products originating from marine species have better amino acid and fatty acid profiles than those based on terrestrial animal and plant sources. The by-product from the aquaculture industry has also been explored in recent years which revealed interesting findings. In our study using hybrid grouper by-products, fins contained a high level of protein (68.12%) and low Na/K ratios (fins: 0.4). Meanwhile, filet and fins had the same PUFA/SFA ratio (0.43) and were higher than bones (0.31) and viscera (0.25). Glycine and proline were the most abundant amino acids, while calcium was the major mineral in the bones. Each part of the fish could be potential new sources of specific nutrient components (Chan et al., 2019).

FIGURE 7.1 Anchovy by-products often discarded with no particular use.

Substantial studies have been carried out to replace fish meal with fishery by-product meal. Most studies were using tuna by-products due to the high by-products quantity which is related to the fish size (Table 7.4). Other fishery by-products such as salmon by-product meal, anchovy by-product meal, and shrimp head meal are also under investigation. There is a high potential for using fishery by-product meal as a source of protein in aquaculture feeds due to the high nutritional values. In some cases, fish by-products are also subjected to hydrolysis treatment to form fish hydrolysate to improve the utilization of the meal. The challenges of using a high amount of fishery by-product meal in feed formulation might be influenced by the presence of toxic heavy metals such as mercury (Kim et al., 2019).

7.3 INGREDIENTS BASED ON PLANT BY-PRODUCTS

7.3.1 RICE BRAN

Rice bran is a by-product of the rice-milling industry which was produced during the conversion of paddy into white rice. Rice bran is available all year round, especially in rice-producing countries, and is often traded as animal feed. Rice bran contains protein (12–17%), lipid (13–23%), fiber (6–14%), carbohydrate (34–54%) and ash (8–18%) (Saunders, 1985). Other than that, rice bran also contains non-starch polysaccharides, phenolic acids, flavonoids, tocopherols, tocotrienols, c-oryzanol, and phytic acid (Aluddin

TABLE 7.4 Replacement of Fish Meal with Fishery By-Product Meal

Target Species	Findings	References
Juvenile olive flounder (*Paralichthys olivaceus*)	Substitution of fishmeal with up to 30 and 20% tuna by-product meal could be made without adverse effects on growth and feed utilization of juvenile olive flounder, respectively.	Kim et al. (2014)
Juvenile Korean rockfish (*Sebastes schlegeli*)	Tuna by-product meal used in this study could replace up to 75% of fishmeal protein in the diet without reducing growth and feed utilization in juvenile Korean rockfish.	Kim et al. (2018)
Red tilapia	Tilapia fed FBPM 50 attained the highest final body weight, weight gain, specific growth rate, feed intake, protein efficiency ratio, and the best feed conversion ratio. All fish fed FBPM-based diets also attained better survival than those fed the control diet.	Fum et al. (2017)
Nile tilapia (*Oreochromis niloticus*), fingerlings	Tilapia fed the tuna by-product meal diet had greater weight gain and feed intake, and better feed conversion ratios than those fed diets containing soybean meal diet.	Hernández et al. (2013)
Spotted rose snapper (*Lutjanus guttatus*)	Fish meal protein can be successfully replaced by tuna by-product meal at 25–30% replacement levels in spotted rose snapper diets but higher replacement levels reduce fish performance.	Hernández et al. (2014)
Pacific white shrimp (*Litopenaeus vannamei*)	Growth performance and feed conversion ratio were not statistically different when salmon by-product meal replaced anchovy meal at up to 100% replacement level.	Guo et al. (2020)
Sex-reversed red tilapia (*Oreochromis niloticus*) x *O. mossambicus*	Shrimp head meal at the level of 50% replacement or 13.84% by weight of diet is a potential protein source in sex-reversed red tilapia diet.	Chimsung et al. (2006)

et al., 2019; Zhuang et al., 2019). The biochemical composition of rice bran is influenced by the plant variety and planting site, grinding process, and the physical characteristics of the grain.

In the aquaculture industry, rice bran is commonly added to fish feeds to supply energy and reduce feed costs. For example, rice bran can substitute soybean meal in the diet for catfish, Clarias gariepinus at up to 80% without significantly compromising growth and feed utilization (Samuel & Shapawi, 2016). Since rice bran is about 60% cheaper than soybean meal, the feed cost will be reduced significantly. Comparable growth performance, survival, and condition factors between tilapia, Oreochromis niloticus fed on rice bran diet, and mixed ingredients were observed indicating the potential of this ingredient (Limbu et al., 2016). Khan (2008) reported reasonably high apparent digestibility coefficients of rice bran in tropical freshwater catfish with values for dry matter, protein, and energy were 85.81%, 81%, and 67.93%, respectively. The main challenge in the use of rice bran is the considerably high-fat content that can promote auto-oxidation and reduce its shelf-life and high concentrations of fiber that can reduce the digestibility of energy and bioavailability of nutrients. Nevertheless, the de-fatting process can be applied to reduce the fat in rice bran and rice bran can be used in combination with other ingredients to balance out the nutritional profile of the formulated feeds.

7.3.2 FRUIT AND VEGETABLE WASTE

Waste from fruit and vegetable accounts for about one-third of the industry's total production. Due to their biodegradable nature and irresponsible consumption, these products are either spoiled when they are still on the farm, during transportation, or wasted by the consumers. Fruit and vegetable waste is also derived from the unutilized parts that are thrown away at several stations such as collection, handling, shipping, and processing, for example, pomace, peels, rind, and seeds. They are rich sources of fiber, bioactive compounds such as carotenoids, enzymes, polyphenols, oils, vitamins, lignin, sugar, and many other components (Arvanitoyannis & Varzakas, 2002; Kumar et al., 2020).

Fruit and vegetable wastes significantly contribute to environmental pollution via different pathways. Reusing the waste from fruit and vegetable waste is an important effort to reduce pollution and increase economic benefits. So far, after several processes and manipulations, fruit and vegetable waste are being considered raw materials in producing fertilizer, beverages, animal feeds, renewable energy, food additives, biosorbents, and microbial media. Meanwhile, several kinds of research have been conducted to evaluate the potential of fruit and vegetable waste as an ingredient in aquaculture feeds, especially citrus fruit peel. It is well documented that citrus peels are a rich source of fiber, vitamins, minerals, antioxidants, and essential oils. In Nile tilapia fingerlings, dietary orange peel can act as a growth promoter for with an optimum level of 2 g kg–1 diet by improving the nutrient absorptive ability of the intestine (Salem & Abdel-Ghany, 2018). When diets were supplemented with 1%–3% fermented lemon peel, improved intestinal health of Asian sea bass was observed. However, it was suggested that 5%

supplementation may lead to low immune response and high oxidative stress since high lipid content was observed in the fish liver (Zhuo et al., 2021). Based on these findings, citrus peels are more suitable to be included in aquaculture feed formulation at low levels and serve as feed additives.

The major constraint in the utilization of vegetable waste and by-products as aquaculture feed ingredients are their high moisture content and presence of pesticide residues. In most cases, vegetable waste is normally subjected to a drying, grinding, and ensiling process before it can be used as an ingredient. Al-Azad & Lal (2018) utilized leafy vegetable waste by bioconverting them with bacteria to improve the nutritional value and included them in the diets of tilapia (Oreochromis niloticus). There were no significant differences in the feed intake and feed conversion ratio among the diet composed of commercial feed with diets added with 5% and 10% of the bio-converted product. Vegetable waste consisting of cabbage, water spinach, spinach, and mustard were turned into silage with 5% formic acid giving the best results when incorporated into the diets for tilapia (Putra et al., 2019).

7.3.3 MICROALGAE CULTURE BY-PRODUCT

Aquaculture farms use different types of culture systems. One of them is the green water culture system which uses microalgae such as Chlorella and Nannochloropsis. Green water serves as direct food for larvae, stabilizing the water quality, providing color contrast inside the tank for better vision of larval fish, microbial control, and stimulating the immune system of larval fish. At a certain point in the production phase, the green water needs to be discharged to maintain the water quality. Instead of discharging the water into the environment, it can be reused by harvesting and concentrating the green water into meal (powder) form. Green water meal contains about 39% crude protein, 1.6% crude lipid, 8.4% crude fiber, and 6% crude ash. Green water is also a good source of carotenoids (Shapawi et al., 2017). When green water meal was used to replace fish meal in the diets formulated for Pacific white shrimps at 10% replacement level, growth performance and feed utilization were not compromised. At the same time, diets based on green water meal produced shrimps with a more intense red color and higher total carotenoid contents (Abdul Basri et al., 2015; Figure 7.2). Nevertheless, like other plant-based ingredients, the challenges in using high inclusion of green water meal in the dies are influenced by the low digestibility and inferior certain essential amino acid content.

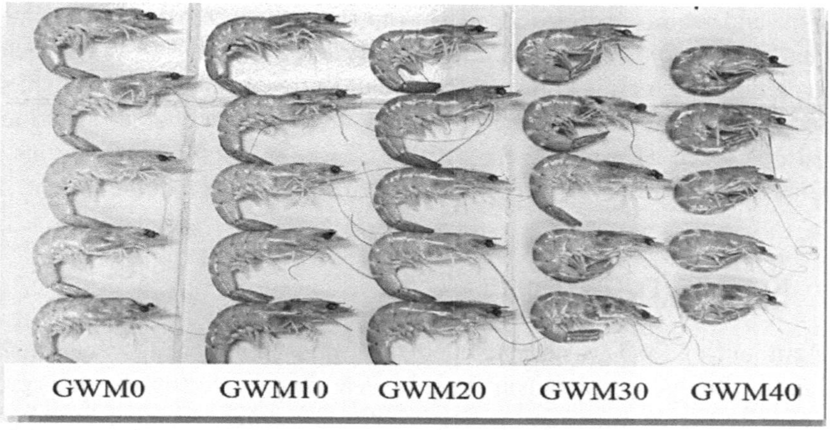

FIGURE 7.2 The increased color intensity of shrimp when fed diets containing green water meal.

Source: Reprinted with permission from Abdul Basri et al., 2015. Copyright © 2014, Springer Science Business Media Dordrecht

7.3.4 BREWERY WASTE

The brewery industry is a large-scale industry producing alcoholic drinks from cereal grains such as barley and hops. The production of alcoholic drinks (beer) can be divided into several processes to convert the grain into simple sugars. The extracted sugars undergo fermentation with the use of yeast to produce low-alcohol beer. This results in the production of a huge amount of organic waste, especially in the form of spent yeast, grain, and hops (Rachwał et al., 2020). Brewery-spent yeast is still underutilized even though it is characterized by high protein, carbohydrate, minerals, and vitamin contents. At present, the animal feed industry is the main user of this by-product with several positive outcomes.

Increasing interest in exploring the use of brewery waste in aquaculture feed is witnessed in recent years. The effect of partial replacement of fishmeal by the brewery industry by-products in isoproteic and isolipidic diets for gilthead sea bream (Sparus aurata) revealed that spent yeast and spent grain are highly digestible by the fish and can be included up to 30% and 15%, respectively, yielding similar growth and filet quality to those fed commercial feed (Estéveza et al., 2021). In Pacific white shrimp, Litopenaeus vannamei, the protein-rich product produced from brewer's spent grain can partially replace fishmeal at up to 50% without negatively affecting shrimp survival, growth performance, feed utilization efficiency, or the protein content and amino acid profile of shrimp (He et al., 2020). Unfortunately, untreated brewery spent grain contains high levels of anti-nutritional factors such as lignin, which is known to reduce digestive performance in fish, and under an unsuitable storage environment, it can be perishable due to high moisture content (Karlsen & Skov, 2022). More in-depth investigations are required to determine the full potential of this high-potential ingredient.

7.4 KITCHEN FOOD WASTE

The amount of food waste generated from households, restaurants, hotels, and airline kitchens is increasing year by year. The waste poses serious problems to the environment and is very costly to manage. Typical food waste from the kitchens includes meat (red meat, poultry, and seafood), bones, bakery products, cereal, fruits, and vegetables. In general, kitchen food waste is supposed to contain good nutritional values as they were targeted for human consumption. However, this waste is normally inconsistent in composition and contains very high moisture and high contaminant contents. Efforts have been made to convert the waste into fertilizer, biogas, and also as a substrate in the black soldier fly breeding program. Unlike other organic by-products that were previously discussed, little attempt to utilize kitchen food waste as aquaculture feed ingredients due to the complicated nature of the by-products. They can either be separated into categories (e.g., vegetables and fruits category) or used as it is (mixture of different kinds). Grass carp fed with food waste feeds achieved a specific growth rate and feed conversion ratio that was similar to commercial feeds. However, the concentrations of heavy metals in food waste feeds were relatively higher than in other feeds (Cheng et al., 2015). On the other hand, heavy metal concentrations in bighead carp, grass carp, and mud carp fed with food waste feeds were all below the international maximum permissible levels in food. This study suggested that health risk assessments indicated that human consumption of the fish fed with food waste-based feeds was safe for consumption (Cheng et al., 2016). More research is needed to strategize on how to improve the utilization of this abundant waste.

7.5 CONCLUSION

Organic by-products are valuable sources of nutrients and biomasses. At the moment, these by-products were still under exploitation and cause a major issue in their collection and disposal. Increasing interest in the use of organic by-products as ingredients in aquaculture feeds is based on their promising performance in other animal feed sectors. So far, animal-based by-products appear to be more promising than those originated from plants in terms of performance when included in the aquaculture feeds. For example, poultry by-products are commercially used as major protein sources in aquaculture feeds. This is due to the fact that aquatic species such as fish and crustacean have limited ability to digest dietary carbohydrate due to the lack of amylase enzyme and requires high protein in their diets. In recent years, plant-based by-products such as brewery waste have gained more attention due to their promising performance and availability. In general, partial replacement of fish meal with organic by-products is highly possible and the advancement in technology has contributed to the improved utilization of organic by-products in aquaculture feeds. Some by-product ingredients have shown the potential to improve growth, pigmentation, and immunity. Deficiency of essential nutrients, presence of anti-nutritional factors, low digestibility, and palatability issues are the main constraints in replacing fish-based dietary ingredients with those from organic waste. Nevertheless, several approaches have been adopted to overcome these issues. The success in using organic by-product-based ingredients in aquaculture feed formulation will help the industry to grow sustainably and consistent with the principles of circular bioeconomy.

KEYWORDS

- **animal by-products**
- **aquaculture feeds**
- **by-products**
- **circular bioeconomy**
- **microalgae**
- **organic by-products**
- **plant by-products**

REFERENCES

Abdul, B. N., Muhamad, S. S. R., Matanjun, P., Mohammad, N. N., & Shapawi, R., (2015). The potential of microalgae meal as an ingredient in the diets of early juvenile Pacific white shrimp, Litopenaeus vannamei. Journal of Applied Phycology, 27, 857–863. doi: 10.1007/s10811-014-0383-6.

Agbebi, O. T., Otubusin, S. O., & Ogunleye, F. O., (2009). Effect of different levels of substitution of fishmeal with blood meal in pelleted feeds on catfish Clarias gariepinus (Burchell, 1822) culture in net cages. European Journal of Scientific Research, 31(1), 6–10.

Al Azad, S., & Lal, M. T. B. M., (2018). Potentiality of bioconverted leafy vegetable waste as aquaculture feed supplement. Advances in Bioscience and Biotechnology, 9, 571–583. https://doi.org/10.4236/abb.2018.911040.

Alauddin, M. D., Rahman, S., Islam, J., Shirakawa, H., Komai, M., & Howlader, M. Z. H., (2019). "Chapter 8 – development of rice bran functional food and evaluation of its healthful properties." In: Ling-Zhi, C., & Xuebing, X., (eds.), Rice Bran and Rice Bran Oil (pp. 183–206). AOCS Press. https://doi.org/10.1016/B978-0-12-812828-2.00008-1.

Arunlertaree, C., & Rakyuttithamkul, E., (2006). Utilization of fermented feather meal as a replacement for fish meal in the diet of hybrid Clarias catfish. Kasetsart Journal – Natural Science, 40(2), 436–448.

Arvanitoyannis, I. S., & Varzakas, T. H., (2002). Management: Treatment methods treated waste. In: Waste Management for the Food Industries. https://doi.org/10.1016/B978-0-12-373654-3.50014-6.

Begum, S., Ikejima, K., Ara, H., & Islam, M. Z., (2006). Solar drying as an option for shrimp processing biowaste in Khulna district-southwest Bangladesh. Journal of Applied Sciences, 6(6), 1302–1306. https://doi.org/10.3923/JAS.2006.1302.1306.

Bishop, C. D., Angus, R. A., & Watts, S. A., (1995). The use of feather meal as a replacement for fish meal in the diet of Oreochromis niloticus fry. Bioresource Technology, 54, 291–295. https://doi.org/10.1016/0960-8524(95)00146-8.

Bureau, D. P., Harris, A. M., & Cho, C. Y., (1999). Apparent digestibility of rendered animal protein ingredients for rainbow trout. Aquaculture, 180, 345–358. doi: 10.1016/S0044-8486(99)00210-0.

Chaklader, M. R., Siddik, M. A. B., & Fotedar, R., (2020). Total replacement of fishmeal with poultry by-product meal affected the growth, muscle quality, histological structure, antioxidant capacity, and immune response of juvenile barramundi, Lates calcarifer. PLoS One, 15(11), e0242079. https://doi.org/10.1371/journal.pone.0242079.

Chan, P. T., Matanjun, P., Shapawi, R., Budiman, C., & Lee, J. S., (2019). Chemical composition of the fillet, fins, bones, and viscera of hybrid grouper (Epinephelus lanceolatus x Epinephelus fuscoguttatus). Journal of Physics: Conference Series, 1358, 012008. doi: 10.1088/1742-6596/1358/1/012008.

Cheng, Z., Lam, C. L., Mo, W. Y., Nie, X. P., Choi, W. M., Man, Y. B., & Wong, M. H., (2016). Food wastes as fish feeds for polyculture of low-trophic-level fish: Bioaccumulation and health risk assessments of heavy metals in the cultured fish. Environmental Science Pollution Research, 23(8), 7195–7203. doi: 10.1007/s11356-016-6484-9.

Cheng, Z., Mo, W. Y., Nie, X. P., Li, K. B., Choi, W. M., Man, Y. B., & Wong, M. H., (2015). The use of food waste-based diets and Napier grass to culture grass carp: Growth performance and contaminants contained in cultured fish. Environmental Science Pollution Research. doi: 10.1007/s11356-530 015-5189-9. 531.

Chimsung, N., Chealoh, N., Pimolrat, P., & Tantikitti, C., (2006). Effects of shrimp head meal in the diets on growth, feed efficiency and pigmentation of sex-reversed red tilapia, Oreochromis niloticus x O. mossambicus. Songklanakarin Journal of Science and Technology, 8(5), 951–964.

Chor, W. K., Lim, L. S., & Shapawi, R., (2013). Evaluation of feather meal as a protein source for African catfish, Clarias gariepinus. Journal of Fisheries and Aquatic Science, 8(6), 697–705. 10.3923/jfas.2013.697.705.

Donkoh, A., Moughan, P. J., & Smith, W. C., (1994). True ileal digestibility of amino acids in meat and bone meal for the growing pig—application of a routine rat digestibility assay. Animal Feed Science, 49(1, 2), 73–86. https://doi.org/10.1016/0377-8401(94)90082-5.

El-Sayed, A. F. M., (1998). Total replacement of fish meal with animal protein sources in Nile Tilapia, Oreochromis niloticus (L.), feeds. Aquaculture Research, 29, 275–280. https://doi.org/10.1046/j.1365-2109.1998.00199.x.

Estéveza, A., Padrell, L., Iñarra, B., Orive, M., & San, M. D., (2021). Brewery by-products (yeast and spent grain) as protein sources in gilthead seabream (Sparus aurata) feeds. Aquaculture, 543, 736921. https://doi.org/10.1016/j.aquaculture.2021.736921.

Ewing, W. N., (1997). The Feed Directory: Commodity Products Guide. Context Products Ltd., Publications Division. Packington, UK.

FAO, (2020). The state of world fisheries and aquaculture 2020. Sustainability in Action (p. 244). Rome, Italy.

Fowler, L. G., (1990). Feather meal as a dietary protein source during parr-smolt transformation in fall chinook salmon. Aquaculture, 89, 301–314. https://doi.org/10.1016/0044-8486(90)90134-9.

Fum, K. Y., Shapawi, R., & Lim, L. S., (2017). Fish by-product meal served as a good protein source in the formulated diets for red tilapia fry. Songklanakarin Journal of Science & Technology, 39(6), 813–817.

Gaylord, T. G., & Rawles, S. D., (2005). The modification of poultry by-product meal for use in hybrid striped bass Morone chrysops X M. saxatilis diets. Journal of the World Aquaculture Society, 36, 363–374. https://doi.org/10.1111/j.1749-7345.2005.tb00340.x.

Glencross, B. D., Booth, M., & Allana, G. L., (2007). Feed is only as good as its ingredients – a review of ingredient evaluation strategies for aquaculture feeds. Aquaculture Nutrition, 13, 17–34. https://doi.org/10.1111/j.1365-2095.2007.00450.x.

Gunben, E. M., Senoo, S., Yong, A., & Shapawi, R., (2014). High potential of poultry by-product meal as a main protein source in the formulated feeds for a commonly cultured grouper in Malaysia (Epinephelus fuscoguttatus). Sains Malaysiana, 43(3), 399–405.

Guo, J., Swanepoel, A., Qiu, X., Reis, J., Rhodes, M., & Davis, D. A., (2020). Use of salmon by-product meals as a replacement for anchovy meal in practical diets for the Pacific white shrimp (Litopenaeus vannamei). Aquaculture Nutrition, 26(2), 490–501. https://doi.org/10.1111/ANU.13011.

Hamed, S. S., Jiddawi, N. S., & Bwathondi, P. O. J., (2017). Effects of blood meal as a substitute for fish meal in the culture of juvenile silver pompano Trachinotus blochii (Lacepède, 1801) in a circulating aquaculture system. Western Indian Ocean Journal of Marine Science, 16(1), 1–11.

Hasan, M. R., Haq, M. S., Das, P. M., & Mowlah, G., (1997). Evaluation of poultry-feather meal as a dietary protein source for Indian major carp, Labeo rohita fry. Aquaculture, 151, 47–54. 10.1016/S0044-8486(96)01498-6.

He, Y., Galagarza, O. A., Wang, H., Taylor, Z. W., Ferguson, C. S., Ogejo, J. A., O'Keefe, S. F., et al., (2020). Protein-rich product recovered from brewer's spent grain can partially replace fishmeal in diets of Pacific white shrimp, Litopenaeus vannamei. Aquaculture Research, 51(8), 3284–3296. https://doi.org/10.1111/are.14664.

Hernández, C., Hardy, R. W., Contreras-Rojas, D., López-Molina, B., González-Rodríguez, B., & Domínguez-Jimenez, P., (2014). Evaluation of tuna by-product meal as a protein source in feeds for juvenile spotted rose snapper Lutjanus guttatus. Aquaculture Nutrition, 20(6), 574–582. https://doi.org/10.1111/ANU.12110.

Hernández, C., Olvera-Novoa, M. A., Voltolina, D., Hardy, R. W., González-Rodríg, B., Dominguez-Jimenez, P., Valverde-Romero, M., & Agramon-Romero, S., (2013). Use of tuna industry waste in diets for Nile tilapia, Oreochromis niloticus, fingerlings: Effect on digestibility and growth performance. Latin America Journal of Aquatic Research, 41(3), 468–478. doi: 103856/vol41-issue3-fulltext-10.

Jeng, A. S., Haraldsen, T. K., Grønlund, A., & Pedersen, P. A., (2006). Meat and bone meal as nitrogen and phosphorus fertilizer to cereals and ryegrass. Nutrient Cycling in Agroecosystems, 76, 183–191. https://doi.org/10.1007/s10705-005-5170-y.

Kandyliari, A., Mallouchos, A., Papandroulakis, N., Golla, J. P., Lam, T. T., Sakellari, A., Karavoltsos, S., et al., (2020). Nutrient composition and fatty acid and protein profiles of selected fish by-products. Foods, 9, 190. doi: 10.3390/foods9020190.

Karlsen, F., & Skov, P. V., (2022). Review – potentials and limitations of utilizing brewer's spent grain as a protein source in aquaculture feeds. Journal of Cleaner Production, 357, 131986. https://doi.org/10.1016/j.jclepro.2022.131986.

Khan, M. S., (2008). Apparent digestibility coefficients for common feed ingredients in formulated diets for tropical catfish, Mystus nemurus (Cuvier & valenciennes). Aquaculture Research, 25, 167–174. https://doi.org/10.1111/j.1365-2109.1994.tb00572.x.

Kikafunda, J. K., & Sserumaga, P., (2005). Production and use of a shelf-stable bovine blood powder for food fortification as a food-based strategy to combat iron deficiency anemia in Sub-Saharan Africa. African Journal of Food Agriculture and Nutritional Development (AJFAND), 5(1), 1–17. 10.18697/ajfand.8.1590.

Kikuchi, K., Furuta, T., & Honda, H., (1994). Utilization of feather meal as a protein source in the diet of juvenile Japanese flounder. Fisheries Science, 60(2), 203–206.

Kim, H. S., Jung, W. G., Myung, S. H., Cho, S. H., & Kim, D. S., (2014). Substitution effects of fishmeal with tuna by-product meal in the diet on growth, body composition, plasma chemistry and amino acid profiles of juvenile olive flounder (Paralichthys olivaceus). Aquaculture, 431, 92–98. https://doi.org/10.1016/J.AQUACULTURE.2014.03.025.

Kim, K. D., Jang, J. W., Kim, K. W., Lee, B. J., Hur, S. W., & Han, H. S., (2018). Tuna by-product meal as a dietary protein source replacing fishmeal in juvenile Korean rockfish Sebastes schlegeli. Fisheries and Aquatic Sciences, 21(1). https://doi.org/10.1186/s41240-018-0107-y.

Kumar, H., Bhardwaj, K., Sharma, R., Nepovimova, E., Kuča, K., Dhanjal, D. S., Verma, R., et al., (2020). Fruit and vegetable peels: Utilization of high-value horticultural waste in novel industrial applications. Molecules, 25(12), 2812. doi: 10.3390/molecules25122812. PMID: 32570836; PMCID: PMC7356603.

Kureshy, N., Davis, D. A., & Arnold, C. R., (2000). Partial replacement of fishmeal with meat-and-bone meal, flash-dried poultry by-product meal, and enzyme-digested poultry by-product meal in practical diets for juvenile red drum. North American Journal of

Aquaculture, 62, 266–272. https://doi.org/10.1577/1548-8454(2000)062<0266:PROFMW >2.0.CO;2.

Li, M. H., Robinson, E. H., & Lim, C. E., (2008). Use of meatpacking by-products in fish diets. In: Lim, C., Webster, C. D., & Lee, C. S., (eds.), Alternative Protein Sources in Aquaculture Diets (pp. 95–112). The Haworth Press. Taylor & Francis Group: New York & London.

Limbu, S. M., Shoko, A. P., Lamtane, M. A., Kishe, M. A., Joram, M. C., Mbonde, A., Mgana, H. F., & Mgaya, Y. D., (2016). Supplemental effects of mixed ingredients and rice bran on the growth performance, survival, and yield of Nile tilapia, Oreochromis niloticus reared in fertilized earthen ponds. SpringerPlus, 5(5), 2–13. doi: 10.1186/s40064-015-1643-x.

Luzier, J. M., Summerfelt, R. C., & Ketola, H. G., (1995). Partial replacement of fish meal with spray dried blood powder to reduce phosphorus concentrations in diets for juvenile rainbow trout Oncorhynchus mykiss. Aquaculture Research, 26, 577–587. https://doi.org/10.1111/j.1365-2109.1995.tb00948.x.

Martínez-Llorens, S., Tomás, V. A., Moñino, A. V., Ader, J. G., Torres, M. P., & Joker, C. M., (2008). Blood and haemoglobin meal as protein sources in diets for gilthead sea bream (Sparus aurata): Effects on growth, nutritive efficiency, and fillet sensory differences. Aquaculture Research, 39(10), 1028–1037. https://doi.org/10.1111/j.1365-2109.2008.01961.x.

Martins, S. N., & Guzmán, E. C., (1994). Effect of drying method of bovine blood on the performance of growing diets for tambaqui (Colossoma macropomum, Cuvier 1818) in experimental culture tanks. Aquaculture, 124(1–4), 335–341. 10.1016/0044-8486(94)90406-5.

Mautinho, S., Martínez-Llorens, S., Tomás-Vidal, A., Jover-Cerdá, M., Oliva-Teles, A., & Peres, H., (2017). Meat and bone meal as partial replacement for fish meal in diets for gilthead seabream (Sparus aurata) juveniles: Growth, feed efficiency, amino acid utilization, and economic efficiency. Aquaculture, 468(1), 271–277. 10.1016/j.aquaculture.2016.10.024.

Millamena, O. M., (2002). Replacement of fish meal by animal by-product meals in a practical diet for grow-out culture of grouper Epinephelus coioides. Aquaculture, 204(1–4), 75–84. 10.1016/S0044-8486(01)00629-9.

National Research Council, (1994). Nutrient Requirements of Poultry (9th revised edn.). National Academic Press.

Nengas, I., Alexis, M. N., & Davis, S. J., (1999). High inclusion levels of poultry meals and related by-products in diets for gilthead sea bream, Sparus aurata L. Aquaculture, 179, 12–23. 10.1016/S0044-8486(99)00148-9.

Ockerman, H. W., & Basu, L., (2014). By-products – inedible. In: Devine, C., & Dikeman, M., (eds.), Encyclopedia of Meat Sciences (2nd edn., pp. 679–696). Woodhead Publishing Series in Food Science, Technology and Nutrition. https://doi.org/10.1016/B978-0-08-100694-8.00022-4.

Poolsawat, L., Yang, H., Sun, Y. F., Li, X. Q., Liang, G. Y., & Leng, X. J., (2021). Effect of replacing fish meal with enzymatic feather meal on growth and feed utilization of tilapia (Oreochromis niloticus × O. aureus). Animal Feed Science and Technology, 274, 114895.

Putra, A. N., Jaenudin, Sofia, F., Mustahal, Syamsunarno, M. B., Hermawan, D., & Herjayanto, M., (2019). The utilization of vegetable waste silage as feed ingredient in diets for tilapia Oreochromis niloticus. IOP Conf. Series: Earth and Environmental Science, 383, 012002. IOP Publishing. doi: 10.1088/1755-1315/383/1/012002.

Rachwał, K., Waśko, A., Gustaw, K., & Polak-Berecka, M., (2020). Utilization of brewery wastes in food industry. PeerJ., 8, e9427. https://doi.org/10.7717/peerj.9427.

Ribeiro, R. A., De Almeida, O. R. O., Batista, S. M. G., Pereira-Filho, M., Ono, E. A., & Roubach, R., (2011). Use of spray-dried blood meal as an alternative protein source in pirarucu (Arapaima gigas) diets. Journal of Applied Aquaculture, 23(3), 238–249. https://doi.org/10.1080/10454438.2011.600635.

Rossi, Jr. W., & Davis, D. A., (2014). Meat and bone meal as an alternative for fish meal in soybean meal-based diets for Florida Pompano, Trachinotus carolinus L. Journal of Aquaculture Society, 45(6), 613–624. https://doi.org/10.1111/jwas.12155.

Rustad, T., (2003). Utilization of marine by-products. Electronic Journal of Environmental, Agricultural and Food Chemistry, 2(4), 458−463.

Salem, M., & Abdel-Ghany, H. M., (2018). Effects of dietary orange peel on growth performance of Nile tilapia (Oreochromis niloticus) fingerlings. Aquaculture Studies, 18(2), 127–134. 10.4194/2618-6381-v18_2_06.

Samuel, D. E., & Shapawi, R., (2016). Effects of different inclusion levels of rice bran in the diets on the growth performance of African catfish (Clarias gariepinus) juveniles. Malaysian Applied Biology, 45(2), 17–22.

Saunders, R. M. (1986). Rice bran: Composition and potential food uses. Food Reviews International, 1, 465–495. doi: 10.1080/87559128509540780.

Shapawi, R., Abdul, B. N., & Muhd, S. S. R., (2017). Green water meal as protein and carotenoid sources in grow-out diets for Pacific White Shrimp, Litopenaeus vannamei. Sains Malaysiana, 46(12), 2281–2289. http://dx.doi.org/10.17576/jsm-2017-4612-04.

Shapawi, R., Ng, W. K., & Mustafa, S., (2007). Replacement of fish meal with poultry by-product meal in diets formulated for the humpback grouper, Cromileptes altivelis. Aquaculture, 273, 118–126. https://doi.org/10.1016/j.aquaculture.2007.09.014.

Steffens, W., (1994). Replacing fish meal with poultry by-product meal in diets for rainbow trout Oncorhynchus mykiss. Aquaculture, 124, 27–34. https://doi.org/10.1016/0044-8486(94)90351-4.

Takagi, S. T., Hosokawa, H., Shimeno, S., & Ukawa, M., (2000). Utilization of poultry by-product meal in a diet for red sea bream Pagrus major. Nippon Suisan Gakkaishii, 66, 428–438.

Wang, Y., Guo, J. L., Bureau, D. P., & Cui, Z. H., (2006). Replacement of fish meal by rendered animal protein ingredients in feeds for cuneate drum (Nibea miichthioides). Aquaculture, 252, 476–483.

Yang, Y., Xie, S., Cui, Y., Lei, W., Zhu, X., Yang, Y., & Yu, Y. 2004b. Effect of fish meal by meat and bone meal and poultry by-product meal in diets on the growth and immune response of Macrobrachium nipponense. Fish and Shellfish Immunology, 17(2), 105–114. doi: 10.1016/j.fsi.2003.11.006.

Yang, Y., Xie, S., Cui, Y., Lei, W., Zhu, X., Yang, Y., & Yu, Y., (2004a). Effect of replacement of dietary fish meal by meat and bone meal and poultry by-product meal on growth and feed utilization of Gibel Carp, Carassius auratus gibelio. Aquaculture Nutrition, 10, 289–294. https://doi.org/10.1111/j.1365-2095.2004.00301.x.

Yang, Y., Xie, S., Cui, Y., Zhu, X., Lei, W., & Yang, Y., (2006). Partial and total replacement of fish meal with poultry by-product meal in diets for Gibel carp, Carassius auratus gibelio Bloch. Aquaculture Research, 37, 40–48. 10.1111/j.1365-2109.2005.01391.x.

Yigit, M., Erdem, M., Koshio, S., Ergun, S., Turker, A., & Karaali, B., (2006). Substituting fish meal with poultry by-product meal in diets for Black Sea Turbot Psetta maeotica. Aquaculture Nutrition, 12, 340–347. https://doi.org/10.1111/j.1365-2095.2006.00409.x.

Yu, Y., (2007). Replacement of fish meal with poultry by-product meal and hydrolyzed feather meal in feeds for finfish. In: By Lim, C., Webster, C. D., & Lee, C. S., (eds.), Alternative Protein Sources in Aquaculture Diets (pp. 51–86). The Haworth Press. Taylor and Francis Group: New York & London.

Zhuang, X., Yin, T., Han, W., & Zhang, X. (2019). Chapter 10 – nutritional ingredients and active compositions of defatted rice bran. In: Ling-Zhi, C., & Xuebing, X., (eds.), Rice Bran and Rice Bran Oil (pp. 247–270). AOCS Press. https://doi.org/10.1016/B978-0-12-812828-2.00010-X.

Zhuo, L. C., Abang, Z. D. N. Z., Yong, A. S. K., Shapawi, R., & Lin, Y. H., (2021). Effects of fermented lemon peel supplementation in diet on growth, immune responses, and intestinal morphology of Asian sea bass, Lates calcarifer, Aquaculture Reports, 21, 100801. https://doi.org/10.1016/j.aqrep.2021.100801.

CHAPTER 8

Use of Pineapple (*Ananas comosus* L.) Waste and Its Application in Different Biotechnological Sectors

A. DANITZA CASAS-RODRÍGUEZ,[1] JUAN A. ASCACIO-VALDÉS,[1] MIRIAM DESIRÉE DÁVILA-MEDINA,[1] LILIANA LONDOÑO-HERNÁNDEZ,[2] and LEONARDO SEPÚLVEDA-TORRE[1]

[1]*Food Research Department, School of Chemistry, Autonomous University of Coahuila, Mexico*

[2]*Biotics Group, School of Basic Sciences, Technology, and Engineering, Universidad Nacional Abierta y a Distancia, Colombia*

ABSTRACT

In recent years, there has been a growing interest in using agro-industrial waste (fruit or vegetable peel) since there has been a great increase in this waste, and it has not been fully utilized. However, it is intended that all these wastes have a more useful life since it has been proven that they contain compounds that can be beneficial to health and can also have various applications in cosmetology, food, and drugs, among others. Several studies have been dedicated to investigating the use of agro-industrial waste, and pineapple has not been the exception as it has been proven to have a myriad of benefits to human health and also seeks to give a use to the waste of this (leaves, peels, and core) as it is known to contain ash, fats, proteins, fiber, carbohydrates, etc., In addition, the residues are a source of bioactive compounds since it has been proven that they contain phenolic compounds, aromatic compounds, alcohols, esters, and the enzyme bromelain, among others. It is known that these have antioxidant, anti-inflammatory, and

Waste Biorefineries: Future Energy, Green Products, and Waste Treatment.
Abu Zahrim Yaser, Siew-Xian Chin, Leonardo Sepúlveda Torre, & A. K. Haghi, (Eds.)

even anticarcinogenic activity, among others. In order to obtain this type of compound, different extraction methods have been studied that are also environmentally friendly, and it has been proven that the residues of this fruit are an alternative in the food industry, even in the pharmaceutical industry, in the paper industry and in the textile industry, however, alternatives for the use of pineapple residues must continue to be sought, in this research we have focused on pineapple residues and the extraction methods that can be used to obtain the aforementioned compounds and what purpose they can have in different areas. In addition, it is intended that agro-industrial waste can have a second chance and be discarded.

8.1 INTRODUCTION

Over the years, fruits and vegetables have become particularly important because they contain compounds that are beneficial to human health, and nowadays, the demand for processed and packaged foods is increasing, which generates a greater processing of fruits and vegetables. Tropical fruits are the ones that have had greater importance due to their nutritional value; among them are avocado, papaya, mango, pineapple, banana, pomegranate, and grape, among others (Banerjee, Ranganathan, Patti, & Arora, 2018). Pineapple is known to be an herbaceous plant that reaches between 1 and 2 m in height and 1–2 in width and is cultivated in tropical regions (Asim et al., 2015). It is known that this fruit is native to Latin America, and since its discovery it has been found in different countries and currently occupies the first places among tropical fruits in world production (Hossain & Rahman, 2011; Zakaria et al., 2021). The main pineapple producing countries are Costa Rica, Brazil, Philippines, China, India, Thailand, Mexico, Indonesia, Nigeria, and Colombia; and in Mexico the production of this fruit increases every year and has three main destinations, within the first is the consumption of fresh fruit which corresponds to 70%, the second is the industrialization of the fruit that is dedicated to the production of juices, jams, nectars, candies, vinegar, cattle feed, among others, which corresponds to 25% and finally the export market which corresponds to between 5% and 7% (Sánchez-Hernández, Huja-Mendoza, & Acevedo-Gómez, 2015). Due to the processing of this fruit and many other tropical fruits, substantial amounts of waste are generated which causes a problem (Banerjee, Ranganathan, Patti, & Arora, 2018). The amount of waste generated is diverse for each type of fruit and within the industrialization of pineapple, around 25 to 50%

of waste is generated considering pineapple peels, core, leaves, and crown of the pineapple (Sánchez-Hernández, Huja-Mendoza, & Acevedo-Gómez, 2015). These wastes contain carbohydrates (fructose, glucose, and sucrose), proteins, cellulose, lignin, pectin, hemicellulose, ash, vitamins (vitamins A, B, C), minerals (calcium, magnesium, potassium, manganese, sodium, copper, phosphorous, zinc, among others), enzymes (bromelain), bioactive compounds (phenols, flavonoids, condensed, and hydrolysable polyphenols) carotenoids, among others (Banerjee, Ranganathan, Patti, & Arora, 2018; Mohd-Ali, Hashim, Abd-Aziz, & Lasekan, 2020; Sun, Zhang, Soler, & Marie-Alphonsine, 2016). To obtain bioactive compounds, there are currently technological alternatives for the determination of bioactive compounds, among them are conventional and emerging technologies, however, they have presented a few advantages and disadvantages, so new alternatives are sought that benefit both economically and for a reliable performance for the bioactive compounds that are sought, in addition to being environmentally friendly. Therefore, biotechnological technologies have been proposed, such as fermentations (in solid and liquid state) that allow the release of these compounds (Ascacio-Valdés et al., 2014).

8.2 ORIGIN OF THE PINEAPPLE (*ANANAS COMOSUS* L.)

Pineapple (Figure 8.1) is a type of herbaceous plant, which belongs to the Bromeliaceae family which is constituted by 3010 species of which are distributed in three subfamilies: Bromelioideae (31 genres), Pitcairnioideae (16), and Tillandsioideae (León-Yánez et al., 2019). The adult plant is between 1 and 2 m tall and wide, the main morphological structures are the stem, leaves, peduncle, multiple fruits, and crown (Figure 8.2). Pineapple is currently cultivated in tropical and subtropical areas and is believed to have originated in South America, in central and southern Brazil, northern Argentina and Paraguay. Its discovery was until 1493 by the Spanish, when Columbus arrived at the American continent and gave it the name of pineapple because they saw that it resembled pinecones, the origin of the name comes from the word pine "pineapple" and comosus which means "tufted" referring to the stem of the fruit (Hassan, Othman, & Siriphanich, 2011). Another notable discovery was by the Portuguese in 1502, as they introduced the fruit to the east and west coasts of Africa. It later reached China in 1594 and South Africa in 1655 (Uriza-Ávila et al., 2018). Throughout the years, several varieties of this fruit have been found, among them Cayenne, which in turn is divided into 7 subgroups: Smooth, Cayenne, *Cayenne lisa*, F180, Sarawak, Nanglae, and Fils de Chalvet; another variety is Queen which is subdivided into 4 subgroups being Queen, Mauritius, Moris, and Mac Gregor; the Spanish variety is subdivided into 6 subgroups which are Singapore, Spanish, Red Spanish, Española Roja, Nangka, and Gandul; and the varieties of Pernambuco and Perolera (Uriza-Ávila et al., 2018). Pineapple has a versatility to adapt to various climatic conditions and it is estimated that the most favorable temperature for its growth is from 18 to 32°C, also 5 cm^3 of water is needed for optimal growth, likewise it is considered that the soil either sandy or loamy is with a pH between 4.5 and 6.5 (Wali, 2019).

FIGURE 8.1 Pineapple (*Ananas comosus* L.).

FIGURE 8.2 Morphological structure of pineapple.

8.3 WORLD PRODUCTION OF PINEAPPLE

At present, pineapple is one of the most consumed fruits worldwide, occupying second place (Usano-Alemany, Palá-Paúl, & Díaz, 2014). As of 2018, the world production of this fruit increased by approximately 8% (Vargas-Serna, Ochoa-Martínez, & Velez-Pasos, 2022) and according to FAOSTAT (FAO, 2022) data up to 2020, the top 10 pineapple producing countries are the Philippines, Costa Rica, Brazil, Indonesia, China, India, Thailand, Nigeria, Mexico, and Colombia (Figure 8.3). Nowadays this fruit is more commercialized in a processed form, whether it is found in cans, juices, jams, etc. (Campos, Ribeiro, Teixeira, Pastrana, & Pintado, 2020; Usano-Alemany, Palá-Paúl, & Díaz, 2014).

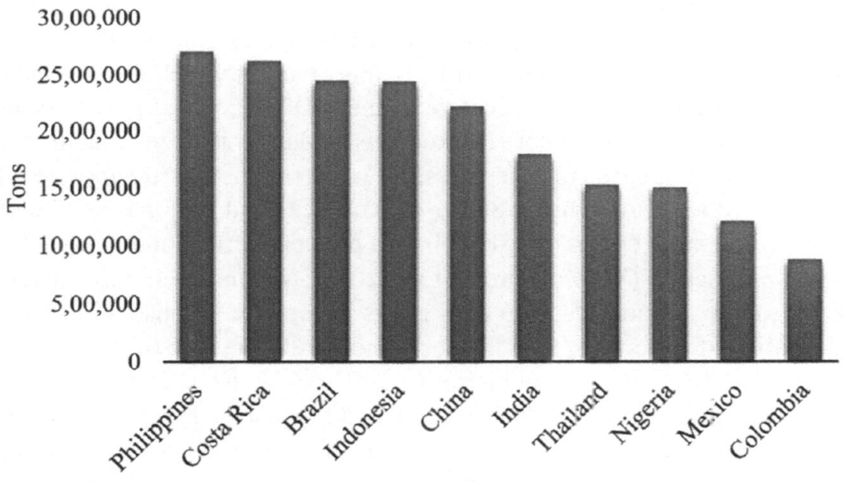

FIGURE 8.3 Pineapple production around the world.

Source: FAO (2022)

8.4 PINEAPPLE IN MEXICO

Pineapple cultivation in Mexico began in the early 20th century. One version suggests that the smooth Cayenne variety was brought from Hawaii in 1903, while another version credits Frank Peters with bringing pineapple seeds to Loma Bonita, Oaxaca in 1908. By 1924, there was a growing demand for pineapples, leading John I. Howe to plant them in the municipality of Isla de Veracruz. The production of smooth Cayenne pineapples increased significantly from the 1940s onwards, eventually occupying 90% of the national surface by 1970. Another introduced variety, the hybrid MD2, also known as Oro, Gold, or pineapple honey, was developed by the Pineapple Research Institute in 1970 (Uriza-Ávila et al., 2018). Currently, Mexico ranks among the top 10 pineapple producers globally, with the smooth Cayenne and MD2 varieties leading the way. The country's five main pineapple-producing states are Veracruz (64.9%), Oaxaca (12.8%), Tabasco (6.2%), Quintana Roo (4.6%), and Jalisco (2.8%) (Secretaria de Agricultura y Desarrollo Rural, 2017). In recent years, pineapple production in Mexico has been on the rise. In 2017, the country produced 945,000 tons of pineapples (Martínez-Castro, Ríos-Castillo, & Castillo-Leal, 2019), and by 2020, the production had reached 1,209,000 tons (Uriza-Ávila et al., 2018). Currently the production of this fruit has three main purposes, in the first place is the consumption of fresh fruit (usually in fruit cocktails) this corresponds to 70%, within the second is the industrialization of this fruit which is dedicated to the production of juices, nectars, candies, vinegar, cookies, popsicles among others, which corresponds to 25% and finally to the export market which corresponds to 5% (Sarabo et al., 2021). In recent years the MD2 hybrid has gained significant importance in the export market, this is due to its size in addition to having a higher degree of sweetness, its main purpose is to supply export markets in the USA (United States of America) and national self-service stores, with a total export of 1,00,000 tons per year corresponding to 10% of annual production (Valdés-García et al., 2021). Within the industrialization for pineapple, around 25–50% of waste is generated, it consists first in the elimination of the crown, then the peels and finally the core of the pineapple (Diaz-Vela, Totosaus, Cruz-Guerrero, & Pérez-Chabela, 2013). In recent years, pineapple and its residues have gained popularity due to its nutritional composition, as it has multiple health benefits, and the use of pineapple waste is being sought for applications in different industrial sectors.

8.5 COMPOSITION OF PINEAPPLE (*ANANAS COMOSUS* L.)

In recent years, this fruit has increased its importance due to its nutritional composition, in addition to its sweet sensory properties and health benefits (Ajayi, Coker, Oyebanjo, Adebanjo, & Ademowo, 2022; Mohd-Ali, Hashim, Abd-Aziz, & Lasekan, 2020). Among the sugars contained in pineapple are fructose, glucose, and sucrose, in addition to vitamins such as A, B1, B2, B3, B6, C, K, E, and folic acid (Sun, Zhang, Soler, & Marie-Alphonsine, 2016; Uriza-Ávila et al., 2018). It contains cellulose, lignin, pectin, hemicellulose, wax, and ash. The main minerals identified in pineapple are calcium, magnesium, potassium, manganese, sodium, copper, phosphorus, and zinc

(Mohd-Ali, Hashim, Abd-Aziz, & Lasekan, 2020). Different aromatic compounds were identified in fresh-cut pineapple juice, the identification of these compounds was carried out by capillary gas chromatography coupled to mass spectrometry (HRGC-MS), among the compounds are ethyl pentanoate, limonene, 1-pentanol, ethyl hexenoate, butyl acetate, 1-butanol, furfural, acetic acid, benzaldehyde, hexanal, linalool, among others (Elss et al., 2005). Different compounds have been identified in pineapple juice concentrate that are unique to this fruit, among them are sinapyl-L-cysteine, N-ϒ-L-glutamyl-s-sinapyl-L-cysteine and s-sinapyl glutathione (Wen & Wrolstad, 2002). Pineapple residues (core, peel, and leaves) are also rich in polysaccharides, cellulose, and hemicellulose (Banerjee, Ranganathan, Patti, & Arora, 2018), esters, terpenes, ketones, and aldehydes have been found in the pineapple core, in the peels fructose, sucrose, and glucose, in addition, they contain cellulose (polysaccharide formed by a chain of β-glucose, hemicellulose (found in different pentoses such as arabinose and xylose) (Chang-Bin et al., 2011; Nadzirah, Zainal, Noriham, Normah, Siti, & Nadya, 2013).

8.6 PINEAPPLE UTILIZATION

Pineapple is consumed as fresh fruit, jams, juices, and more; however, pineapple processing generates residues that are not used (Mohd-Ali, Hashim, Abd-Aziz, & Lasekan, 2020). By-products of pineapple industrialization include livestock feed, production of alcoholic beverages, obtaining the enzyme bromelain, the pineapple heart has come to be used for the manufacture of juices, the fiber obtained from pineapple peels has come to be used in the textile industry Cordenusi et al. (2010) Table 8.1 shows the possible applications of pineapple residues.

8.6.1 HEALTH BENEFITS FROM PINEAPPLE (ANANAS COMOSUS L.)

Like other fruits, pineapple contains different phytochemicals and bioactive compounds that help improve metabolism and human health and is one of the most useful fruits for obtaining high value-added compounds (Mohd-Ali, Hashim, Abd-Aziz, & Lasekan, 2020). It is known that pineapple is rich in bromelain, which is a natural anti-inflammatory, helping in various diseases such as sore throat, arthritis, or acute sinusitis; besides, it contains vitamin C, helping to fight bacterial or viral infections, and this fruit helps in arterial regulation and heart rate (Chaudhary et al., 2019). The different biological activities of bromelain have been studied and it is proven by Seenak, Kumphune, &

TABLE 8.1 Applications of Pineapple Residues

Part of the Pineapple	Application	References
Crown	Textile, paper, animal feed production	Prado & Spinacé (2019)
Crown	Bromelain extraction	Campos, Ribeiro, Teixeira, Pastrana, & Pintado (2020); Prado & Spinacé (2019)
Sheet–fiber	Textiles	Asim et al. (2015)
Sheet–fiber	Medicine, cosmetics, and biopolymers coating	Mohd-Ali, Hashim, Abd-Aziz, & Lasekan (2020)
Pulp and peel	Production of natural pigments	Shahanaz-Hamidin, Abdullah, Mohd-Nor, & Hadibarata (2022)
Peel	Bioethanol production	Casabar, Ramaraj, Tipnee, & Unpaprom (2020); Chintagunta, Ray, & Banerhee (2017)
Peel	Manufacture of hydrogels	Dai, Huang, Zhang, Ma, Huang, Wu, & Zhang (2020)
Peel	Edible film manufacturing	Rodsamran & Sothornvit (2019)
Pulp and peel	Juice production	Difonzo, Vollmer, Caponio, Pasqualone, Carle, & Steingass (2019)
Crown–cellulose	Production of recyclable packaging	Choquecachua-Mamani et al. (2020)

Malakul (2021) where they evaluated the effect of pineapple consumption for lipid reduction, cardiac oxidative stress and inflammation in rats fed a high cholesterol diet and concluded that daily pineapple consumption decreases lipid profiles and cardiac risk. In addition to being shown to have anti-cancer and wound healing activities (Banerjee, Ranganathan, Patti, & Arora, 2018; Uriza-Ávila et al., 2018). Among other benefits pineapple has been found to be effective as a diuretic and for the elimination of intestinal worms, among others it helps prevent ulcers and improves fat excretion, likewise it helps for constipation and gastrointestinal function (Chaumpluk, Chaiprasart, & Vilaivan, 2018; Hossain & Rahman, 2011). It also has certain benefits such as regulating emotional stability (Asim et al., 2015). Pineapple residues have also been shown to be anti-arthritic, hypolipidemic, anti-diabetic (Ajayi, Coker, Oyebanjo, Adebanjo, & Ademowo, 2022).

8.7 METHODS OF EXTRACTING BIOACTIVE COMPOUNDS FROM PINEAPPLE WASTE

The extraction of bioactive compounds has become important nowadays because, as it is well known, they have benefits for human health since they help us to prevent or cure certain diseases, For the extraction of this type of compounds different methods have been used such as ultrasound-assisted extraction (UAE), microwave-assisted extraction (MAE), Soxhlet method, among others, Hossain & Rahman (2011) studied the polyphenolic content and its antioxidant activity, different extractions were carried out giving as the best result the methanol extract giving a result of 51.1% for the phenolic content, in addition to its antioxidant activity was equivalent to polyphenols, in a more recent study by Sarabo et al. (2021) applied four pretreatments with the objective of studying the changes in the physicochemical properties of the slices, using hot blanching (pineapple slices were boiled at 80°C for 10 min.), hot syrup (20 g of sugar were mixed in 100 mL of boiling water until the sugar was completely dissolved, the slices were immersed in the syrup at 80°C for 30 min.), ascorbic acid soaking (a 2% solution was used and soaked for 20 min. at room temperature) and citric acid (a 1% solution was used and soaked for 20 min. at room temperature). They were able to determine that the hot syrup pretreatment showed a significantly higher moisture content, while the citric acid pretreatment has a low pH value and has a high total soluble sugar value, likewise the slices in the ascorbic acid treatment have a stronger activity than the other pretreatments and finally the hot blanching had a higher bromelain activity (0.11 U/ml). In Table 8.2, we can observe in a more concentrated way the extraction methods used to obtain and characterize the bioactive compounds.

TABLE 8.2 Extraction Methods Used to Obtain Bioactive Compounds

Part of the Pineapple	Extraction Method	Compounds Found	References
Sheets	Soxhlet extraction	• Azaserine • Perseitol • Betamethasone • Mitoxantrone • Avocadyne acetate • Cyclohexasiloxane • Campesterol	Kargutkar & Brijesh (2018)
Peel	Liquid-liquid extraction	• Gallic acid • Chlorogenic acid • Caffeic acid • Ferulic acid • Myricetin • Isoferulic acid	Lourenço et al. (2021)

TABLE 8.2 *(Continued)*

Part of the Pineapple	Extraction Method	Compounds Found	References
Peel, center, and crown	Ultrasonic extraction	• Fructose • Alanine • Arginine • Valine • Ferulic acid • Gallic acid • Epicatechin • Pectin	Azizan et al. (2020); Shivamathi et al. (2022)
Peel	Boiling extraction	• Flavonoids • Phenolic acid • Aldehydes • Carboxyl	Das, Patra, Debnath, Ansari, & Shin (2019)
Peel	Microwave-assisted extraction	• Pectin • Ribose • Xylose • Galacturonic acid	Rodsamran & Sothornvit (2019); Zakaria, Rahman, Abang-Zaidel, Dailin, & Jusoh (2021)
Peel	Ion exchange	• Pectin	Yanbo & Bing (2020)
Peel and center	Solid phase microextraction	• Acetic acid • 1-octanol • p-xylene • D-limonene • Alfa murolene • Hexanoic acid • Methyl acetate	Valdés-García et al. (2021)
Peel	Reflux extraction	• Gallic acid • Catechin • Epicatechin • Ferulic acid	Li, Shen, Liu, Liu, Liang, Yan, & Chen (2014)
Peel and center	Autohydrolysis	• Gallic acid • Catechin • Epicatechin • Ferulic acid	Sepúlveda, Romani, Aguilar, & Teixeira (2018)
Center and shell	Green chemistry: extraction using a juice machine	• Chlorogenic acid • Caffeic acid • Ferulic acid	Campos, Ribeiro, Teixeira, Pastrana, & Pintado (2020)

As mentioned above, several extraction methods have been used for these bioactive compounds; however, new alternatives are still being sought, such as solid-state fermentation (SSF), although it was already a known method, it could be considered a new one since it has been little studied; this bioprocess consists of growing a microorganism on a solid substrate with a source of nitrogen and mineral salts under certain conditions of pH, temperature, and humidity (Borrás-Sandoval, & Torres-Vidales, 2016). SSF has become of great interest nowadays due to its simplicity and proximity to the natural growth conditions of microorganisms (fungi or yeasts) (Borrás, Valiño, Elías, Martínez, Sanabria, & Becerra, 2020). Several types of agro-industrial substrates have been used in SSF; however, pineapple by-products have not been the exception. It has been proven that using Rhizopus oryzae lactic acid is produced under favorable conditions of 80% humidity, pH of 6.5, inoculum of 1×104 spores/g and a temperature of 27°C obtaining 0.0236 g of lactic acid/g of pineapple residues (Awasthi et al., 2022).

Another bioprocess considered is submerged fermentation (SmF)/liquid fermentation (LF), which consists of growing the microorganism in a liquid medium under certain conditions such as pH, inoculum, and oxygen, in addition to considering a balance of matter for the elements that participate in the growth of the microorganism, this is significant since the microorganism consumes the substrate that gives it energy, in this case organic residues can be considered (Prieto-Contreras & Grevechova, 2006). Prabha & Rangaiah (2014) used SmF to obtain citric acid from pineapple peel from Aspergillus Niger MTCC281 and determined that the best conditions were at a period of 180 h, initial pH of 2.5, a sugar concentration of 150 g/Lu at 30°C, obtaining 8.58 g/L, in addition, they used three types of alcohols (methanol, ethanol, butane) at different concentrations, which served as stimulants. And methanol at a concentration of 1% produced the highest amount of citric acid (14.71 g/L). Kavuthodi & Sebastian (2018) used pineapple stem extract where they determined pectinase using SmF by Bacillus subtilis BKDS1, the optimum conditions for pectinase production were $CaCO_3$ (1.10 g/L), $CaCl_2$ (0.92 g/L), yeast extract (6.40 g/L), inoculum (1.5%) and pineapple stem extract (12.5%) at pH 7, achieving a production of 1508.5 U/ml. It has been proven that pineapple waste is an optimal material to produce probiotics, using *Lactobacillus* spp. (Pyar, Liong, & Peh, 2014).

8.7.1 OBTAINING BROMELAIN FROM PINEAPPLE WASTE

Bromelain is one of the components of pineapple that can be beneficial to human health, this is a proteolytic enzyme that is constituted by amino acids, generally bromelain is found throughout the pineapple fruit, however, where it has been found most is in the stem of the fruit, it has been shown that the consumption of this enzyme has different benefits and doses between

750 and 100 mg/day have been recommended (Cantú-Perches, Hernández-Salazar, Medina-Lozano, & Martínez-Báes, 2019) and undoubtedly has had several applications in the pharmaceutical industry, since it is known to have anticancer activity (being an important antitumor agent) helps fight breast, colorectal, pancreatic, and liver cancer, it is known to decreases proliferation, invasion, and migration of cholangiocarcinoma cells. It reduces tumor size, promotes cell apoptosis (Artem-Ataide et al., 2017; Colleti, Li, Marengo, Adinolfi, & Cravatto, 2021). It also has anti-inflammatory activity, this activity is partly due to the inhibition of bradykinin, leading to a significant reduction in pain, in addition, by stimulating the proliferation of tenocytes it helps to promote healing, regulates blood levels of bradykinin (Aiyegbusi, Duru, Awelimobor, Noronha, & Okanlawon, 2010; Aiyegbusi, Duru, Anunobi, Noronha, & Okanlawon, 2011), antimicrobial, antibiotic potentiation, anthelmintic potential, immunomodulatory effect, antinociceptive effect, anti-asthmatic, antidiarrheal effect, has perioperative activity, meaning it relieves postoperative pain seven days after surgery, reduces post-traumatic pain (Liu, Zhao, Wang, Zhao, & Ma, 2019) also helps for dermatological disorders (Chakraborty et al., 2021). Bromelain has also been shown to help with food allergies, asthma, and dermatitis by reducing CD44 expression in cell adhesion and metastasis models, presenting inhibition of allergic sensitization (Secor et al., 2013). Currently, several studies have been conducted to obtain bromelain, Misran et al. (2019) conducted a research where they extracted bromelain from Morris variety pineapple, they were 20 pineapples with crown and 20 without crown, the weight of each of the waste such as pulp (45–52%), peel (28–29%), heart (13–14%), crown (7%) and stem (5–6%) was determined, in order to characterize the bromelain extract in each of the waste. The results showed that the pH was 3–4, total soluble solids of 1–6° Brix, volume of the extract at 850–6,000 ml, total protein of 21–98 mg/ml and bromelain activity was 95–251 CDU/ml, under these conditions showed that pineapple peels and pulp were the ones that obtained the highest bromelain activity. Bromelain has had significant effects in the pharmaceutical industry, Abbas et al. (2021) aimed in their study to evaluate the potential of bromelain to combat acne as it is known to have antimicrobial properties, for the study pineapple was used (the fruit, stem, peel, core, and crown) and from each of the residues bromelain enzyme was extracted and purified, pineapple peels obtained higher enzyme activity (3.417 U/μg). In addition, the antimicrobial activity was studied, resulting in the purified bromelain from the peels having the greatest inhibitory effect on *S. aureus* and *P. acne*; with these results, a formulation was made for

washing the face, compared with commercial products, it was concluded that the formulation based on pineapple peels was more effective compared to the commercial ones. Bromelain furthermore has been used as a food additive as meat tenderizers (Banerjee, Ranganathan, Patti, & Arora, 2018). Mala et al. (2021) conducted a study where ultrasound-assisted (UAE) and MAE conditions were evaluated to obtain bromelain enzyme, the extractions were carried finished using distilled water and it was determined that the best extraction conditions were 99.96%, 26.83 min and 20.96 ml/g for EUA and 100 W, 8.99 min and 1:8 g/ml for EAM, which indicated that high yields of bromelain can be obtained, which means that pineapple residues have a high potential for applications in the pharmaceutical or food industry.

8.7.2 PRODUCTION OF BIOFUELS FROM PINEAPPLE PEELS (ANANAS COMOSUS L.)

Agro-industrial wastes continue to increase every day and, as we have seen, they can also cause alterations to the environment, which is why different forms of utilization continue to be studied to improve the environment and obtain biofuels from this biomass. Recently, bioethanol is emerging as a potentially sustainable energy resource, as this can present different environmental and economic advantages since unlike petroleum these sources are obtained from natural raw materials such as agro-industrial waste, which can be valorized since they are generated in enormous quantities, these by-products can be treated with fermentation processes to obtain various biofuels (Malagón-Micán et al., 2017). And in recent years the demand for these products increased and bioethanol can be produced from organic matter thanks to its high fermentable sugar content (Beigbeder, De Medeiros, & Lavoie, 2021). Conesa et al. (2016) aimed to study the microwave extraction conditions as a pretreatment for bioethanol production from pineapple peels and core, concluding that pretreatments using microwave and adequate exposure time improve the bioethanol production yield. Chintagunta et al. (2017) conducted a work in which they performed bioethanol production from pineapple leaves, conducted in a simultaneous way of saccharification and fermentation, resulting in 7.12% (v/v). In another study, Casabar, Unpaprom, & Ramaraj (2019) aimed to produce bioethanol from pineapple residues (peel) inoculated with 2% *S. cerevisiae* and at the end of the experiments it was determined that using pineapple residues produced a yield of 5.98 ± 1.01 g/L at 48 h of fermentation. Dahunsi (2019) conducted a study where he aimed to obtain biogas from pineapple peels, pretreatments such as sulfuric acid and alkaline hydrogen peroxide were performed, It was concluded that the best pretreatment was using H_2O_2 since the biogas production was higher (67%) compared to the H_2SO_4 pretreatment (51%), likewise the thermal energy was obtained resulting in 583 kWh t^{-1} TS for H_2O_2 and 1,436 kWh t^{-1} TS for H_2SO_4, demonstrating that using H_2O_2 is economically stable. Currently, new alternatives for bioethanol production are being sought and the use of *Trichoderma harzianum* has been of interest due to its production of hydrolytic and cellulolytic enzymes, as proven by Casabar et al. (2020) where they evaluated the effect of *T. harzianum* on the sonicated pineapple peel, giving as a result that at a shorter time of 30 min. of sonication a higher production of fermented sugars is obtained giving as a result 567.6 ± 58.4 g/L compared to a longer time as it was at 48 h giving a result of only 197.6 ± 9.9 g/L. Recently Salafia et al. (2022) conducted a research study where they performed a bioconversion of cell wall sugars from pineapple waste (peel and core) to bioethanol using, by simultaneous fermentations using Saccharomyces cerevisiae ATCC 4126, different cell wall sugars were detected which were glucose, xylose, uronic acid, arabinose, and galactose (9.84, 8.16, 3.18, 2.46, 1.58%, respectively) and in smaller amounts mannose, rhamnose, and fructose, which makes it a good substrate for bioethanol production since its highest concentration was 3.9% (30.77 g/L) reached in a period of 24 h.

8.7.3 OBTAINING CELLULOSE FROM PINEAPPLE WASTES

Among other applications is the obtaining of cellulose, which is a polymer composed exclusively of glucose molecules synthesized by plants, its main use is in the manufacture of paper, cardboard, artificial wood, natural fibers, artificial silks, or celluloids (Robles, Saucedo-Corona, Delgado, Sanjuán, & Turrado, 2014). Sibaly & Jeetah (2017) used pineapple leaves from Mauritius located in Mapou, with the objective of determining the feasibility of using fibers for paper production, in the study they used samples mixed with pineapple leaf fiber and sugar cane in different proportions, likewise waste-paper was used with pineapple fiber. The papers obtained were subjected to mechanical and physical tests to check if the use of this residue is viable. The most absorbent paper was found to contain 100% pineapple (1.19 s), in addition to a high rate of attraction and rupture compared to the other papers (6.5 Nm/g and 0.84 kPam^2/s, respectively), the paper that resisted abrasion was the one containing sugarcane bagasse and pineapple fiber (40:60) and the paper that resisted wrinkling was the one containing pineapple residue paper (80: 20), with these results they demonstrated that pineapple leaves for the manufacture of paper is a viable form, in addition they obtained a compatibility with sugarcane bagasse (which as is well known is another agro-industrial waste), this can give a twist to the paper industry since in addition to protecting the environment can be given a second chance to waste generated by the food industry, this is proven by Presenda-Gerónimo et al. (2020) conducted a research where the objective was to obtain high quality cellulose from pineapple residues (crowns) where a pre-hydrolysis treatment combined with the Soda-Anthraquinone pulping method and elemental chlorine free bleaching was performed, where it was demonstrated to obtain cellulose free of residues being 93% and it is proposed as an alternative for the manufacture of biomaterials. Among other applications we can mention the manufacture of paper, since it is known that the main raw material is wood and over the years, we have seen that this generates a negative impact on the environment, therefore new environmentally friendly alternatives are sought to obtain paper and one of the ways is through the use of agro-industrial waste, since these contain lignocellulosic fibers that act as a possible resource for cellulose.

8.7.4 PINEAPPLE WINE AND VINEGAR PRODUCTION

Another application of pineapple waste is the production of pineapple wine and vinegar, its high sugar content helps the fermentation process to produce wine, it is considered a viable alternative to grapes, this is proven by Kodagoda & Marapana (2017) as they aimed to develop a non-alcoholic wine from pineapple waste (external and internal peels and the core of the pineapple), the results obtained from the proximate analysis and fermentation for 56 days it can be concluded that the development of non-alcoholic wines is an alternative to the one we already know. Another contribution made by Dellacassa et al. (2017) carried out a process of pineapple vinification by inoculated and spontaneous fermentation using Saccharomyces cerevisiae strain, the results they obtained showed that the production of pineapple wine is of good quality and could be pleasant for white wine consumers, since it contains 61 volatile compounds and 18 aromatic compounds, they also reported two new lactones and two terpenes. Qi et al. (2017) studied the fermentation process for obtaining pineapple wine, the quantitative and sensory evaluation indicated that pineapple wine is still one of the alternatives for its production, in this research 68 aromatic compounds were detected, in addition to 34 esters and 13 alcohols. Roda et al. (2017) aimed to investigate whether the alcoholic fermentation process was suitable for wine production and to subsequently carry out an acetic fermentation for vinegar production, the work showed that the alcohol yield was approximately 7% and a 5% yield for acetic acid and among the main components were acetoin, 2,3-butanediol, ethyl acetate, and 2-phenylethanol, the work showed that the processing of waste could be effective to obtain alternative value-added products. Huang et al. (2017) reported that vinegar production could be increased by 10% value-added using pineapple bran (another by-product of pineapple waste), it has been even proven that pineapple bran contains vitamins and amino acids with rich nutritional value. In a more recent publication by Boondaeng et al. (2022) where they evaluated the fermentation process for the elaboration of an alcoholic beverage, where they obtained a high alcohol content of 10.71%, in addition to presenting an antioxidant activity similar to that of red and white wines, it can be said that even today pineapple residues are an alternative for the elaboration of wines, giving them a second chance, since it has been proven that they contain compounds that help the fermentation process and generate compounds for the benefit of human health.

8.7.4.1 PRODUCTION OF TEPACHE FROM PINEAPPLE PEELS

It is one of Mexico's traditional beverages that has been consumed since pre-Hispanic times. It is known to be a fermented beverage prepared from pineapple peel (sometimes the pulp), brown sugar, water, piloncillo, cinnamon, and even beer can be added, depending on how it is prepared in each region of Mexico, and the fermentation process is carried out at room temperature (Bamforth, 2014; Corona-González et al., 2013). Among the microorganisms that contribute to the fermentation of tepache are the bacteria *Bacillus megaterium* and *Bacillus subtillis*, and among the yeasts are *Saccharomyces cerevisiae, Pichia membranifaciens*, *Candida boidinii* and *Candida incospicua* (Mapes & Basurto, 2016). In addition, it has been demonstrated the presence of tibicos, which are microcolonies, formed of symbiotic bacteria and yeasts, they are characterized by being compact, whitish, yellowish, translucent, or opalescent, these develop in a liquid environment enriched with sugars, among the tibicos are *Saccharomyces radaisii*, *Bacillus mexicanus*, *Bacillus subtilis*, *Bacillus graveolens, E. coli*, *Proteus vulgaris*, *Proteus membranifaciens* and *S. cerevisiae* (Robledo-Márquez et al., 2021; Romero-Luna, Hernández-Sánchez, & Dávila-Ortiz, 2017). The study conducted by Corona-González et al. (2013) aimed to evaluate the fermentation conditions since in the elaboration of tepache there is no adequate control since it is made by hand, they evaluated the sugar concentration, pH, temperature, and fermentation time, and determined that the best conditions were 22°C at 10% (m/v) sugar with piloncillo, for 72 h, with a pH of 5, giving as a result for ethanol 7.30 (g/L), for lactic acid 5.25 (g/L), acetic acid 4.16 (g/L), sucrose 67.63 (g/L), glucose 4.48 (g/L) and fructose 1.09 (g/L) and the yeast identified were *Candida apicola*, *Crytococus skinneri*, *Hanseniaspora*, *Saccharomyces* sp. and *Saccharomyces cerevisiae*.

8.7.5 EXTRACTION OF ESSENTIAL OILS FROM PINEAPPLE PEEL

Essential oils are complex mixtures of volatile compounds, usually terpenes; these oils are synthesized, obtained from vegetable raw materials, usually found in plants and flowers, and currently essential oils are sought from agro-industrial residues of fruits and vegetables, these oils are widely used in the food, cosmetic, pharmaceutical, and perfumery industries (Said-Al, Hikal, & Mahmound, 2017; Usano-Alemany, Palá-Paúl, & Díaz, 2014). Chang-Bin et al. (2011) conducted a work in which volatile aromatic compounds were identified from pineapple residues (center and pulp) of the Smooth Cayenne variety, they performed a microextraction and from the oil 54 compounds were identified, among the main ones were, terpenes, ketones, and aldehydes, methyl octanoate, ethyl hexane, ethyl butane, ethyl acetoxy butyrate, among others, the main compounds responsible for its characteristic odor were: Ethyl 2-methylbutanoate, ethyl hexanoate and 2,5-dimethyl-4-hydroxy-3(2H)-furanone (DMHF). Machado & Aparecida (2011) determined that the main compounds that give it its characteristic flavor are ethyl hexanoate and decanal.

8.8 CONCLUSIONS AND FUTURE PERSPECTIVES

Over the years, the industrialization of pineapple has led to an increase in agro-industrial waste. However, this waste can be used to develop value-added products as it contains high levels of beneficial nutrients and bioactive compounds. To extract these compounds, various methods, both conventional and emerging, have been used. While these methods have been effective, there is a growing interest in more environmentally friendly and efficient technologies such as biotechnologies and green chemistry. Implementing these technologies can help obtain valuable compounds that can be applied to human and animal nutrition, as well as medicines. Additionally, enzymes such as bromelain have been extracted from pineapple residues, which have been found to have multiple biological activities. By studying these activities, we can better understand how these enzymes work and how they can be used in various industries such as paper, cosmetics, and fabrics. This review emphasizes the importance of utilizing fruit and vegetable waste to create value-added products and reduce waste.

KEYWORDS

- ***Ananas comosus***
- **bioactive compounds**
- **biofuels**
- **pineapple utilization**
- **pineapple waste**
- **tepache**
- **vinegar production**

REFERENCES

Abbas, S., Shanbhag, T. T., & Kothare, A., (2021). Applications of bromelain from pineapple waste towards acne. *Saudi J. Biol. Sci., 28*, 1001–1009.

Aiyegbusi, A. I., Duru, F. I. O., Anunobi, C. C., Noronha, C. C., & Okanlawon, A. O., (2011). Bromelain in the early phase of healing in acute crush Achilles tendon injury. *Phytother. Res., 25*, 49–52.

Aiyegbusi, A. I., Duru, F. I. O., Awelimobor, D., Noronha, C. C., & Okanlawon, A. O., (2010). The role of aqueous extract of pineapple fruit parts on the healing of acute crush tendon injury. *Nig. Q J. Hosp. Me., 20*, 223–227.

Ajayi, A. M., Coker, A. I., Oyebanjo, O. T., Adebanjo, I. M., & Ademowo, O. G., (2022). *Ananas comosus* (L) Merril (pineapple) fruit peel extract demonstrates antimalarial, anti-nociceptive and anti-inflammatory activities in experimental models. *J. Ethnopharmacol., 282*, 114576.

Artem-Ataide, J., Mendes De, C. N., Araújo-Rebelo, M., Vinícius-Chaud, M., Grotto, D., Gerenutt, M., Rai, M., et al., (2017). Bacterial nanocellulose loaded with bromelain:

Assessment of antimicrobial, antioxidant, and physical-chemical properties. *Sci. Report, 7*, 18031.

Ascacio-Valdés, J. A., Buenrostro, J. J., De La Cruz, R., Sepúlveda, L., Aguilera, A. F., Prado, A., Contreras, J. J., et al., (2014). Fungal biodegradation of pomegranate ellagitannins. *J. Basic Microbiol., 54*, 28–34.

Asim, M., Abdan, K., Jawaid, M., Nasir, M., Dashtizadeh, Z., Ishak, M. R., & Hoque, M. E., (2015). A review on pineapple leaves fiber and its composites. *Int. J. Polym. Sci.*, 950567.

Awasthi, M. K., Wan-Azelee, N. I., Mazila-Ramli, A. N., Rashid, S. A., Abdul-Manas, N. H., Dailin, D. J., Illias, R. M., et al., (2022). Microbial biotechnology approaches for conversion of pineapple waste in to emerging source of healthy food for sustainable environment. *Int. J. Food Microbiol., 373*, 109714.

Azizan, A., Lee, A. X., Abdul-Hamid, N. A., Maulidiani, M., Mediani, A., Abdul-Ghafar, S. Z., Zulaikha-Zolkeflee, N. K., & Abas, F., (2020). Potentially bioactive metabolites from pineapple waste extracts and their antioxidant and α-Glucosidase inhibitory activities by ^{1}H NMR. *Foods, 11*, 173.

Bamforth, C. W., (2014). Fermented beverages. *Encyclopedia of Agriculture and Food Systems* (pp. 124–136). Academic Press, Elsevier.

Banerjee, S., Ranganathan, V., Patti, A., & Arora, A., (2018). Valorization of pineapple wastes for food and therapeutic applications. *Trends Food Sci. Technol., 28*, 60–70.

Beigbeder, J. B., De Medeiros, D. J. M., & Lavoie, J. M., (2021). Optimization of yeast, sugar, and nutrient concentrations for high ethanol production rate using industrial sugar beet molasses and response surface methodology. *Fermentation, 7*, 86.

Boondaeng, A., Kasemsumran, S., Ngowsuwam, K., Vaithanomsat, P., Apiwatanapiwat, W., Trakunjae, C., Janchai, P., et al., (2022). Fermentation condition and quality evaluation of pineapple fruit wine. *Fermentation, 8*, 11.

Borrás, L. M., Valiño, E. C., Elías, A., Martínez, J. J., Sanabria, A. M., & Becerra, M. L., (2020). Solid-state fermentation of *Solanum tuberosum* postharvest residues and a microbial preparation. *Cuba J. Agric. Sci., 54*, 525–533.

Borrás-Sandoval, L. M., & Torres-Vidales, G., (2016). Animal feed production by solid-state fermentation. *Orinoquia., 20*, 47–54.

Campos, D. A., Coscueta, E. R., Valetti, N. W., Pastrana-Castro, L. M., Teixeira, J. A., Picó, G. A., & Pintado, M. M., (2019). Optimization of bromelain isolation from pineapple byproducts by polysaccharide complex formation. *Food Hydrocoll., 87*, 792–804.

Campos, D. A., Ribeiro, T. B., Teixeira, J. A., Pastrana, L., & Pintado, M. M., (2020). Integral valorization of pineapple (*Ananas comosus* L.) by-products through a green chemistry approach towards added value ingredients. *Foods, 9*, 60.

Cantú-Perches, M. L., Hernández-Salazar, M., Medina-Lozano, I. G., & Martínez-Báes, A. Z., (2019). Aplicaciones terapéuticas y composición química de *Ananas comosus* (L.) Merr. (Piña). Un studio de revision. *Planta, 15*, 15–20.

Casabar, J. T., Ramaraj, R., Tipnee, S., & Unpaprom, Y., (2020). Enhancement of hydrolysis with *Tricoderma harzianum* for bioethanol production of sonicated pineapple fruit peel. *Fuel, 279*, 118437.

Casabar, J. T., Unpaprom, Y., & Ramaraj, R., (2019). Fermentation of pineapple fruit peel wastes for bioethanol production. *Biomass Convers. Biorefin., 9*, 761–765.

Chakraborty, A. J., Mitra, S., Tallei, T. E., Tareq, A. M., Nainu, F., Cicia, D., Dhama, K., et al., (2021). Bromelain a potential bioactive compound: A comprehensive overview from a pharmacological perspective. *Life (Basel), 11*, 317.

Chang-Bin, W., Sheng-Hui, L., Yu-Ge, L., Ling-Ling, L., Wen-Xiu, Y., & Guang-Ming, S., (2011). Characteristic aroma compounds from different pineapple parts. *Molecules, 16*, 5104–5112.

Chaudhary, V., Kumar, V., Sunil, Vaishali, Singh, K., Kumar, R., & Kumar, V., (2019). Pineapple (*Ananas comosus*) product processing: A review. *J Pharmacogn Phytochem., 8*, 4642–4652.

Chaumpluk, P., Chaiprasart, P., & Vilaivan, T., (2018). Postharvest non-destructive determination of fruits: A model on fruit maturity assay via biosensor based on colorimetric change of gold nanoparticles. *Acta Horticulturae, 945*, 205–212.

Chintagunta, A. D., Ray, S., & Banerhee, R., (2017). An integrated bioprocess for bioethanol and bio manure production from pineapple leaf waste. *J. Clean. Prod., 165*, 1508–1516.

Choquecachua-Mamani, D., Otero-Nole, K. S., Chaparro-Montoya, E. E., Mayta-Huiza, D. A., Pastrana-Alta, R. Y., & Aguila-Vitorino, H., (2020). Minimizing organic waste generated by pineapple crown: A simple process to obtain cellulose for the preparation of recyclable containers. *Recycling, 5*, 24.

Colleti, A., Li, S., Marengo, M., Adinolfi, S., & Cravatto, G., (2021). Recent advances and insights into bromelain processing, pharmacokinetics, and therapeutic uses. *Appl. Sci., 11*, 8428.

Conesa, C., Seguí, L., Laguarda-Miró, N., & Fito, P., (2016). Microwaves as a pretreatment for enhancing enzymatic hydrolysis of pineapple industrial waste for bioethanol production. *Food Bioprod. Process., 100*, 203–213.

Cordenunsi, B., Saura-Calixto, F., Diaz-Rubio, M. E., Zuleta, A., Tiné, M. A., Buckeridge, M. S., Da Silva, G. B., et al., (2010). Carbohydrate composition of ripe pineapple (cv. perola) and the glycemic response in humans. *Food Sci. Technol., 30*, 282–288.

Corona-González, R. I., Ramos-Ibarra, J. R., Gutiérrez-González, P., Pelayo-Ortiz, G. M., Guatemala-Morales, G. M., & Arriola-Guevara, E., (2013). The use of response surface methodology to evaluate the fermentation conditions in the production of tepache. *Rev. Mex. Ing. Quí., 12*, 19–28.

Dahunsi, S. O., (2019). Liquefaction of pineapple peel: Pretreatment and process optimization. *Energy, 185*, 1017–1031.

Dai, H., Huang, Y., Zhang, H., Ma, L., Huang, H., Wu, J., & Zhang, Y., (2020). Direct fabrication of hierarchically processed pineapple peel hydrogels for efficient Congo red adsorption. *Carbohyd. Polymers., 230*, 115599.

Das, G., Patra, J. K., Debnath, T., Ansari, A., & Shin, H. S., (2019). Investigation of antioxidant, antibacterial, antidiabetic, and cytotoxicity potential of silver nanoparticles synthesized using the outer peel extract of *Ananas comosus* (L). *PLoS One, 14*, e0220950.

Dellacassa, E., Trenchs, O., Feriña, L., Debernardis, F., Perez, G., Boido, E., & Carrau, F., (2017). Pineapple (*Ananas comosus* L. Merr) wine production in Angola: Characterization of volatile aroma compounds and yeast native flora. *Int. J. Food Microbiol., 241*, 161–167.

Diaz-Vela, J., Totosaus, A., Cruz-Guerrero, A. E., & Pérez-Chabela, M. L., (2013). In vitro evaluation of the fermentation of added-value agro-industrial by-products: Cactus pear (*Opuntia ficus-indica* L.) peel and pineapple (*Ananas comosus*) peel as functional ingredients. *Int. J. Food Sci. Technol., 48*, 1460–1467.

Difonzo, G., Vollmer, K., Caponio, F., Pasqualone, A., Carle, R., & Steingass, C. B., (2019). Characterization and classification of pineapple (*Ananas comosus* [L.] Merr.) juice from pulp and peel. *Food Control., 96*, 260–270.

Elss, S., Preston, C., Hertzig, C., Heckel, F., Richling, E., & Schreier, P., (2005). Aroma profiles of pineapple fruit (*Ananas comosus* [L.] Merr.) and pineapple products. *LWT – Food Sci. Technol., 38*, 263–274.

FAO (Food and Agriculture Organization of the United Nations) Production Indices, (2022). Retrieved from: FAOSTAT.

Gutiérrez-Sarmiento, W., Peña-Ocaña, B. A., Lam-Gutiérrez, A., Guzmán-Albores, J. M., Jasso-Chávez, R., & Ruíz-Valdiviezo, V. M., (2022). Microbial community structure, physicochemical characteristic, and predictive functionalities of the Mexican tepache fermented beverage. *Microbiol. Res., 260*, 127045.

Hassan, A., Othman, Z., & Siriphanich, J., (2011). Pineapple (*Ananas comosus* L. Merr). In: Yahia, E. M., (ed.), *Postharvest Biology and Technology of Tropical and Subtropical Fruits* (pp. 194–217). Woodhead Publishing, Elsevier.

Hossain, M. A., & Rahman, S. M. M., (2011). Total phenolics, flavonoids and antioxidant activity of tropical fruit pineapple. *Food Res. Inter., 44*, 672–676.

Hossain, M. F., (2016). World pineapple production: An overview. *African J. Food Agric. Nutr. Dev., 16*, 11443–11456.

Huang, H., Liao, L., Lin, L., Wang, X., Gong, X., Zhang, F., & Gong, J., (2017). Progress in the study of pineapple bran vinegar. *Adv. Eng. Res., 125*, 145–149.

Kargutkar, S., & Brijesh, S., (2018). Anti- inflammatory evaluation and characterization of leaf extract f *Ananas comosus*. *Inflammopharmacology, 26*, 469–477.

Kavuthodi, B., & Sebastian, D., (2018). Biotechnological valorization of pineapple stem for pectinase production by *Bacillus subtilis* BKDS1: Media formulation and statistical optimization for submerged fermentation. *Biocatal. Agric. Biotechnol., 16*, 715–722.

Kodagoda, K., & Marapana, R., (2017). Development of non-alcoholic wines from the wastes of Mauritius pineapple variety and its physicochemical properties. *J. Pharma. Phytochem.*, *6*, 492–497.

León-Yánez, S., Valencia, R., Pitmam, N., Endara, L., Ulloa, U. C., & Navarrete, H., (2019). *Libro Rojo de Plantas Endémicas del Ecuador*. Publicaciones del Herbario QCA, Pontificia Universidad Católica del Ecuador, Quito. https://bioweb.bio/floraweb/librorojo (accessed on 15 November 2023).

Li, T., Shen, P., Liu, W., Liu, C., Liang, R., Yan, N., & Chen, J., (2014). Major polyphenolics in pineapple peels and their antioxidant interactions. *Int. J. Food Prop., 17*, 1805–1817.

Liu, S., Zhao, H., Wang, Y., Zhao, H., & Ma, C., (2019). Oral bromelain for the control of facial swelling, trismus, and pain after mandibular third molar surgery: A systematic review and meta-analysis. *J. Oral Maxillofac. Surg., 77*, 1566–1574.

Lourenço, S. C., Campos, D. A., Gómez-García, R., Pintado, M., Oliveira, M. C., Santos, D. I., Corrêa-Filho, L. C., et al., (2021). Optimization of natural antioxidants extraction from pineapple peel and their stabilization by spray drying. *Foods, 10*, 1255.

Machado, M. M., & Aparecida, S. M., (2011). Aroma retention in drying with normal and modified atmosphere: Development of a study system. *Food Sci. Technol., 31*, 295–302.

Mala, T., Sadiq, M. B., & Anal, A. K., (2021). Comparative extraction of bromelain and bioactive peptides from pineapple byproducts by ultrasonic- and microwave-assisted extractions. *J. Food Process Eng., 44*, e13709.

Malagón-Micán, M. L., Paéz, A. I., Lache-Muñoz, A., Santos-Aguilar, J., & Zabala-García, D. A., (2017). Bioethanol production from different mixes of organic wastes in a food company. *Revista de Investigación., 10*, 47–59.

Mapes, C., & Basurto, F., (2016). Biodiversity and edible plants of Mexico. In: Lira, R., Casas, A., & Blancas, J., (eds.), *Ethnobotany of Mexico: Interactions of People and Plants in Mesoamerica* (pp. 83–131). New York, Springer.

Martínez-Castro, C. J., Ríos-Castillo, M., & Castillo-Leal, M., (2019). Levels of technological adoption in pineapple production units in Loma Bonita, Oaxaca, Mexico. *Revista de Geografía Agrícola*, 232–257.

Misran, E., Idris, A., Sarip, S. H. S., & Ya'akob, H., (2019). Properties of bromelain extract from different parts of the pineapple variety Morris. *Biocatal. Agric. Biotechnol., 18*, 101095.

Mohd-Ali, M., Hashim, N., Abd-Aziz, S., & Lasekan, O., (2020). Pineapple (*Ananas comosus*): A comprehensive review of nutritional values, volatile compounds, health benefits and potential food products. *Food Res. Int., 137*, 109675.

Müller, A., Barat, S., Chen, X., Bui, K. C., Bozko, P., Malek, N. P., & Plentz, R. R., (2016). Comparative study of antitumor effects of bromlain and papaína in human colangiocarcinoma cell lines. *Int. J. Oncol., 48*, 2025–2034.

Nadzirah, K. Z., Zainal, S., Noriham, A., Normah, I., Siti, R. A. M., & Nadya, H., (2013). Physic-chemical properties of pineapple variety N36 harvested and stored at different maturity stages. *Int. Food Res. J., 20*, 225–303.

Prabha, M. S., & Rangaiah, G. S., (2014). Citric acid production using *Ananas comosus* and its waste with the effect of alcohols. *Int. J. Curr. Microbiol. Appl. Sci., 3*, 747–754.

Prado, K. S., & Spinacé, M. A. S., (2019). Isolation and characterization of cellulose nanocrystals from pineapple crown waste and their potential uses. *Int. J. Biol. Macromol., 122*, 410–416.

Presenda-Gerónimo, A., Bolio-López, G. I., Azmar-Barrios, J. A., Ramírez-Casillas, R., Sarracino-Martínez, O., Hernández-Villegas, M. M., & De La Cruz-Burelo, P., (2020). Isolation of cellulose from pineapple waste (*Ananas comosus* L. Merril). *Agro. Productividad., 13*, 77–82.

Prieto-Contreras, L., & Grevechova, R., (2006). Transport phenomena in the production of pectolytic enzymes from *Aspergillus foetidus* under submerged fermentation conditions. *Épsilon., 7*, 65–77.

Pyar, H., Liong, M. T., & Peh, K. K., (2014). Potentials of pineapple waste as growth medium for *Lactobacillus* species. *Int. J. Pharm Pharm Sci., 6*, 142–145.

Qi, N., Ma, L., Li, L., Gong, X., & Ye, J., (2017). Production and quality evaluation of pineapple fruit wine. *IOP Confer Ser.: Earth Environ. Sci., 100*, 012028.

Robledo-Márquez, K., Ramírez, V., González-Córdova, A. F., Ramírez-Rodríguez, García-Ortega, L., & Trujillo, J., (2021). Research opportunities: Traditional fermented beverages in Mexico. Cultural, microbiological, chemical, and functional aspects. *Food Res Int., 147*, 110482.

Robles, M. N., Saucedo-Corona, A. R., Delgado, F., Sanjuán, D., & Turrado, S. J., (2014). Effect of cellulose microfibers on paper with high contents of recycled fiber. *Rev. Mex. De cienc. Forestales., 5*, 70–79.

Roda, A., Lucini, L., Torchio, F., Doroni, R., De Faveri, D. M., & Lambri, M., (2017). Metabolite profiling and voltiles of pineapple wine and vinegar obtained from pineapple waste. *Food Chem., 229*, 734–742.

Rodsamran, P., & Sothornvit, R., (2019). Preparation and characterization of pectin fraction from pineapple peel as a natural plasticizer and material for biopolymer film. *Food Bioprod. Process., 118*, 198–206.

Romero-Luna, H. E., Hernández-Sánchez, H., & Dávila-Ortiz, G., (2017). Traditional fermented beverages from Mexico as a potential probiotic source. *Ann Microbiol., 67*, 577–586.

Said-Al, A. H. A. A., Hikal, W. M., & Mahmound, A. A., (2017). Essential oils: Biosynthesis, chemistry, and biological functions. *J. Chem. Pharm. Res., 9*, 190–200.

Salafia, F., Ferracane, A., & Tropea, A., (2022). Pineapple waste cell wall sugar fermentation by *Saccharomyces cerevisiae* for second generation bioethanol production. *Fermentation, 8*, 1–10.

Sánchez-Hernández, M. A., Huja-Mendoza, S., & Acevedo-Gómez, R., (2015). Producción de piña Cayena Lisa y MD2 (*Ananas comosus* L.) en condiciones de Loma Bonita, Oaxaca. In: Figueroa, E., Godinez, L., & Pérez, F., (eds.), *Ciencias de la Biología y Agronomía, Handbook T-I* (pp. 100–110).

Sarabo, Z., Hanafi, N., Azzuan-Rosli, M., Abd-Rashid, S. N., Mohd-Ropi, N. A., Hasham, R., Roji-Sarmidi, M., et al., (2021). Effect of different pre-treatments on the physicochemical properties of freeze-dried *Ananas comosus* L. *Materials Today: Proceedings, 42*, 229–233.

Secor, E. R., Szczepanek, S. M., Castater, C. A., Adami, A. J., Matson, A. P., Rafti, E. T., Guernsey, L., et al., (2013). Bromelain inhibits allergic sensitization and murine asthma via modulation of dendritic cells. *Evid Based Complement Alternat Med.*, 702196.

Secretaria de Agricultura y Desarrollo Rural, (2017). *Se Afianza Méxcio Dentro de Los 10 Productores Mundiales de Piña*. Retrieved from: https://www.gob.mx/agricultura/prensa/se-afianza-mexico-dentro-de-los-10-productores-mundiales-de-pina (accessed on 15 November 2023).

Seenak, P., Kumphune, S., Malakul, W., Chotima, R., & Nernpermpisooth, (2021). Pineapple consumption reduced cardiac oxidative stress and inflammation in high cholesterol diet-fed rats. *Nutr. Metab. (Lond.), 18*, 36.

Sepúlveda, L., Romani, A., Aguilar, C. N., & Teixeira, J., (2018). Valorization of pineapple waste for the extraction of bioactive compounds and glycosides using autohydrolysis. *Innov. Food Sci. Emerg. Technol., 47*, 38–45.

Shahanaz-Hamidin, N. A., Abdullah, S., Mohd-Nor, F. H., & Hadibarata, T., (2022). Isolation and identification of natural green and yellow pigments from pineapple pulp and peel. *Materials today: Proceedings, 63*, S406–S410.

Shivamathi, C. S., Gunaseelan, S., Soosai, M. R., Vignesh, N. S., Varalakshmi, P., Kumar, R. S., Karthikumar, S., et al., (2022). Process optimization and characterization of pectin derived from underexploited pineapple peel biowaste as a value-added product. *Food Hydrocoll., 123*, 107141.

Sibaly, S., & Jeeath, P., (2017). Production of paper from pineapple leaves. *J. Environ. Chem. Eng., 5*, 5978–5986.

Sun, G. M., Zhang, X. M., Soler, A., & Marie-Alphonsine, P. A., (2016). Nutritional composition of pineapple (*Ananas comosus* (L.) Merr.). In: Simmonds, M. S. J., & Preedy, V. R., (eds.). *Nutritional Composition of Fruit Cultivars* (pp. 609–637). Academic Press, Elsevier.

Uriza-Ávila, D. E., Torres-Ávila, A., Aguilar-Ávila, J., Santoyo-Cortés, V. H., Zetina-Lezama, R., & Rebolledo-Martínez, A., (2018). La piña mexicana frente al reto de a innovación. *Avances y Retos en la Gestión de la Innovación* (pp. 1–481). Colección Trópico Húmedo, Chapingo, Estado de México. México: UACh.

Usano-Alemany, J., Palá-Paúl, J., & Díaz, S., (2014). Aceite esenciales: Conceptos básicos y actividad antibacteriana. *Reduca (Biología), 7*, 60–70.

Valdés-García, A., Domingo-Martínez, A. I., Ponce-Landete, M., Prats-Moya, M. S., & Beltrán-Sanahuja, A., (2021). Potential of industrial pineapple (*Ananas comosus* (L.) Merrill by-products as aromatic and antioxidant sources. *Antioxidants (Basel)*, *10*, 1767.

Vargas-Serna, C. L., Ochoa-Martínez, C. I., & Velez-Pasos, C., (2022). Microwave-assisted extraction of phenolic compounds from pineapple peel using deep eutectic solvents. *Horticulture, 8*(9), 791.

Wali, N., (2019). Pineapple (*Ananas comosus*). In: Nabavi, M., & Silva, A. S., (eds.), *Nonvitamin and Nonmineral Nutritional Supplements* (pp. 367–373). Academic Press, Elsevier.

Wen, L., & Wrolstad, R., (2002). Phenolic composition of authentic pineapple juice. *J. Food Sci., 67*, 155–161.

Yanbo, G., & Bing, J., (2020). Kinetic analysis and optimum design of extracting pectin from pineapple peel by ion exchange. *IOP Conf. Ser.: Earth Environ. Sci., 546*, 052048.

Yuris, A., & Siow, L. F., (2014). A comparative study of the antioxidant properties of three pineapple (*Ananas comosus* L.) varieties. *J Food Stud.*, *3*, 40–56.

Zakaria, N. A., Rahman, R. A., Abang-Zaidel, D. N., Dailin, D. J., & Jusoh, M., (2021). Microwave-assisted extraction of pectin from pineapple peel. *Mal. J. Fund. Appl. Sci., 17*, 33–38.

PART III
POTENTIAL AND PERSPECTIVES

CHAPTER 9

Waste Biorefineries: Future Energy, Green Products, and Waste Treatment

NARENDER SINGH and SONIA KHANNA

Advanced Polymeric Materials Research Laboratory, Department of Chemistry and Biochemistry, Sharda University, Greater Noida, Uttar Pradesh, India

ABSTRACT

Energy recovery and generating valuable products from different waste resources grip a remarkable capacity in the feasible and regular waste management position to keep up the idea of "reusing" economies with proper framework and to diminish the confront of waste-originated issues of public health in conjunction with the public clean environment. In today's world, waste disposal of any material to landfills is the generally used method, especially in developing countries, due to the lack of good infrastructure and insufficient budgets alongside the improper facilities to maintain efficient and technical universal standards. As a result, the dumpsites or polluted landfills have become important sources of greenhouse gases (GHGs) release, unclean water and dirty soil, annoying odors, dangerous contamination, and disease escalation vectors. It is worth mentioning that proper treatment and the well management of waste could always be a good source of energy, fuels, and value-added products.

This chapter indicates and explores the aspects going throughout the world with proper measure of concept of biorefineries, technical, and practical frameworks and future growth of the biorefineries technologies in a defined manner. Biomass production and converting it to bioenergy products

Waste Biorefineries: Future Energy, Green Products, and Waste Treatment.
Abu Zahrim Yaser, Siew-Xian Chin, Leonardo Sepúlveda Torre, & A. K. Haghi, (Eds.)

with the various conversion processes and in the end achieving the valuable biobased products. The biobased products help the whole world to understand an assessment cycle, i.e., original make to use to reuse to recycle.

9.1 INTRODUCTION

The present global population is 7.97 billion people as of September 2022, still the number increasing day by day and expected to reach 10 billion people by 2057 (https://www.worldometers.info/world-population) as shown in Figure 9.1. High annual rates of population expansion are creating urgent problems with regard to global issues including security of food, pollution conditions and environmental issues which affect and slow down the sustainable development goals (SDGs). Environmental considerations such as pollution, climate variability, global warming, waste disposal, and the loss of natural resources, have increased drastically in recent years and are largely due to 'human beings' destructive unchecked actions on our planet Earth (Hoornweg, Bhada-Tata & Kennedy, 2013). Humans are evolving with the newest developments to fulfill their requirements and using the resources of petroleum and fossil fuels in a disturbed manner to produce energy, chemicals, products, and synthetic

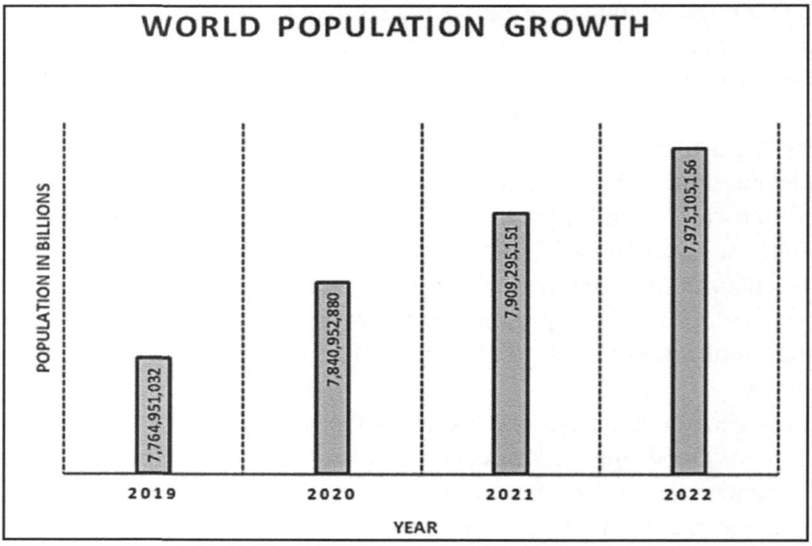

FIGURE 9.1 World population growth trend from 2019 to 2022.

Source: Reprinted from macrotrends. https://www.macrotrends.net/countries/WLD/world/population-growth-rate.

materials, which causes the shortage of non-renewable natural resources and also increases the percentage of greenhouse gasses (GHGs) emission; all these factors play a major role in spoiling the environment drastically. Therefore, sustainable development is essential, and especially biorefineries in a circular bioeconomy attain impeccable and encouraging paths (Severo, Siqueira, Depra, Maroneze, Zepka, & Jacob-Lopes, 2019).

In this framework, biorefineries have gained attention as a flexible and potential substitute for negative-valued petroleum and fossil fuel-based waste refineries where a variety of products, including biofuel, industrial biochemicals, and biomaterials including commercially important biopolymers are produced by biomass of uneatable feedstock and biogenic waste as raw materials (Clark & Deswarte, 2015; Venkata, 2014a). Figure 9.2 depicts the assessment tooling for the circular bioeconomy of biorefineries.

In upshot, these global issues want urgent and impactable solution so that the human race can overcome the problem of waste spreading and waste treatment (United Nations Foundation, 2020) and circular bioeconomy can play an important role to tackle these issues.

The term circular bioeconomy, which can be considered as bio-based reused economy, is an essential and useful concept of bioeconomy and circular economy, it represents the running use of biomass from biological resources into a working and systematic way for development of economy.

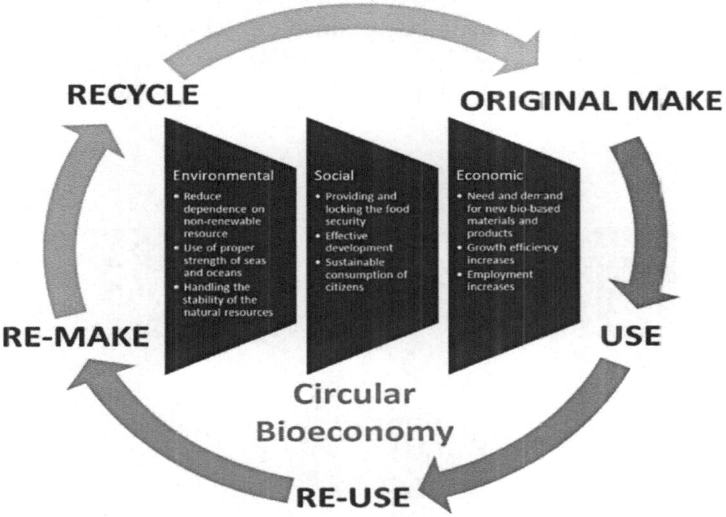

FIGURE 9.2 Assessment tool for circular bioeconomy.

This circular bioeconomy provides an instinctive utilization of biomass, which includes wastes and influent for the reusable production, and suppress the environment contamination (Satchatippavarn et al., 2016; Sadhukhan et al., 2016). The benefit of circular bioeconomy includes – (i) less GHGs footprints; (ii) valorization of waste products and materials from various sources; (iii) improved and useful resource with eco-efficiency; and (iv) reduced environment contamination. Initially the concept of bioeconomy was projected in the late 1990s by Enriquez in 1998 indicating the importance of the genome and genomics applications. In the beginning of 20th century, bioeconomy gained lot of attraction and forced the Europe to transform it as a motion policy wherein awareness about the sustainability developed (Enriquez, 1998; Birner, 2018; Fund, El-Chichakli, Patermann, & Dieckhoff, 2015; Strategy, 2018). The circular bioeconomy adopts the circular economy structural process, utilizing biomass as a vital factor for the generation of various bio-products, bioenergy, biofuels, and biochemicals in a biorefinery (Carrez et al., 2017; Carus & Dammer, 2018).

9.2 BIOREFINERY AND BIOMASS PRODUCTION

Biorefining is significantly important and primary path to facilitate the ideas and strategies of the circular bioeconomy that bind and holds the chain between fresh, raw resources, mineral, water, and carbon.

The term 'biorefinery' means the sustainable process that effectively utilize the biomass-based resources for the production of numerous commercialized products and metabolites for example: lipids, carbohydrates, proteins,

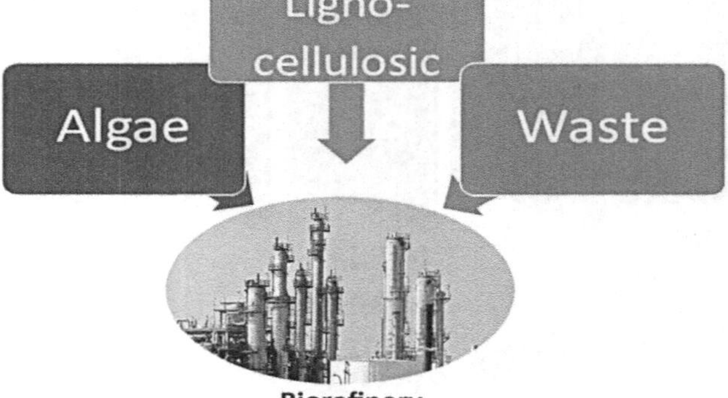

FIGURE 9.3 Major types of biorefineries.

bioactive compounds, biochemicals, and biomaterials (Mohan et al., 2016; Mohan, Modestra, Amulya, Butti, & Velvizhi, 2016; https://task42.ieabioenergy.com/publications/global-biorefinery-status-report-2022). Biorefineries receives much more interest as it provides a proper and applicable channel for waste management approach (Dahiya et al., 2018; Mishra, Roy, & Mohanty, 2019; Hoornweg, Bhada-Tata, & Kennedy, 2013). This concept designed and utilized to process a greater number of bio-mass raw materials such as lignocelluloses, algae, and different kinds of wastes as shown in Figure 9.3

9.2.1 LIGNOCELLULOSIC BIOREFINERY

Lignocellulosic biomass refers to second generation biomass feedstocks and provides an alternative approach to first generation biomass feedstocks which tackles with the land needs of foodstuffs. They are real source with workable method of biomass feedstock, considering the broad channel of plant options and their good accessibility in tropical climates. About 1.3 billion tons of lignocellulosic biomass per year produced worldly, from which only 3% are consumed for bioenergy, biochemicals, and bio-products not related to food (Baruah, Nath, Sharma, Kumar, Deka, Baruah, & Kalita, 2018). Generally, lignocellulosic biomass is obtained from coconut husk, empty fruit bunch, corn stover, barely straw, straw, rice, sugarcane bagasse, wheat, sorghum stalks and wood. It contains lignin hemicellulose and conversion of cellulose to number of multi-products by the lignocellulosic fractionation in a biorefinery (Zhang, 2008; De Bhowmick, Sarmah, & Sen, 2018). Lignocellulosic biorefinery creates a platform to produce the sustainable development of up-scalable bio-products alongside with biofuels production. Both recycling biomass and its wastes with the use of different way of conversion technologies with feasible routes.

9.2.1.1 LIGNOCELLULOSIC BIOMASS PRODUCTION

Lignocellulosic biomass is a good source of carbon for the production of sugars and other useful chemicals. It is a renewable resource which includes a variety of sources such as energy crops, residues generated from forests and grass, residues, and by-products generated from agricultural area (Patinvoh, Osadolor, Chandolias, Horváth, & Taherzadeh, 2017; Zheng, Zhao, Xu, & Li, 2014). It mainly consists of three polymers – (i) cellulose; (ii) hemicellulose; and (iii) lignin. Cellulose is the primary carbohydrate building up the lignocellulosic cell wall which forms longer and larger fibrils with long and aligned microfibrils (Isikgor & Becer, 2015). Hemicellulose are polysaccharides in plant cell wall. They are branched, amorphous, and single chained polysaccharides which attach to adjacent cellulose fibrils through non-covalent cross links (Somerville, 2006; Wyman, 2013). Lignin is a complex cross-linked macromolecule. It binds the lignocellulose matrix together and toughens the cell walls (Zakzeski, Bruijnincx, Jongerius, & Weckhuysen, 2010; Abomohra, 2019). Effect of pretreatment on lignocellulosic biomass shown in Figure 9.4.

Thermochemical conversion is one of the most crucial methods for biomass conversion to energy and fuels. Thermochemical conversion can be defined as the controlled heating or oxidation of feedstocks to generate energy products and/or heat. This covers several distinct techniques, primarily including pyrolysis, liquefaction, gasification, and combustion. The thermochemical conversion techniques differ by their range of conditions and degree of control of both oxidation and heating (DeFoort & Leach, 2014). Pyrolysis can be described as the initial step of both gasification and combustion processes (Babu, 2008). Pyrolysis processes can be significantly affected by the introduction of catalysts. Generally, these aim to reduce the oxygen content of the pyrolysis oil products. Whilst pyrolysis still occurs through thermal depolymerization followed by subsequent decompositions into volatile compounds, acid-catalyzed deoxygenation reactions, including decarboxylation, decarbonylation, and dehydration, result in greater formation of hydrocarbon products (Yaman, 2004).

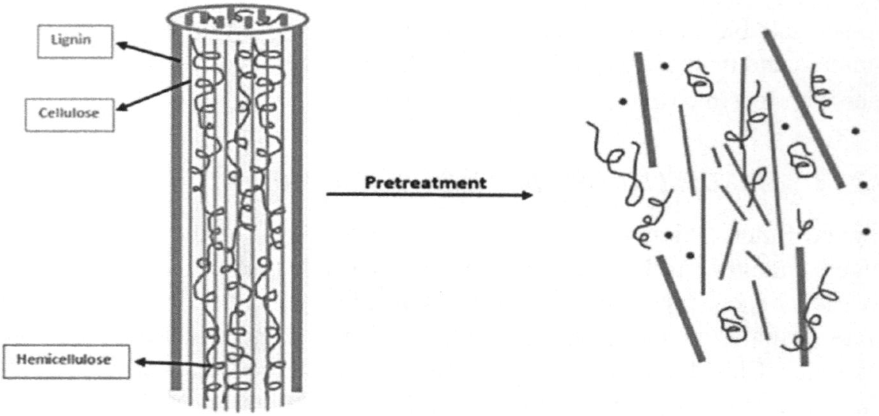

FIGURE 9.4 Effect of pretreatment on lignocellulosic biomass.

1. **Mechanical Pretreatment:** This of lignocellulosic biomass involves milling, i.e., lignocellulosic biomass cuts down in smaller pieces. The aim of this mechanical treatment is to lower the particle size and crystallinity which affects grow of certain surface and decrease the degree of polymerization (Palmowski & Müller, 2000; Park, Jones, Koo, Chen, Tucker, Yu, & Park, 2016). Milling causes the shearing of the biomass.
2. **Thermal Pretreatment:** It involves the heating of lignocellulosic biomass. If the temperature increases above 150–180°C, parts of the lignocellulosic biomass, the hemicelluloses at first and then the lignin, will begin to solubilize (Bobleter, 1994; Garrote, Dominguez, & Parajo, 1999). From the hemicellulose backbone formation and the branching groups, we can easily determine the thermal, acid, and alkali stability of the hemicellulose. On the other hand, from the two governing components of hemicelluloses (xylan and glucomannan), the xylans are thermally reported to have the minimal stability, but the variation with the glucomannans is only small. Above 180°C an exothermal reaction (probably solubilization) of the hemicellulose starts (Beall & Eickner, 1970; Domansky & Rendos, 1962). This temperature of 180°C could possibly explain the temperature that an exothermal reaction of the hemicellulose starts. This is obviously due to the thermal reactivity of lignocellulosic biomass, which related significantly on its composition.
3. **Acid Pretreatment:** In this pretreatment lignocellulosic biomass react with strong and dilute acids. The main reaction can be expected during acid pretreatment is the hydrolysis of hemicellulose, especially xylan as glucomannan is relatively acid stable. Solubilized hemicelluloses (oligomers) could be considered as:
   - Furfural;
   - Hydrolytic reactions producing monomers;
   - HMF;
   - And other (volatile) products in acidic environments (Woiciechowski et al., 2020; Fengel & Wegener, 1984; Ramos, 2003; Clauser, Felissia, Area, Vallejos, & Gutierrez, 2020; Zhang, Qi, Liu, Su, Wu, & He, 2010).
4. **Alkaline Pretreatment:** In this alkaline pretreatment the first response occur are solvation and saphonication. This in fact will result in a swollen state of the biomass and accomplish more

available for enzymes and bacteria. With 'strong' alkali concentrations dissolution, 'peeling' of end-groups, alkaline hydrolysis and degradation and decomposition of dissolved polysaccharides can be expected to develop. Dropping of polysaccharides is mostly due to the peeling and hydrolytic reactions (Fengel & Wegener, 1984). This peeling lead for later conversion, but, because lower molecular compounds are formed as a result, the chance on degradation and reduction of carbon, in the form of carbon dioxide (CO2), also increases. Although lignocellulosic material reacts with the aqueous solution of sodium sulfide and sodium hydroxide at 145–170°C with the concentration of 15–20% and (6–7 bar) pressure for 2–3 hours. In such kind of environment, the cellulose fibers are isolated, and the remaining liquid fraction known as black liquor (Woiciechowski et al., 2020 Brännvall, & Bäckström, 2016; Fernández-Rodríguez, Erdocia, Hernández-Ramos, Alriols, & Labidi, 2019).

5. **Oxidative Pretreatment:** This oxidative pretreatment involves the inclusion of an oxidizing compound, such as hydrogen peroxide or peracetic acid, to the biomass that is suspended in water. The main objective here is to remove the hemicellulose and lignin to increase the accessibility of the cellulose. Meanwhile, at the time of oxidative pretreatment, different reactions can be considered:
   - Displacement of side chains;
   - Electrophilic substitution;
   - Cleavage of alkyl aryl ether linkages; or
   - The oxidative cleavage of aromatic nuclei (Baeza & Freer, 2001).

9.2.2 ALGAE BIOREFINERY

Algae biomass is referred as advanced third generation biomass feedstocks which delivers profits such as lower land requirement and higher biomass production and its yield, this is quite high as compared with lignocellulosic biomass. Algae biomasses are classified into two main categories – (i) macroalgae; and (ii) microalgae (Torres, Kraan, & Domínguez, 2019; Chew, Yap, Show, Suan, Juan, Ling, & Chang, 2017).

Microalgae are photosynthetic microorganisms which effectively uses solar energy to biomass which comprises of important biological compounds (Leu & Boussiba, 2014). Microalgae can be cultivated in multiple reactor systems including photobioreactors that could be designed in vertical manner

and fabricated to give lesser land requirements (Ubando, Cuello, El-Halwagi, Culaba, Promentilla, & Tan, 2016) as compared to conventional methods of cultivation. Many biorefinery pathways have been developed for microalgae biomass to sustainably produce various microalgal-based products (Barreiro, Samorì, Terranella, Hornung, Kruse, & Prins, 2014). Macroalgae are majorly belongs to marine microorganisms Seaweeds which are majorly grown offshore and are mostly found on coastal regions near the edge of substantial water bodies such as sea and ocean (Lehahn, Ingle, & Golberg, 2016). Biorefinery based on macroalgae was developed and started for the production to obtain high-valued bio-products from the known seaweeds (Sadhukhan, Ng, & Martinez-Hernandez, 2016; Ingle, Vitkin, Robin, Yakhini, Mishori, & Golberg, 2018).

9.2.2.1 ALGAE BIOMASS PRODUCTION

Algae biomasses are converted to bio-based products such as biochar, bio-oils, syngas, and biofuels with the help of biochemical method, chemical method and thermochemical method. To avoid or stop issues like expenses uses in harvesting are too high and dewatering occurs with stimulated speed, thermochemical method being investigated by researchers to obtain solid, liquid, and gaseous products from algal biomass (Chen, Lin, Huang, & Chang, 2015; Chan, Cheah, How, Loy, Shahbaz, Singh, & Ngan, 2019; Chiaramonti, Prussi, Buffi, Casini, & Rizzo, 2015).

Algae biomass involves three main conversions:

i. Chemical conversion;
ii. Biochemical conversion; and
iii. Thermochemical conversion.

9.2.2.1.1 Chemical Conversion

Chemical conversion involves the transesterification or esterification of algae-based oil or lipid, depending on the feed composition. The chemical conversion method mainly uses the lipids (microalgae oil) as its raw material with the 90–98% of triglyceride, content of moisture and parts of other residual materials. If the lipids contain more quantity of free fatty acids or moisture, then a transesterification reaction takes place instead of an esterification reaction to avoid saponification (Tubino, Junior, & Bauerfeldt, 2016; Endalew, Kiros, & Zanzi, 2011; Li, Niu, Lu, & Li, 2016; Solis, Alejo, & Kiros, 2016).

In Transesterification process, microalgae lipid reacts with the mono-alcohol in the presence of catalyst through the transesterification for the formation of the biodiesel. These reactions are fall under equilibrium reactions needed higher alcohol to bio-oil ratio to shift the equilibrium to the higher side of reaction (Singh & Singh, 2010). The transesterification reaction is divided into three types depends on the catalyst used in the reaction, i.e., (i) acid-based transesterification; (ii) base-based transesterification; and (iii) enzyme-based transesterification. These types can also be further divided into two categories, i.e., homogeneous transesterification and heterogeneous transesterification as shown in Figure 9.5. Biofuel based diesel which is renewable are derived from biological resources via hydro-treating process (Davis et al., 2012), whereas biodiesel which is derived with the help of transesterification process are fatty acids methyl esters (Amin, 2019; Takisawa, Kanemoto, Kartikawati, & Kitamura, 2014). Not only the general catalytic method of transesterification is in use, non-catalytic method of transesterification in which they are using supercritical and subcritical methanol is used for the production of microalgal algae biomass (Levine, Pinnarat, & Savage, 2010; Patil et al., 2011). Microalgal biomass which are rich in carbohydrates are extracted from oil/FAME are highly suitable for the production of bioethanol (Batlle, Mastropietro, & Rodilla, 2020; Chng, Chan, & Lee, 2016).

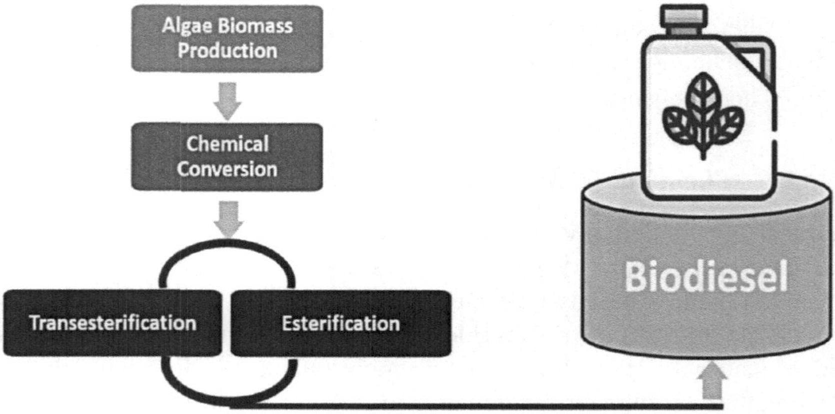

FIGURE 9.5 Chemical conversion of algae biomass production in biodiesel.

processes, i.e., (i) gasification; (ii) pyrolysis; and (iii) torrefaction as shown in Figure 9.7. The gasification involves the partial oxidation of wet biomass into combustible gas with low heat of value thus can be directly used for burning (Duman, Uddin, & Yanik, 2014; Biller & Ross, 2012; Singh & Olsen, 2011; Kröger, & Müller-Langer, 2012).

The reaction occurs in this process in the presence of catalyst at the elevated temperature (600–1,000°C). CO and H2 are the major products while CO2, N2, and CH4 are the minor products coming from the gasification chamber. However, the major task for this gasification process is to avoid the formation of tar which contains the mixture of the complex compound in the end-product because it creates the fouling in pipe, turbines, filters or the materials which are used for heat transfer (Adnan, Muraza, Razzak, Hossain, & De Lasa, 2017; Adnan, Hidayat, Ajumobi, Adamu, Muraza, & Hossain, 2018).

The pyrolysis process is an intermediate reaction which conducted at the temperature around 400–600°C which converts the algae biomass to liquid oils, charcoal, and syngas. However, there are some practical challenges connected to the pyrolysis fuel, i.e., acidity value is high, dissolved chemicals, viscosity is high and higher moisture content. Therefore, it requires an upstream oil upgradation area for the liquid oil purification (Demirbas, 2007; Hornung, Apfelbacher, & Sagi, 2011). Maddi et al. performed pyrolysis with algal and lignocellulosic feedstocks under the same reactor conditions and found that algal cultures deficient in lipids especially nuisance algae which are obtained from natural blooms are converted to liquid fuels by thermochemical conversion (Maddi, Viamajala, & Varanasi, 2011).

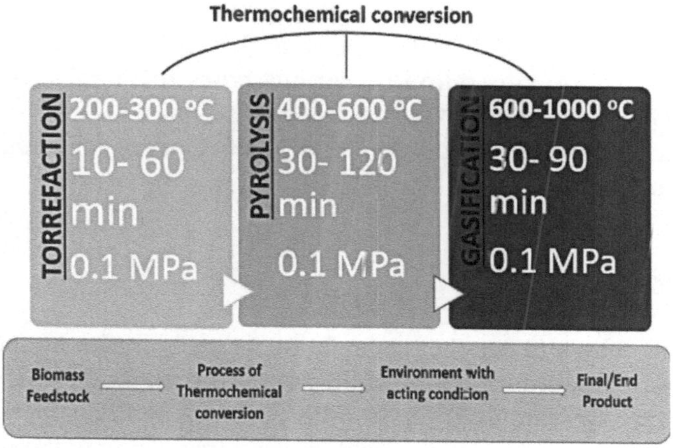

FIGURE 9.7 Types of thermochemical conversion.

Vasudev et al. performed kinetic and thermodynamic study of microalgae Spirulina via pyrolysis. The activation energies were found to be 109–340 kJ/mol and a novel kinetic model was proposed which tells the evolution of weight loss over a wide range of heating rates. In the end, spontaneity of reaction hindered by the remaining lipids and biochar (Vasudev, Ku, & Lin, 2020).

Biomass pyrolysis is a favorable renewable and reused source of fuels and petrochemical substitutes. It might assist in making up for the gradual consumption of fossil fuel supplies (Fahmy, Fahmy, Mobarak, El-Sakhawy, & Abou-Zeid, 2020; Chen, Meng, Han, Lan, & Zhang, 2019; Mao, Zou, Chen, Du, & Zuo, 2015; Hoang et al., 2021).

9.2.3 WASTE BIOREFINERY

In the present time, the growing population on the planet Earth is considered a major issue, but the growing population also releases waste in the form of solids (unwanted material, used stuff), liquids (unwanted liquids, uncleaned liquid), and a mixture of both (compounds, semi-solids, and liquids), this we need to consider as a major issue as well. Several scientists in the related field have investigated different technologies of valorization in order to handle, recycle, and reuse the waste generated by the population. Many biorefineries are now developed to convert this waste into a rich source of renewable feedstock, making value-added products and energy. It can be classified in food, biomass (agricultural), industrial, and municipal waste (Bhaskar, Pandey, Mohan, Lee, & Khanal, 2018; Nizami et al., 2017; Yadav, Dash, & Sen, 2019; Yaashikaa, Kumar, & Varjani, 2022). Valorization of various types of biomasses shown in Figure 9.8.

Different bioproducts produced with different types of conversion process as shown in Table 9.1.

9.2.3.1 FOOD WASTE

Food waste comes under the category of waste biorefinery and is main in the organic fraction of public solid waste (Barampouti, Mai, Malamis, Moustakas, & Loizidou, 2019). Food processing and distribution places majorly produces the food waste, and they are increasing with the growing population on the planet. It should be noted that the composting process processes, i.e., (i) gasification; (ii) pyrolysis; and (iii) torrefaction as shown in Figure 9.7. The gasification involves the partial oxidation of wet biomass into combustible gas with low heat of value thus can be directly used for burning (Duman, Uddin, & Yanik, 2014; Biller & Ross, 2012; Singh & Olsen, 2011; Kröger, & Müller-Langer, 2012).

The reaction occurs in this process in the presence of catalyst at the elevated temperature (600–1,000°C). CO and H2 are the major products while CO2, N2, and CH4 are the minor products coming from the gasification chamber. However, the major task for this gasification process is to avoid the formation of tar which contains the mixture of the complex compound in the end-product because it creates the fouling in pipe, turbines, filters or the materials which are used for heat transfer (Adnan, Muraza, Razzak, Hossain, & De Lasa, 2017; Adnan, Hidayat, Ajumobi, Adamu, Muraza, & Hossain, 2018).

The pyrolysis process is an intermediate reaction which conducted at the temperature around 400–600°C which converts the algae biomass to liquid oils, charcoal, and syngas. However, there are some practical challenges connected to the pyrolysis fuel, i.e., acidity value is high, dissolved chemicals, viscosity is high and higher moisture content. Therefore, it requires an upstream oil upgradation area for the liquid oil purification (Demirbas, 2007; Hornung, Apfelbacher, & Sagi, 2011). Maddi et al. performed pyrolysis with algal and lignocellulosic feedstocks under the same reactor conditions and found that algal cultures deficient in lipids especially nuisance algae which are obtained from natural blooms are converted to liquid fuels by thermochemical conversion (Maddi, Viamajala, & Varanasi, 2011).

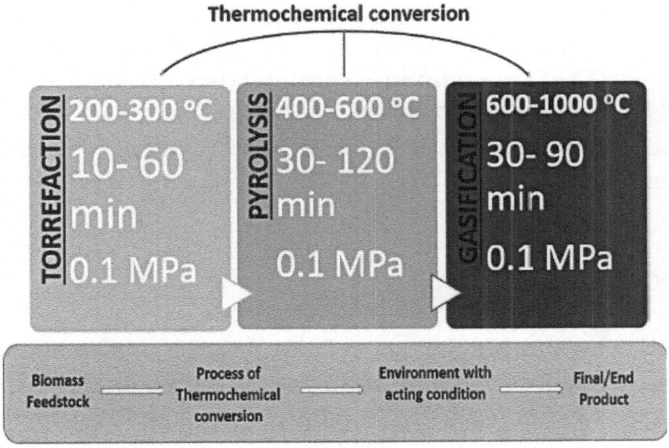

FIGURE 9.7 Types of thermochemical conversion.

Vasudev et al. performed kinetic and thermodynamic study of microalgae Spirulina via pyrolysis. The activation energies were found to be 109–340 kJ/mol and a novel kinetic model was proposed which tells the evolution of weight loss over a wide range of heating rates. In the end, spontaneity of reaction hindered by the remaining lipids and biochar (Vasudev, Ku, & Lin, 2020).

Biomass pyrolysis is a favorable renewable and reused source of fuels and petrochemical substitutes. It might assist in making up for the gradual consumption of fossil fuel supplies (Fahmy, Fahmy, Mobarak, El-Sakhawy, & Abou-Zeid, 2020; Chen, Meng, Han, Lan, & Zhang, 2019; Mao, Zou, Chen, Du, & Zuo, 2015; Hoang et al., 2021).

9.2.3 WASTE BIOREFINERY

In the present time, the growing population on the planet Earth is considered a major issue, but the growing population also releases waste in the form of solids (unwanted material, used stuff), liquids (unwanted liquids, uncleaned liquid), and a mixture of both (compounds, semi-solids, and liquids), this we need to consider as a major issue as well. Several scientists in the related field have investigated different technologies of valorization in order to handle, recycle, and reuse the waste generated by the population. Many biorefineries are now developed to convert this waste into a rich source of renewable feedstock, making value-added products and energy. It can be classified in food, biomass (agricultural), industrial, and municipal waste (Bhaskar, Pandey, Mohan, Lee, & Khanal, 2018; Nizami et al., 2017; Yadav, Dash, & Sen, 2019; Yaashikaa, Kumar, & Varjani, 2022). Valorization of various types of biomasses shown in Figure 9.8.

Different bioproducts produced with different types of conversion process as shown in Table 9.1.

9.2.3.1 FOOD WASTE

Food waste comes under the category of waste biorefinery and is main in the organic fraction of public solid waste (Barampouti, Mai, Malamis, Moustakas, & Loizidou, 2019). Food processing and distribution places majorly produces the food waste, and they are increasing with the growing population on the planet. It should be noted that the composting process

The fermentation process involves the production of bioethanol, which contains sugars, cellulose, and starch. In the typical fermentation method, biomass first converted to sugar in the presence of enzymes or acids, after that it further react by yeast and finally to bioethanol with the help of the distillation column for separation. Bioethanol can be used at is in same form or can blend with gasoline to form gasohol (De Farias, & Bertucco, 2016). Likewise, it can be used as an octane enhancer or gasoline improver. Microalgae are highly potential feedstock for the production of bioethanol as they own high concentrations of carbohydrates (about 12–51%) in the form of starch and cellulose (Doma, Abdo, Hemdan, & Ali, 2018; De Farias, Meneghello, & Bertucco, 2018; Pejin, Mojović, Vučurović, Pejin, Denčić, & Rakin, 2009; Jalilian, Najafpour, & Khajouei, 2020).

The use of fossil fuels in the disturbed and rough manner has resulted in the release of harmful GHGs, leading to change in climate conditions and increases the effect of global warming (Antar, Lyu, Nazari, Shah, Zhou, & Smith, 2021; Lu, Alam, Liu, Xu, & Saldivar, 2020; Martin-Gamboa, Marques, Freire, Arroja, & Dias, 2020). Ongoing reduction in the fossil fuels points towards the search for substituent fuels. Hydrogen has the capacity to set as a substituent for it, as hydrogen act as strong energy carrier and combustion of hydrogen only releases the water molecule with no harmful effect on the environment (Nagarajan, Dong, Chen, Lee, & Chang, 2021; Dębowski, Zieliński, Świca, & Kazimierowicz, 2021). Hydrogen can be use in the generation of power in fuel cells as it provides alternate energy carrier. Production of biohydrogen can be done by the microbial methods and the use of microalgae for hydrogen production is great due to the carbon mitigation effect. Photo fermentation and dark fermentation are suitable for the production of anaerobic hydrogen. Microalgae can produce of photo-biological H2 as it has both metabolic and enzyme characteristics (Ghirardi et al., 2000; Shuba & Kifle, 2018; Khetkorn et al., 2017). Melis and Happe reported that maximum yield of 200 kg H2 ha–1d–1 is produced by using green algae based two stage photosynthesis process (Melis & Happe, 2001), as the reversible nature of the photosynthetic process hydrogen can jump between the end product and reactant side depending on the proton reaction with hydrogen.

9.2.2.1.3 *Thermochemical Conversion*

Thermochemical conversion uses the thermal energy and convert the biomass to usable biofuel. Thermochemical conversion involves three main

9.2.2.1.2 *Biochemical Conversion*

The conversion of microalgae biomass production through biochemical pathway is very important and usable method. It could be divided into 3 groups, i.e., (i) anaerobic digestion; (ii) fermentation process for bioethanol production; and (iii) photobiological H2 production or biohydrogen production as shown in Figure 9.6. Fermentation process for the production of bioethanol is the best and popular method as it converts biomass to useful biofuels in the presence of microorganism or enzymes. The biochemical process is best fit for the organic waste which are having higher moisture content (more than 55% content) and also fit for the wet algae biomass (McKendry, 2002).

In anaerobic digestion, the biomass containing organic matter gets converted to biogas (methane (CH4)), NH3 and CO2 with trace amount of some uninvited gases like H2S (Singh & Olsen, 2011; Membere & Sallis, 2018). A typical anaerobic digestion process decreases the emission of harmful gases and oxygen demand value in wastewater. It can produce valuable by-products like fertilizer and compost, but anaerobic digestion has some disadvantages which includes high installation cost, requires long time running operations with high maintenance and needed large area farms for anaerobic digestion process. During the initial stage, the organic matters are converted to the sugar soluble products followed by the conversion of bacteria of these soluble sugars (which includes volatile fatty acids, alcohols, acetic acid, H2 and CO2) into CH4 and CO2 (Lü, Ji, Shao, & He, 2013; Lakaniemi, Hulatt, Thomas, Tuovinen, & Puhakka, 2011). Algae biomass production is dependent on species with lipids content about 7–21%, proteins content about 6–51% and carbohydrates content about 5–22%. Higher the lipid content, greater the yield of CH4 production as compared to carbohydrates and proteins of the cell wall (Li, Chen, & Wu, 2019; D'Silva et al., 2021).

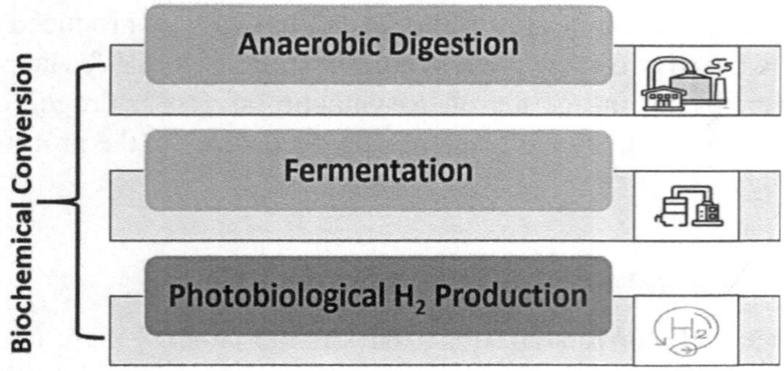

FIGURE 9.6 Methods for biochemical conversion.

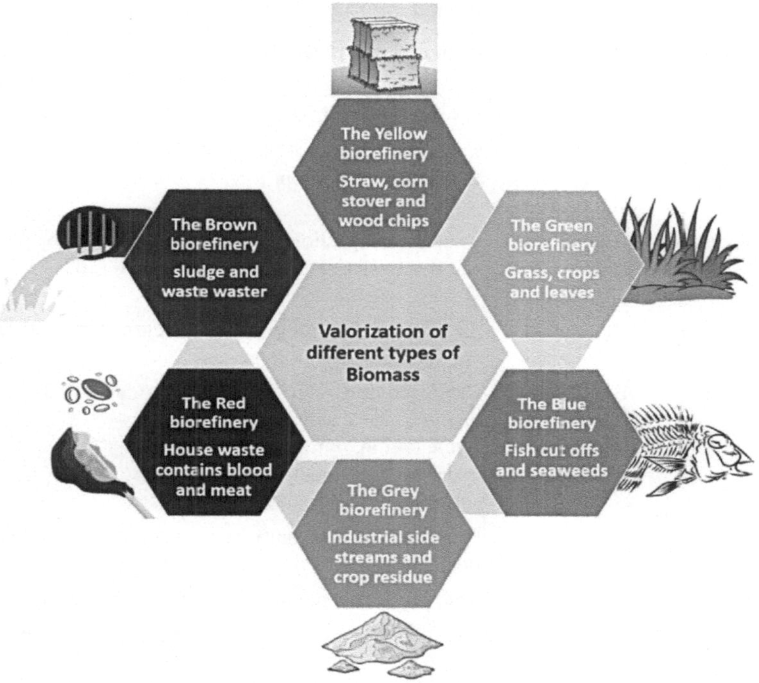

FIGURE 9.8 Valorization of different types of biomasses.

alongside with landfilling and also the production of animal feed are the most developed handling methods to food waste treatment (Capson-Tojo, Rouez, Crest, Steyer, Delgenès, & Escudié, 2016). However, these methods are costly and produce rather low value-added products, apart from investigated environmental concerns. Recent studies and reports argue that food waste can be used as alimentation or fuel sources. As an example, food waste utilization deals with compounds extraction from pretreatment by using hydrolysis and the production of bio-based chemicals (Li, Ng, Chen, Chiang, & Chao, 2018). Enzymatic and thermal conversion processes are other biotechnologies that using and developing bioenergy products from food waste (Banerjee, Ranganathan, Patti, & Arora, 2018; Gnansounou & Pandey, 2017; Giampietro, 2019). Nevertheless, many studies proving the feasibility of optimization strategies are still inappreciable. Anyhow, profitability alongside the importance of marketability, and economies of scale needed to be considered (Cristóbal, Caldeira, Corrado, & Sala, 2018).

TABLE 9.1 The Different Bioenergy Products Using Different Conversion Process

Categories of Biomass	Biobased Product	Conversion Process Used	Source of Information
Algal (microalgae)	Biochar	Pyrolysis experiment	Hilali et al. (2019)
Waste (orange peel)	Essential oils, polyphenols, and pectins	Conventional and solar hydro-distillation extraction.	Hilali et al. (2019)
Waste (olive tree)	Biomethane and bioethanol	Anaerobic digestion process and Fermentation process	Najafi, Castro, & Karimi (2021)
Waste (olive tree)	Bioethanol	Steam explosion pretreatment and then fermentation process	Negro et al. (2014)
Waste (coffee grounds)	Biodiesel, bio-oil, and bioethanol	Catalyzed transesterification, in-situ transesterification, fermentation, pyrolysis, hydrothermal liquefaction, and anaerobic digestion.	Karmee (2018)
Waste (food)	Biodiesel	Esterification and transesterification process.	Carmona-Cabello et al. (2018)
Lignocellulosic (olive tree)	Bioethanol	Acid pretreatment and fermentation.	Solarte-Toro, Chacón-Pérez, & Cardona-Alzate (2018)
Algal (macroalgae laminaria and fucus)	Biogas	Anerobic digestion process-extracted residues.	Tedesco & Stokes (2017)

9.2.3.2 SOLID WASTE

Solid wastes, especially in extremely developed countries, always have been an important issue for the people and governing authorities due to their bad effect on the environment and health. The model of the circular economy has also greatly changed the way of handling and treatment of the wastes from the conventional disposal and recycling practices into usable utilization method towards zero-waste economy (Mohan, Butti, Amulya, Dahiya, & Modestra, 2016). The composition of municipal solid waste is greatly related on the source, way of collection, and social surrounding activities of a particular region of an area (Maina, Kachrimanidou, & Koutinas, 2017). The development of many different substances in municipal solid waste streams shown serious interest for creation of new policies to effectively overcome the present issue with dynamic regenerating the energy in various field or separation of these substances and convert them into positive environmental impacts or potential income. As an example, the organic content available in municipal solid waste has been an encouraging free energy source. As a matter of fact, it does not provoke supplementary costs comparing to other biomass materials that require their own cultivation and harvesting (Iaquaniello, Centi, Salladini, Palo, & Perathoner, 2018). Incineration provides a technical conversion method to energy production and heat generation alongside the chemicals production from municipal solid waste (Hytönen & Stuart, 2011). As reported by Morf et al. (2013); and Rana et al. (2020), There have been efforts to effectively quantify and recover valuable metals, especially from electronic wastes. It should be noted that the effective recycling of desired substances in municipal solid waste is also limited because of structural impurities (Sadhukhan, Ng, & Martinez-Hernandez, 2016; Sadhukhan et al., 2018). As reported by Rakotovao, Gobert, & Dermine-Brullot (2018), functional process design and integration to produce multiple bioproducts have been verified to exhibit viability in terms of techno-economic aspects.

9.2.3.3 BIOMASS (AGRICULTURAL WASTE)

Under biorefinery context, the wastes and remains from agricultural products and the different food processing technics such as: olive mills, corn stover, bagasse, grapeseed, fruit and vegetable scraps, and manure have also been investigated for their potential to produce biochemicals and biofuels.

According to Awasthi et al. (2019); and Bosco, Casale, Gribaudo, Mollea, & Malucelli (2017), the production of low-value compounds such as fertilizers have been the current applications of these feedstocks According to Lucarini et al. (2018), making use of other biochemicals found in these materials has motivated researchers due to consider compelling economic benefits in their studies. As many of them are lignocellulosic biomass in nature, they can also be considered for conversion into value-added bioproducts. For such reasons, similar methods and technics and are being developed to lignocellulosic feedstocks regarding fractionation, pretreatment, and isolation of bioactive compounds. They contain (Zuin & Ramin, 2018) different types of solvent, ultrasound-assisted, microwave-assisted, and supercritical fluid extractions (Zuin & Ramin, 2018). According to Ferreira, Brancoli, Agnihotri, Bolton, & Taherzadeh (2018), for sustainability, several of these processes are being integrated with commercial establishments such as in first-generation biofuel.

9.2.3.4 INDUSTRIAL WASTE

Rise of demand for any main industrial product leads to a remarkable increase in the amount of by-products by the final process and wastewaters, which requires for recycling or treatment. For case in point, higher amount of glycerol with top level of impurities produced by biodiesel plants which provide it to be of lower value (Manara & Zabaniotou, 2016). The metals and minerals industry also produce great amounts of wastewater in their processes (Maina, Kachrimanidou, & Koutinas, 2017). On the other hand, current policies lined up with the idea of a circular economy strongly motivate these industries to develop functionalized waste treatment strategies and methods. In fact, according to Lotti, Carretti, Berti, Martina, Lubello, & Malpei (2019), the techniques for the effective recovery of needed substances from by-products and wastewater also draw great amount of attention as they develop opportunities for exploration of other potential applications. In various areas of industrial wastes such as glycerol (D'Angelo et al., 2018), beet pulp (Alexandri et al., 2019) and paper (Adu, Jolly, & Thakur, 2018), the bioactive compounds extraction and biomaterials production have been successfully observed. Arranging of industrial symbiosis (whereas one industry waste could be considered as raw material for 2nd industry) could also be one of the practical options that could lower their dependence on natural resources (Strategy, 2018). The awareness of a bioeconomy has also encouraged openings for co-production that would let the private sector to benefit from mixed product while rectifying their problems in spreading the waste and refining its environmental performance. Although economic and logistical issues have always been an important complication from their actual implementation, further technological improvements to optimize systems and processes are expected (Kaur, Uisan, Ong, & Lin, 2018).

9.3 GREEN PRODUCTS FROM THE WASTE BIOREFINERIES

The various technologies have been developed to minimize the waste issue caused by the factories, industries, and other innovation sectors. The waste biorefineries are playing a major role to reducing the problems into solution, using the waste biorefineries various types of waste converted into valuable products which are used in day-to-day life and minimizing the harmful effect on the environment (Wicker, Kumar, Khan, & Bhatnagar, 2021). Earth by 2057 reach the 10 billion population with increase in value of waste generated, to maintain the ecological value we need to setup new policies with running capacity of technologies. However, waste biorefinery uses the different approaches for the conversion of biomass into bioenergy.

Different types of biobased products are as follows:

- Bioethanol;
- Biomethanol;
- Biodiesel;
- Biochemicals;
- Biomaterials;
- Volatile fatty acids;
- Energy;
- Biogas;
- Biopolymers;
- Bioelectricity;
- Biofertilizer;
- Biopolymers.

9.3.1 BIOETHANOL

In order to meet the enormous demand, the global ethanol production demand surpassed 100 billion liters in 2015 and is constantly rising. Consequently, it is necessary to switch from producing ethanol from fossil fuels to renewable based resources. Currently, sugarcane and corn are used for the production of bioethanol, sugarcane, and corn are also consumed as food and animal feed for their survival, thus creating problem in the form of competition as they used for both food and fuel.

In the worldwide, bioethanol produced to fulfill the need for the society. Every country focuses in to produce bioethanol as per the requirement, Americas produced about 86 billion liters of bioethanol in 2018 (https://www.statista.com/statistics/271471/bioethanol-production-in-selected-countries/) which made it as the top ranker in the production of bioethanol as shown in Figure 9.9.

The application of the waste, produced during bioethanol production form bagasse, could ecologically and economically considered as great challenges. Nowadays, bioethanol production by liquid residuals of pentoses utilized by the microorganism at very low concentration (Kaparaju, Serrano, Thomsen, Kongjan, & Angelidaki, 2009). As pointed out by Andrade, Rivera,

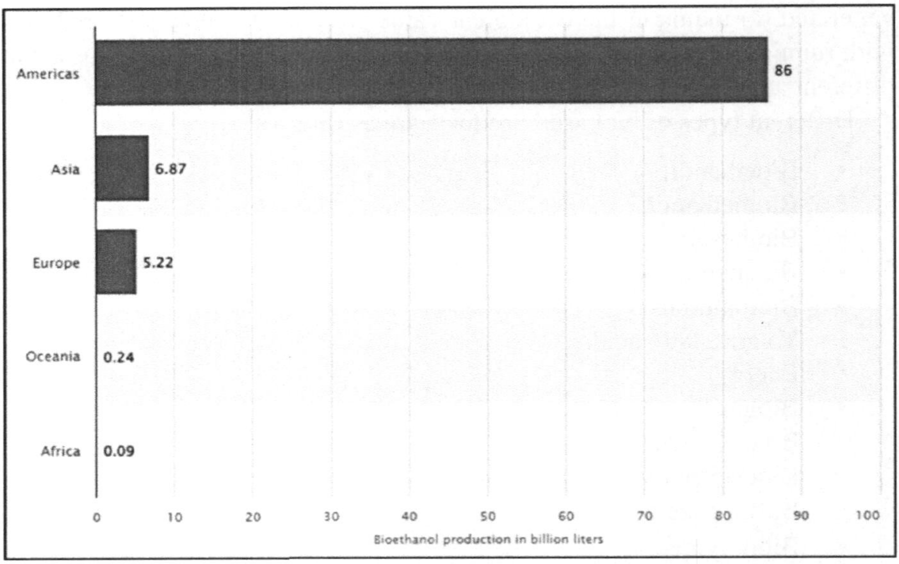

FIGURE 9.9 Production of bioethanol worldwide in year 2018.

Source: Reprinted from Statista (2020). https://www.statista.com/statistics/271471/bioethanol-production-in-selected-countries/.

Atala, Maugeri, Maciel, & Costa (2009), there are new techniques to reuse vinasse, such as: utilization for animal feed, thermal concentration alongside recycling to form partially dilute molasses.

These wastes are used for the production of biogas to remove the organic matter from effluents. Meanwhile, as mentioned by Liu, Liu, Zeng, & Angelidaki (2006), the effluent obtained during biogas formation could be used as fertilizer for farming land it worth mentioning that based on the report by Sassner, Mårtensson, Galbe, & Zacchi (2008), lignin presented in liquid residue could be employed to precipitation followed by heat and energy production (Sassner, Mårtensson, Galbe, & Zacchi, 2008).

9.3.2 BIOBUTANOL

Butyl alcohol is known as bio-butanol is commonly used as solvent in the chemical process, extractant, supplements in the health sector, eluent, precursor for other practical compounds, etc. Conversely, by blend it with the gasoline it can be used as biofuel for transportation. The properties of biobutanol are low vapor pressure, no sensitivity to water, reduced volatility, toxicity level is low with low flammability. Biobutanol is of particular interest due to its ability to directly replace gasoline, thus considered a drop-in-fuel. Starch-rich crops such as maize, wheat, rice, cassava, grasses, trees and energy crops and algae are widely used substrates for bio-butanol production and based on the substrate used it is designated as first, second, and third generation bio-butanol (Stoeberl, Werkmeister, Faulstich, & Russ, 2011).

Domestic food waste is the rich source of carbon hence used for the production of biobutanol (Huang, Singh, & Qureshi, 2015) but still there is a need to analyze the potential of food waste for bio-butanol production.

9.3.3 BIODIESEL

Biodiesel is simply green energy produced from renewable resources. Chemically, biodiesel is a mixture of methyl esters and long-chain fatty acids. Traditionally, biodiesel is obtained from a variety of feedstock, including waste cooking oils, vegetable oils, and animal fat. The limited supply of these feedstock restricts the further expansion of biodiesel industry. However, fungi (especially molds) and oleaginous yeasts can potentially be considered as organic waste as feedstock for the synthesis of lipids. Different types of inexpensive carbon sources viz. molasses (Alvarez et al., 1992), flour extracts, grape must radish brine, hydrolysates of agricultural residue, distillery wastewater, municipal wastewater (Chi, Zheng, Jiang, & Chen, 2011), food and feed waste, cheese whey (Castanha, Mariano, Morais, Scramin, & Monteiro, 2014), food/municipal wastewater, etc., could be considered as primary feedstock for cultivation of oleaginous yeast. Meanwhile, some yeast strains are reported (Ratledge & Wynn, 2002) to accumulate intracellular lipids as high as 70% of their biomass dry weight (such as Rhodosporidium sp., Rhodotorula sp., and Lipomyces sp.).

In a study by Chandra et al. highest FAME yield (19.02% of biomass) and biomass production (2.05 g L–1 in 5 days) were observed with 5 g/l of VFA. Also, a high lipid content of 34% with acetate followed by 18% with butyrate and 3.5% with propionate under mixotrophic condition were reported (Chandra, Arora, Rohit, & Mohan, 2015).

In another study, Han et al. reported Chlorella ellipsoidea and Scenedesmus quadricauda were tested to grow on effluent from anaerobic digestion of kitchen waste. They have well observed lipid productivity (Han, Pei, Hu, Jiang, Ma, Zhang, & Han, 2015). Meanwhile, it should be concluded that the acidogenic effluent from biohydrogen production process can serve as a potential carbon source for lipid biosynthesis by integrating with heterotrophic cultivation of microalgae.

9.3.4 BIOELECTRICITY

According to Rani, Sharma, & Kumar (2019), microbial fuel cell (MFC) is a sustainable and modern technique that is able to convert the chemical energy of waste directly to bio-electrical energy. The microorganisms in the anode function as biocatalyst, producing reducing equivalents (electrons and protons) that help in electrochemical oxidation of organic substrate. The presence of a terminal electron acceptor (TEA) in the cathode acts as an electron driving force, helping in transfer of electrons (e-) through the external circuit, where they can be harnessed for bioelectricity production, whereas the protons (H+) are delivered to the cathode using a membrane separating the anode and cathode (Butti et al., 2016).

9.3.5 Biogas

Biogas is produced from different methods and environments. Its composition includes several factors such as the nature of the substrate used to produce biogas, and the process design (Yentekakis & Goula, 2017). Europe is producing in large number of Mtoe in comparison of other parts of the world as shown in Figure 9.10. Crops include energy crops, crop residues, and sequential crops.

1 Mtoe = 11.63 terawatt-hours (TWh) = 41.9 petajoules (PJ) (https://www.iea.org/data-and-statistics/charts/biogas-production-by-region-and-by- feedstock-type-2018).

Anaerobic digestion is known as the biological conversion process in an oxygen-free environment and is carried out in four steps including hydrolysis, acidogenesis, acetogenesis, and methanogenesis. Different types of waste materials such as food waste, agricultural and industrial wastes, municipal solid waste, wastewater, and crops can be used as feedstock for the anaerobic digestion process. CO2 and the CH4 are the main constituents of the biogas. Biogas is mainly composed of CH4, carbon dioxide (CO2), nitrogen (N2), moisture and traces of hydrogen sulfide (H2S). CH4 production is perhaps the most explored technology for the generation of bioenergy using food waste as substrate with anaerobic fermentation. CH4 production from food waste is dependent on the various process parameters viz. design of reactor, type of inoculum (Deepanraj, Sivasubramanian, & Jayaraj, 2017), temperature condition, pH, ratio of carbon/nitrogen, rate of organic load and volatile fatty acid.

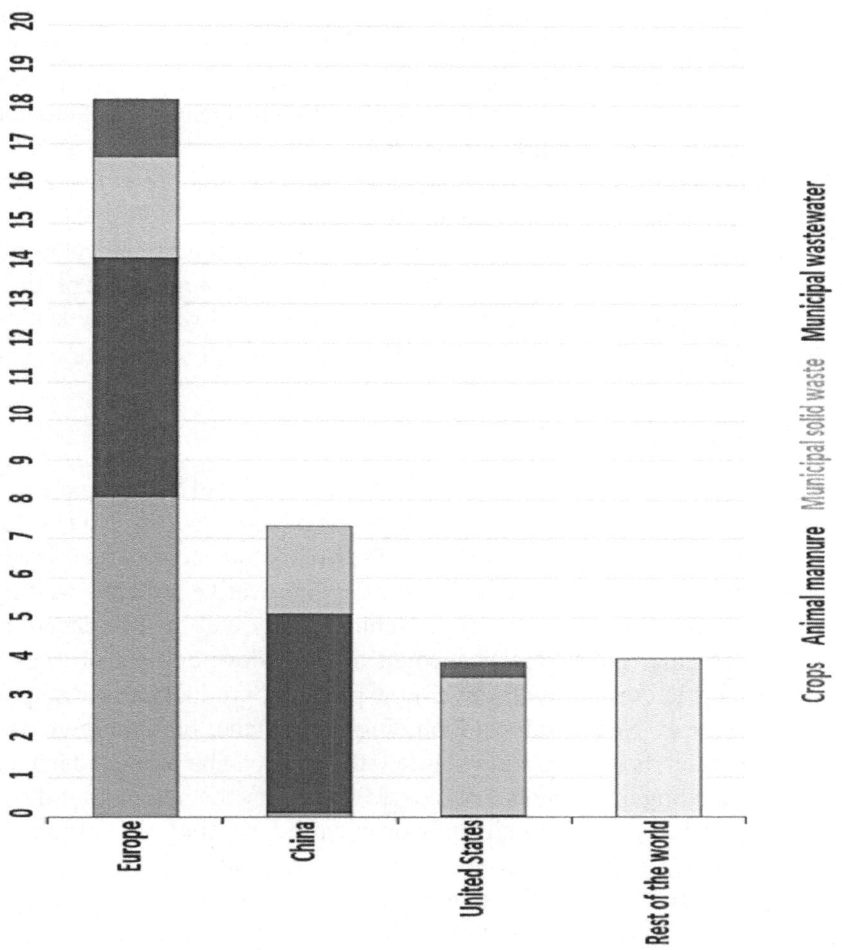

FIGURE 9.10 Production of biogas by region and by type of feedstock in (Mtoe), 2018.

Source: Reprinted with permission from IEA (2020). https://creativecommons.org/licenses/by/4.0/

9.3.6 BIOHYDROGEN

Biohydrogen (H2) is considered as a sustainable green fuel and its production from waste attracted attention of the researchers worldwide. Biologically, H2 can be produced using photo-fermentation, bio-photolysis, and anaerobic fermentation or integrating these processes. H2 is produced by the action of obligatory H2 producing acetogenic bacteria. Biological H2 production was majorly studied using a variety of waste viz. composite vegetable-based market waste, food waste (Sarkar & Mohan, 2017), wastewater, vegetable waste, co-supplemented domestic sewage supplementation, etc. The potential of food waste for H2 production is well established (Sarkar & Venkata, 2016) with semi-pilot and pilot scale studies.

The photo-fermentation and anaerobic fermentation can be either performed in two-phase operation or single stage operation. Mixing dark and photo-fermentation was found to be the ideal route for the highest possible theoretical biohydrogen yield. The developments in the biological H2 production from variety of biogenic waste specifically with food waste may lead to worldwide scaling up of pilot plants and H2 used as major transportation fuel.

9.4 FUTURE PERSPECTIVE

While carrying simultaneously various waste substrates and microorganisms, a nonstop discovery alongside with development of new processes and up-to-date products takes place. Meanwhile, to the prosperous commercial green products supplied recently, there are "rounds" of innovative products waiting to spread the same development level. Naturally, this will be influenced by economical parameters. A greater account of underdevelopment processes is the key driver to constantly discover new potential products, for example, the biogas process. New products including biohythane, biohydrogen, and different type of carboxylates and volatile fatty acids can be produced from a better understanding of the biogas processes or classify the potential of those products as replacement to fossil fuels or even the existing green biofuels and biochemicals, hydrogen is observed as a side product from anaerobic digestion, even if CH4, the main component of biogas, is desired. It should be noted that the optimization of hydrogen production is considered as another potential green fuel combustion from which water originates (Sarkar, Butti, & Mohan, 2018). Similar to biomethane production, food waste has also been the main-choice substrate for production of biohydrogen using dark fermentation. Being a transitional by-product, efforts have been concentrated on codigestion techniques and on specific enrichment of microbial communities for increasing biohydrogen production.

The design and strategy of biorefineries, various measures with policies, and characteristics must be measured and verified attentively before the commercialization of integrated processes of the whole model. Matters connected to feedstock selection, transportation of feedstock to various areas, stability factor, and acceptance by society should be solved in future time.

9.5 CONCLUSION

The exploitation of petroleum-based fuel products has created an alarming situation and called for reliance on non-detrimental renewable energy sources. Biofuels are produced mainly from plants, animals, and algae. Plants provide starch, cellulosic, and non-cellulosic biomolecules have been extensively incorporated in production cycle of bioethanol and biobutanol. Fats and oils in animals, plants, and algae are raw material for biogas and biodiesel production. The formation and practical innovation of biorefineries significantly contributed to the mechanism of transition of energy areas from petrochemical to biochemical energy. It is mandatory to develop the effective and highly productive biorefineries with feasible handling manner so that no harmful effect on our environment can be spread and reducing the less climatic condition. The clean technology and ecofriendly process can be attained based on integrated biorefineries. Though some commercial scale biorefineries are in developing stage, a significant and continuous development will involve the sustainability approach involved by governing authorities, which covers particular guidelines and policies indicating the proper channel for the growth for research and its promotion in biorefineries, consistent industrial strategy framework settings for biobased productions, framework settings of the market environments, etc.

KEYWORDS

- **agricultural waste**
- **biobutanol**
- **bioelectricity**
- **bioethanol**
- **biomass**
- **green products**
- **waste biorefineries**

REFERENCES

Abomohra, A. E. F., (2019). Biomass for Bioenergy: Recent Trends and Future Challenges.

Adnan, M. A., Hidayat, A., Ajumobi, O. O., Adamu, S., Muraza, O., & Hossain, M. M., (2018). Fluidizable Fe–Co/Ce–ZrO2 catalysts for steam reforming of toluene as a tar surrogate in biomass gasification. Energy & Fuels, 32(12), 12833–12842.

Adnan, M. A., Muraza, O., Razzak, S. A., Hossain, M. M., & De Lasa, H. I., (2017). Iron oxide over silica-doped alumina catalyst for catalytic steam reforming of toluene as a surrogate tar biomass species. Energy & Fuels, 31(7), 7471–7481.

Adu, C., Jolly, M., & Thakur, V. K., (2018). Exploring new horizons for paper recycling: A review of biomaterials and biorefinery feedstocks derived from wastepaper. Current Opinion in Green and Sustainable Chemistry, 13, 21–26.

Alexandri, M., Schneider, R., Papapostolou, H., Ladakis, D., Koutinas, A., & Venus, J., (2019). Restructuring the conventional sugar beet industry into a novel biorefinery: Fractionation and bioconversion of sugar beet pulp into succinic acid and value-added coproducts. ACS Sustainable Chemistry & Engineering, 7(7), 6569–6579.

Alvarez, R. M., Rodriguez, B., Romano, J. M., Diaz, A. O., Gomez, E., Miro, D., & Garcia, J. L., (1992). Lipid accumulation in Rhodotorula glutinis on sugar cane molasses in single-stage continuous culture. World Journal of Microbiology and Biotechnology, 8(2), 214, 215.

Amin, A., (2019). Review of diesel production from renewable resources: Catalysis, process kinetics and technologies. Ain Shams Engineering Journal, 10(4), 821–839.

Andrade, R. R., Rivera, E. A. C., Atala, D. I. P., Maugeri, F. F., Maciel, F. R., & Costa, A. C., (2009). Development of extractive processes and robust mathematical model for bioethanol production. Bioethanol: Production. Benefits and Economics (pp. 75–92). Nova Science Publishers, New York.

Antar, M., Lyu, D., Nazari, M., Shah, A., Zhou, X., & Smith, D. L., (2021). Biomass for a sustainable bioeconomy: An overview of world biomass production and utilization. Renewable and Sustainable Energy Reviews, 139, 110691.

Awasthi, M. K., Sarsaiya, S., Wainaina, S., Rajendran, K., Kumar, S., Quan, W., & Taherzadeh, M. J., (2019). A critical review of organic manure biorefinery models toward sustainable circular bioeconomy: Technological challenges, advancements, innovations, and future perspectives. Renewable and Sustainable Energy Reviews, 111, 115–131.

Babu, B. V., (2008). Biomass pyrolysis: A state-of-the-art review. Biofuels, Bioproducts and Biorefining: Innovation for a Sustainable Economy, 2(5), 393–414.

Baeza, J., & Freer, J., (2001). Wood and Cellulosic Chemistry (p. 275).

Banerjee, S., Ranganathan, V., Patti, A., & Arora, A., (2018). Valorization of pineapple wastes for food and therapeutic applications. Trends in Food Science & Technology, 82, 60–70.

Barampouti, E. M., Mai, S., Malamis, D., Moustakas, K., & Loizidou, M., (2019). Liquid biofuels from the organic fraction of municipal solid waste: A review. Renewable and Sustainable Energy Reviews, 110, 298–314.

Barreiro, D. L., Samorì, C., Terranella, G., Hornung, U., Kruse, A., & Prins, W., (2014). Assessing microalgae biorefinery routes for the production of biofuels via hydrothermal liquefaction. Bioresource Technology, 174, 256–265.

Baruah, J., Nath, B. K., Sharma, R., Kumar, S., Deka, R. C., Baruah, D. C., & Kalita, E., (2018). Recent trends in the pretreatment of lignocellulosic biomass for value-added products. Frontiers in Energy Research, 6, 141.

Batlle, C., Mastropietro, P., & Rodilla, P., (2020). Redesigning residual cost allocation in electricity tariffs: A proposal to balance efficiency, equity, and cost recovery. Renewable Energy, 155, 257–266.

Beall, F. C., & Eickner, H. W., (1970). Thermal Degradation of Wood Components: A Review of the Literature. 1–30.

Bhaskar, T., Pandey, A., Mohan, S. V., Lee, D. J., & Khanal, S. K., (2018). Waste Biorefinery: Potential and Perspectives. Elsevier. 1st Edition, pp. 1–890.

Biller, P., & Ross, A. B., (2012). Hydrothermal processing of algal biomass for the production of biofuels and chemicals. Biofuels, 3(5), 603–623.

Birner, R., (2018). Bioeconomy concepts. In: Bioeconomy (pp. 17–38). Springer, Cham.

Bobleter, O., (1994). Hydrothermal degradation of polymers derived from plants. Progress in Polymer Science, 19(5), 797–841.

Bosco, F., Casale, A., Gribaudo, G., Mollea, C., & Malucelli, G., (2017). Nucleic acids from agro-industrial wastes: A green recovery method for fire retardant applications. Industrial Crops and Products, 108, 208–218.

Brännvall, E., & Bäckström, M., (2016). Improved impregnation efficiency and pulp yield of softwood kraft pulp by high effective alkali charge in the impregnation stage. Holzforschung, 70(11), 1031–1037.

Butti, S. K., Velvizhi, G., Sulonen, M. L., Haavisto, J. M., Koroglu, E. O., Cetinkaya, A. Y., & Mohan, S. V., (2016). Microbial electrochemical technologies with the perspective of harnessing bioenergy: Maneuvering towards upscaling. Renewable and Sustainable Energy Reviews, 53, 462–476.

Capson-Tojo, G., Rouez, M., Crest, M., Steyer, J. P., Delgenès, J. P., & Escudié, R., (2016). Food waste valorization via anaerobic processes: A review. Reviews in Environmental Science and Bio/Technology, 15(3), 499–547.

Carmona-Cabello, M., Leiva-Candia, D., Castro-Cantarero, J. L., Pinzi, S., & Dorado, M. P., (2018). Valorization of food waste from restaurants by transesterification of the lipid fraction. Fuel, 215, 492–498.

Carrez, D., Carus, M., Griffon, M., Jilkova, J., Juhász, A., Lange, L., & Consuelo, V. O., (2017). Expert Group Report-Review of the EU Bioeconomy Strategy and its Action Plan.

Carus, M., & Dammer, L., (2018). The circular bioeconomy—Concepts, opportunities, and limitations. Industrial Biotechnology, 14(2), 83–91.

Castanha, R. F., Mariano, A. P., Morais, L. A. S. D., Scramin, S., & Monteiro, R. T. R., (2014). Optimization of lipids production by Cryptococcus laurentii 11 using cheese whey with molasses. Brazilian Journal of Microbiology, 45, 379–387.

Chan, Y. H., Cheah, K. W., How, B. S., Loy, A. C. M., Shahbaz, M., Singh, H. K. G., & Ngan, S. L., (2019). An overview of biomass thermochemical conversion technologies in Malaysia. Science of The Total Environment, 680, 105–123.

Chandra, R., Arora, S., Rohit, M. V., & Mohan, S. V., (2015). Lipid metabolism in response to individual short chain fatty acids during mixotrophic mode of microalgal cultivation: Influence on biodiesel saturation and protein profile. Bioresource Technology, 188, 169–176.

Chen, W. H., Lin, B. J., Huang, M. Y., & Chang, J. S., (2015). Thermochemical conversion of microalgal biomass into biofuels: A review. Bioresource Technology, 184, 314–327.

Chen, W., Meng, J., Han, X., Lan, Y., & Zhang, W., (2019). Past, present, and future of biochar. Biochar, 1(1), 75–87.

Cheng, P., Li, Y., Wang, C., Guo, J., Zhou, C., Zhang, R., & Ruan, R., (2022). Integrated marine microalgae biorefineries for improved bioactive compounds: A review. Science of The Total Environment, 152895.

Chew, K. W., Yap, J. Y., Show, P. L., Suan, N. H., Juan, J. C., Ling, T. C., & Chang, J. S., (2017). Microalgae biorefinery: High value products perspectives. Bioresource Technology, 229, 53–62.

Chi, Z., Zheng, Y., Jiang, A., & Chen, S., (2011). Lipid production by culturing oleaginous yeast and algae with food waste and municipal wastewater in an integrated process. Applied Biochemistry and Biotechnology, 165(2), 442–453.

Chiaramonti, D., Prussi, M., Buffi, M., Casini, D., & Rizzo, A. M., (2015). Thermochemical conversion of microalgae: Challenges and opportunities. Energy Procedia, 75, 819–826.

Chng, L. M., Chan, D. J., & Lee, K. T., (2016). Sustainable production of bioethanol using lipid-extracted biomass from Scenedesmus dimorphus. Journal of Cleaner Production, 130, 68–73.

Clark, J. H., & Deswarte, F., (2015). Introduction to Chemicals from Biomass. John Wiley & Sons.

Clauser, N. M., Felissia, F. E., Area, M. C., Vallejos, M. E., & Gutierrez, S., (2020). An energy-saving re-use strategy for the treatment of lignocellulosic biomass applied to the production of levulinc acid. Journal of Cleaner Production, 257, 120549.

Cristóbal, J., Caldeira, C., Corrado, S., & Sala, S., (2018). Techno-economic and profitability analysis of food waste biorefineries at European level. Bioresource Technology, 259, 244–252.

D'Angelo, S. C., Dall'Ara, A., Mondelli, C., Perez-Ramirez, J., & Papadokonstantakis, S., (2018). Techno-economic analysis of a glycerol biorefinery. ACS Sustainable Chemistry & Engineering, 6(12), 16563–16572.

D'Silva, T. C., Isha, A., Chandra, R., Vijay, V. K., Subbarao, P. M. V., Kumar, R., & Kovács, K. L., (2021). Enhancing methane production in anaerobic digestion through hydrogen assisted pathways–a state-of-the-art review. Renewable and Sustainable Energy Reviews, 151, 111536.

Dahiya, S., Kumar, A. N., Sravan, J. S., Chatterjee, S., Sarkar, O., & Mohan, S. V., (2018). Food waste biorefinery: Sustainable strategy for circular bioeconomy. Bioresource Technology, 248, 2–12.

Davis, R., Fishman, D., Frank, E. D., Wigmosta, M. S., Aden, A., Coleman, A. M., & Wang, M. Q., (2012). Renewable Diesel from Algal Lipids: An Integrated Baseline for Cost, Emissions, and Resource Potential from a Harmonized Model (No. NREL/TP-5100–55431; ANL/ESD/12-4; PNNL-21437). National Renewable Energy Lab. (NREL), Golden, CO (United States).

De Bhowmick, G., Sarmah, A. K., & Sen, R., (2018). Lignocellulosic biorefinery as a model for sustainable development of biofuels and value-added products. Bioresource Technology, 247, 1144–1154.

De Farias, S. C. E., & Bertucco, A., (2016). Bioethanol from microalgae and cyanobacteria: A review and technological outlook. Process Biochemistry, 51(11), 1833–1842.

De Farias, S. C. E., Meneghello, D., & Bertucco, A., (2018). A systematic study regarding hydrolysis and ethanol fermentation from microalgal biomass. Biocatalysis and Agricultural Biotechnology, 14, 172–182.

Dębowski, M., Zieliński, M., Świca, I., & Kazimierowicz, J., (2021). Algae biomass as a potential source of liquid fuels. Phycology, 1(2), 105–118.

Deepanraj, B., Sivasubramanian, V., & Jayaraj, S., (2017). Multi-response optimization of process parameters in biogas production from food waste using Taguchi–Grey relational analysis. Energy Conversion and Management, 141, 429–438.

DeFoort, W., & Leach, J. E., (2014). Biomass for thermochemical conversion: Targets and challenges. Variation in Cell Wall Composition and Bioenergy Potential of Rice Straw, 12.

Demirbas, A., (2007). Biodiesel from sunflower oil in supercritical methanol with calcium oxide. Energy Conversion and Management, 48(3), 937–941.

Doma, H. S., Abdo, S. M., Hemdan, B. A., & Ali, G. H., (2018). Research article enhancing biomass, energy and value-added compounds yield from pilot scale pond system. Journal of Environmental Science and Technology, 11, 199–208.

Domansky, R., & Rendos, F., (1962). On the pyrolysis of wood and its components. Holz Roh-Werkst, 20, 473–476.

Duman, G., Uddin, M. A., & Yanik, J., (2014). Hydrogen production from algal biomass via steam gasification. Bioresource Technology, 166, 24–30.

Endalew, A. K., Kiros, Y., & Zanzi, R., (2011). Inorganic heterogeneous catalysts for biodiesel production from vegetable oils. Biomass and Bioenergy, 35(9), 3787–3809.

Enriquez, J., (1998). Genomics and the world's economy. Science, 281(5379), 925–926.

Fahmy, T. Y., Fahmy, Y., Mobarak, F., El-Sakhawy, M., & Abou-Zeid, R. E., (2020). Biomass pyrolysis: Past, present, and future. Environment, Development and Sustainability, 22(1), 17–32.

Fengel, D., & Wegener, G., (1984). Wood: Chemistry, ultrastructure. Reactions, 613, 1960–1982.

Fernández-Rodríguez, J., Erdocia, X., Hernández-Ramos, F., Alriols, M. G., & Labidi, J., (2019). Lignin separation and fractionation by ultrafiltration. In: Separation of Functional Molecules in Food by Membrane Technology (pp. 229–265). Academic Press.

Ferreira, J. A., Brancoli, P., Agnihotri, S., Bolton, K., & Taherzadeh, M. J., (2018). A review of integration strategies of lignocelluloses and other wastes in 1st generation bioethanol processes. Process Biochemistry, 75, 173–186.

Fund, C., El-Chichakli, B., Patermann, C., & Dieckhoff, P., (2015). Bioeconomy policy (Part II). Synopsis of National Strategies Around the World.

Garrote, G. D. H. P., Dominguez, H., & Parajo, J. C., (1999). Hydrothermal processing of lignocellulosic materials. Holz als Roh- und Werkstoff, 57(3), 191–202.

Ghirardi, M. L., Zhang, L., Lee, J. W., Flynn, T., Seibert, M., Greenbaum, E., & Melis, A., (2000). Microalgae: A green source of renewable H2. Trends in Biotechnology, 18(12), 506–511.

Giampietro, M., (2019). On the circular bioeconomy and decoupling: Implications for sustainable growth. Ecological Economics, 162, 143–156.

Gnansounou, E., & Pandey, A., (2017). Classification of Biorefineries Taking into Account Sustainability Potentials and Flexibility. Elsevier.

Han, L., Pei, H., Hu, W., Jiang, L., Ma, G., Zhang, S., & Han, F., (2015). Integrated campus sewage treatment and biomass production by Scenedesmus quadricauda SDEC-13. Bioresource Technology, 175, 262–268.

Hilali, S., Fabiano-Tixier, A. S., Ruiz, K., Hejjaj, A., Ait, N. F., Idlimam, A., & Chemat, F., (2019). Green extraction of essential oils, polyphenols, and pectins from orange peel employing solar energy: Toward a zero-waste biorefinery. ACS Sustainable Chemistry & Engineering, 7(13), 11815–11822.

Hoang, A. T., Ong, H. C., Fattah, I. R., Chong, C. T., Cheng, C. K., Sakthivel, R., & Ok, Y. S., (2021). Progress on the lignocellulosic biomass pyrolysis for biofuel production toward environmental sustainability. Fuel Processing Technology, 223, 106997.

Hoornweg, D., Bhada-Tata, P., & Kennedy, C., (2013). Environment: Waste production must peak this century. Nature, 502(7473), 615–617.

Hornung, A., Apfelbacher, A., & Sagi, S., (2011). Intermediate Pyrolysis: A Sustainable Biomass-toEnergy Concept—Biothermal Valorization of Biomass BtVB Process. Journal of Scientific & Industrial Research, 70, 664–667.

https://task42.ieabioenergy.com/publications/global-biorefinery-status-report-2022/ (accessed on 15 November 2023).

https://www.iea.org/data-and-statistics/charts/biogas-production-by-region-and-by-feedstock-type-2018 (accessed on 15 November 2023).

https://www.macrotrends.net/countries/WLD/world/population-growth-rate (accessed on 15 November 2023).

https://www.statista.com/statistics/271471/bioethanol-production-in-selected-countries/ (accessed on 15 November 2023).

https://www.worldometers.info/world-population/ (accessed on 15 November 2023).

Huang, H., Singh, V., & Qureshi, N., (2015). Butanol production from food waste: A novel process for producing sustainable energy and reducing environmental pollution. Biotechnology for Biofuels, 8(1), 1–12.

Hytönen, E., & Stuart, P., (2011). Techno-economic assessment and risk analysis of biorefinery processes. In: Computer Aided Chemical Engineering (Vol. 29, pp. 1376–1380). Elsevier.

Iaquaniello, G., Centi, G., Salladini, A., Palo, E., & Perathoner, S., (2018). Waste to chemicals for a circular economy. Chemistry–A European Journal, 24(46), 11831–11839.

Ingle, K., Vitkin, E., Robin, A., Yakhini, Z., Mishori, D., & Golberg, A., (2018). Macroalgae biorefinery from Kappaphycus alvarezii: Conversion modeling and performance prediction for India and Philippines as examples. BioEnergy Research, 11(1), 22–32.

Isikgor, F. H., & Becer, C. R., (2015). Lignocellulosic biomass: A sustainable platform for the production of bio-based chemicals and polymers. Polymer Chemistry, 6(25), 4497–4559.

Jalilian, N., Najafpour, G. D., & Khajouei, M., (2020). Macro and micro algae in pollution control and biofuel production–a review. ChemBioEng Reviews, 7(1), 18–33.

Kaparaju, P., Serrano, M., Thomsen, A. B., Kongjan, P., & Angelidaki, I., (2009). Bioethanol, biohydrogen and biogas production from wheat straw in a biorefinery concept. Bioresource Technology, 100(9), 2562–2568.

Karmee, S. K., (2018). A spent coffee grounds based biorefinery for the production of biofuels, biopolymers, antioxidants and biocomposites. Waste Management, 72, 240–254.

Kaur, G., Uisan, K., Ong, K. L., & Lin, C. S. K., (2018). Recent trends in green and sustainable chemistry & waste valorization: Rethinking plastics in a circular economy. Current Opinion in Green and Sustainable Chemistry, 9, 30–39.

Khetkorn, W., Rastogi, R. P., Incharoensakdi, A., Lindblad, P., Madamwar, D., Pandey, A., & Larroche, C., (2017). Microalgal hydrogen production–A review. Bioresource Technology, 243, 1194–1206.

Kostas, E. T., Adams, J. M., Ruiz, H. A., Durán-Jiménez, G., & Lye, G. J., (2021). Macroalgal biorefinery concepts for the circular bioeconomy: A review on biotechnological developments and future perspectives. Renewable and Sustainable Energy Reviews, 151, 111553.

Kröger, M., & Müller-Langer, F., (2012). Review on possible algal-biofuel production processes. Biofuels, 3(3), 333–349.

Kumar, B., & Verma, P., (2021). Biomass-based biorefineries: An important architype towards a circular economy. Fuel, 288, 119622.

Lakaniemi, A. M., Hulatt, C. J., Thomas, D. N., Tuovinen, O. H., & Puhakka, J. A., (2011). Biogenic hydrogen and methane production from Chlorella vulgaris and Dunaliella tertiolectabiomass. Biotechnology for Biofuels, 4(1), 1–12.

Lehahn, Y., Ingle, K. N., & Golberg, A., (2016). Global potential of offshore and shallow waters macroalgal biorefineries to provide for food, chemicals, and energy: Feasibility and sustainability. Algal Research, 17, 150–160.

Leu, S., & Boussiba, S., (2014). Advances in the production of high-value products by microalgae. Industrial Biotechnology, 10(3), 169–183.

Levine, R. B., Pinnarat, T., & Savage, P. E., (2010). Biodiesel production from wet algal biomass through in situ lipid hydrolysis and supercritical transesterification. Energy & Fuels, 24(9), 5235–5243.

Li, H., Niu, S., Lu, C., & Li, J., (2016). Calcium oxide functionalized with strontium as heterogeneous transesterification catalyst for biodiesel production. Fuel, 176, 63–71.

Li, S. Y., Ng, I. S., Chen, P. T., Chiang, C. J., & Chao, Y. P., (2018). Biorefining of protein waste for production of sustainable fuels and chemicals. Biotechnology for Biofuels, 11(1), 1–15.

Li, Y., Chen, Y., & Wu, J., (2019). Enhancement of methane production in anaerobic digestion process: A review. Applied Energy, 240, 120–137.

Liu, D., Liu, D., Zeng, R. J., & Angelidaki, I., (2006). Hydrogen and methane production from household solid waste in the two-stage fermentation process. Water Research, 40(11), 2230–2236.

Lotti, T., Carretti, E., Berti, D., Martina, M. R., Lubello, C., & Malpei, F., (2019). Extraction, recovery, and characterization of structural extracellular polymeric substances from anammox granular sludge. Journal of Environmental Management, 236, 649–656.

Lü, F., Ji, J., Shao, L., & He, P., (2013). Bacterial bioaugmentation for improving methane and hydrogen production from microalgae. Biotechnology for Biofuels, 6(1), 1–11.

Lu, W., Alam, M. A., Liu, S., Xu, J., & Saldivar, R. P., (2020). Critical processes and variables in microalgae biomass production coupled with bioremediation of nutrients and CO2 from livestock farms: A review. Science of the Total Environment, 716, 135247.

Lucarini, M., Durazzo, A., Romani, A., Campo, M., Lombardi-Boccia, G., & Cecchini, F., (2018). Bio-based compounds from grape seeds: A biorefinery approach. Molecules, 23(8), 1888.

Maddi, B., Viamajala, S., & Varanasi, S., (2011). Comparative study of pyrolysis of algal biomass from natural lake blooms with lignocellulosic biomass. Bioresource Technology, 102(23), 11018–11026.

Maina, S., Kachrimanidou, V., & Koutinas, A., (2017). A roadmap towards a circular and sustainable bioeconomy through waste valorization. Current Opinion in Green and Sustainable Chemistry, 8, 18–23.

Manara, P., & Zabaniotou, A., (2016). Co-valorization of crude glycerol waste streams with conventional and/or renewable fuels for power generation and industrial symbiosis perspectives. Waste and Biomass Valorization, 7(1), 135–150.

Mao, G., Zou, H., Chen, G., Du, H., & Zuo, J., (2015). Past, current, and future of biomass energy research: A bibliometric analysis. Renewable and Sustainable Energy Reviews, 52, 1823–1833.

Martin-Gamboa, M., Marques, P., Freire, F., Arroja, L., & Dias, A. C., (2020). Life cycle assessment of biomass pellets: A review of methodological choices and results. Renewable and Sustainable Energy Reviews, 133, 110278.

McKendry, P., (2002). Energy production from biomass (part 2): Conversion technologies. Bioresource Technology, 83(1), 47–54.

Melis, A., & Happe, T., (2001). Hydrogen production. Green algae as a source of energy. Plant Physiology, 127(3), 740–748.
Membere, E., & Sallis, P., (2018). Effect of temperature on kinetics of biogas production from macroalgae. Bioresource Technology, 263, 410–417.
Mishra, S., Roy, M., & Mohanty, K., (2019). Microalgal bioenergy production under zero-waste biorefinery approach: Recent advances and future perspectives. Bioresource Technology, 292, 122008.
Mohan, S. V., Butti, S. K., Amulya, K., Dahiya, S., & Modestra, J. A., (2016). Waste biorefinery: A new paradigm for a sustainable bioelectro economy. Trends in Biotechnology, 34(11), 852–855.
Mohan, S. V., Modestra, J. A., Amulya, K., Butti, S. K., & Velvizhi, G., (2016). A circular bioeconomy with biobased products from CO2 sequestration. Trends in Biotechnology, 34(6), 506–519.
Mohan, S. V., Nikhil, G. N., Chiranjeevi, P., Reddy, C. N., Rohit, M. V., Kumar, A. N., & Sarkar, O., (2016). Waste biorefinery models towards sustainable circular bioeconomy: Critical review and future perspectives. Bioresource Technology, 215, 2–12.
Morf, L. S., Gloor, R., Haag, O., Haupt, M., Skutan, S., Di Lorenzo, F., & Böni, D., (2013). Precious metals and rare earth elements in municipal solid waste–sources and fate in a Swiss incineration plant. Waste Management, 33(3), 634–644.
Nagarajan, D., Dong, C. D., Chen, C. Y., Lee, D. J., & Chang, J. S., (2021). Biohydrogen production from microalgae—Major bottlenecks and future research perspectives. Biotechnology Journal, 16(5), 2000124.
Najafi, E., Castro, E., & Karimi, K., (2021). Biorefining for olive wastes management and efficient bioenergy production. Energy Conversion and Management, 244, 114467.
Negro, M. J., Alvarez, C., Ballesteros, I., Romero, I., Ballesteros, M., Castro, E., & Oliva, J. M., (2014). Ethanol production from glucose and xylose obtained from steam exploded water-extracted olive tree pruning using phosphoric acid as catalyst. Bioresource Technology, 153, 101–107.
Nishshanka, G. K. S. H., Liyanaarachchi, V. C., Premaratne, M., Nimarshana, P. H. V., Ariyadasa, T. U., & Kornaros, M., (2021). Wastewater-based microalgal biorefineries for the production of astaxanthin and co-products: Current status, challenges and future perspectives. Bioresource Technology, 342, 126018.
Nizami, A. S., Rehan, M., Waqas, M., Naqvi, M., Ouda, O. K., Shahzad, K., & Pant, D., (2017). Waste biorefineries: Enabling circular economies in developing countries. Bioresource Technology, 241, 1101–1117.
Palmowski, L. M., & Müller, J. A., (2000). Influence of the size reduction of organic waste on their anaerobic digestion. Water Science and Technology, 41(3), 155–162.
Park, J., Jones, B., Koo, B., Chen, X., Tucker, M., Yu, J. H., & Park, S., (2016). Use of mechanical refining to improve the production of low-cost sugars from lignocellulosic biomass. Bioresource Technology, 199, 59–67.
Patil, P. D., Gude, V. G., Mannarswamy, A., Deng, S., Cooke, P., Munson-McGee, S., & Nirmalakhandan, N., (2011). Optimization of direct conversion of wet algae to biodiesel under supercritical methanol conditions. Bioresource Technology, 102(1), 118–122.
Patinvoh, R. J., Osadolor, O. A., Chandolias, K., Horváth, I. S., & Taherzadeh, M. J., (2017). Innovative pretreatment strategies for biogas production. Bioresource Technology, 224, 13–24.

Pejin, D., Mojović, L. J., Vučurović, V., Pejin, J., Denčić, S., & Rakin, M., (2009). Fermentation of wheat and triticale hydrolysates: A comparative study. Fuel, 88(9), 1625–1628.

Rakotovao, M., Gobert, J., & Dermine-Brullot, S., (2018). Developing a socio-economic framework for the assessment of rural biorefinery projects. In: 26th European Biomass Conference and Exhibition.

Ramos, L. P., (2003). The chemistry involved in the steam treatment of lignocellulosic materials. Química Nova, 26, 863–871.

Rana, S., Mishra, P., Ab Wahid, Z., Thakur, S., Pant, D., & Singh, L., (2020). Microbe-mediated sustainable bio-recovery of gold from low-grade precious solid waste: A microbiological overview. Journal of Environmental Sciences, 89, 47–64.

Rani, R., Sharma, D., & Kumar, S., (2019). Optimization of operating conditions of miniaturize single chambered microbial fuel cell using NiWO4/graphene oxide modified anode for performance improvement and microbial communities dynamics. Bioresource Technology, 285, 121337.

Ratledge, C., & Wynn, J. P., (2002). The biochemistry and molecular biology of lipid accumulation in oleaginous microorganisms. Advances in Applied Microbiology, 51, 1–52.

Reshmy, R., Philip, E., Madhavan, A., Sirohi, R., Pugazhendhi, A., Binod, P., & Sindhu, R., (2022). Lignocellulose in future biorefineries: Strategies for cost-effective production of biomaterials and bioenergy. Bioresource Technology, 344, 126241.

Sadhukhan, J., Lloyd, J. R., Scott, K., Premier, G. C., Eileen, H. Y., Curtis, T., & Head, I. M., (2016). A critical review of integration analysis of microbial electrosynthesis (MES) systems with waste biorefineries for the production of biofuel and chemical from reuse of CO2. Renewable and Sustainable Energy Reviews, 56, 116–132.

Sadhukhan, J., Martinez-Hernandez, E., Murphy, R. J., Ng, D. K., Hassim, M. H., Ng, K. S., & Andiappan, V., (2018). Role of bioenergy, biorefinery and bioeconomy in sustainable development: Strategic pathways for Malaysia. Renewable and Sustainable Energy Reviews, 81, 1966–1987.

Sadhukhan, J., Ng, K. S., & Martinez-Hernandez, E., (2016). Novel integrated mechanical biological chemical treatment (MBCT) systems for the production of levulinic acid from fraction of municipal solid waste: A comprehensive techno-economic analysis. Bioresource Technology, 215, 131–143.

Sarkar, O., & Mohan, S. V., (2017). Pre-aeration of food waste to augment acidogenic process at higher organic load: Valorizing biohydrogen, volatile fatty acids and biohythane. Bioresource Technology, 242, 68–76.

Sarkar, O., & Venkata, M. S., (2016). Deciphering acidogenic process towards biohydrogen, biohythane, and short chain fatty acids production: Multi-output optimization strategy. Biofuel Research Journal, 3(3), 458–469.

Sarkar, O., Butti, S. K., & Mohan, S. V., (2018). Acidogenic biorefinery: Food waste valorization to biogas and platform chemicals. Waste Biorefinery, 203–218.

Sassner, P., Mårtensson, C. G., Galbe, M., & Zacchi, G., (2008). Steam pretreatment of H2SO4-impregnated Salix for the production of bioethanol. Bioresource Technology, 99(1), 137–145.m.

Satchatippavarn, S., Martinez-Hernandez, E., Hang, M. Y. L. P., Leach, M., & Yang, A., (2016). Urban biorefinery for waste processing. Chemical Engineering Research and Design, 107, 81–90.

Severo, I. A., Siqueira, S. F., Depra, M. C., Maroneze, M. M., Zepka, L. Q., & Jacob-Lopes, E., (2019). Biodiesel facilities: What can we address to make biorefineries commercially competitive? Renewable and Sustainable Energy Reviews, 112, 686–705.

Shahid, M. K., Batool, A., Kashif, A., Nawaz, M. H., Aslam, M., Iqbal, N., & Choi, Y., (2021). Biofuels and biorefineries: Development, application and future perspectives emphasizing the environmental and economic aspects. Journal of Environmental Management, 297, 113268.

Sharma, P., Srinivas, G. L. K., Varjani, S., & Kumar, S., (2021). Emerging microalgae-based technologies in biorefinery and risk assessment issues: Bioeconomy for sustainable development. Science of The Total Environment, 152417.

Shuba, E. S., & Kifle, D., (2018). Microalgae to biofuels: 'promising' alternative and renewable energy, review. Renewable and Sustainable Energy Reviews, 81, 743–755.

Singh, A., & Olsen, S. I., (2011). A critical review of biochemical conversion, sustainability, and life cycle assessment of algal biofuels. Applied Energy, 88(10), 3548–3555.

Singh, S. P., & Singh, D., (2010). Biodiesel production through the use of different sources and characterization of oils and their esters as the substitute of diesel: A review. Renewable and Sustainable Energy Reviews, 14(1), 200–216.

Solarte-Toro, J. C., Chacón-Pérez, Y., & Cardona-Alzate, C. A., (2018). Evaluation of biogas and syngas as energy vectors for heat and power generation using lignocellulosic biomass as raw material. Electronic Journal of Biotechnology, 33, 52–62.

Solis, J. L., Alejo, L., & Kiros, Y., (2016). Calcium and tin oxides for heterogeneous transesterification of Babasssu oil (Attalea speciosa). Journal of Environmental Chemical Engineering, 4(4), 4870–4877.

Somerville, C., (2006). Cellulose synthesis in higher plants. Annual Review of Cell and Developmental Biology, 22(1), 53–78.

Stoeberl, M., Werkmeister, R., Faulstich, M., & Russ, W., (2011). Biobutanol from food wastes–fermentative production, use as biofuel the influence on the emissions. Procedia Food Science, 1, 1867–1874.

Strategy, U. B., (2018). A Sustainable Bioeconomy for Europe: Strengthening the Connection Between Economy, Society, and the Environment. European Commission – 2018. https://ec.europa.eu/research/bioeconomy/pdf/ec_bioeconomy_strategy_2018.pdf (accessed on 15 November 2023).

Strategy, U. B., (2018). A Sustainable Bioeconomy for Europe: Strengthening the Connection Between Economy, Society. and the Environment. European Commission.–2018.–URL: https://ec.europa.eu/research/bioeconomy/pdf/ec_bioeconomy_strategy_2018.pdf (accessed on 15 November 2023).

Takisawa, K., Kanemoto, K., Kartikawati, M., & Kitamura, Y., (2014). Overview of biodiesel production from microalgae. Journal of Developments in Sustainable Agriculture, 9(2), 120–128.

Tedesco, S., & Stokes, J., (2017). Valorization to biogas of macroalgal waste streams: A circular approach to bioproducts and bioenergy in Ireland. Chemical Papers, 71(4), 721–728.

Torres, M. D., Kraan, S., & Domínguez, H., (2019). Seaweed biorefinery. Reviews in Environmental Science and Bio/Technology, 18(2), 335–388.

Tubino, M., Junior, J. G. R., & Bauerfeldt, G. F., (2016). Biodiesel synthesis: A study of the triglyceride methanolysis reaction with alkaline catalysts. Catalysis Communications, 75, 6–12.

Ubando, A. T., Cuello, J. L., El-Halwagi, M. M., Culaba, A. B., Promentilla, M. A. B., & Tan, R. R., (2016). Application of stochastic analytic hierarchy process for evaluating algal cultivation systems for sustainable biofuel production. Clean Technologies and Environmental Policy, 18(5), 1281–1294.

United Nations Foundation, (2020). 5 Global Issues to Watch in 2020. https://unfoundation.org/blog/post/5-global-issues-to-watch-in-2020/ (accessed on 15 November 2023).

Vandenberghe, L. P. S., Valladares-Diestra, K. K., Bittencourt, G. A., Torres, L. Z., Vieira, S., Karp, S. G., & Soccol, C. R., (2022). Beyond sugar and ethanol: The future of sugarcane biorefineries in Brazil. Renewable and Sustainable Energy Reviews, 167, 112721.

Vasudev, V., Ku, X., & Lin, J., (2020). Pyrolysis of algal biomass: Determination of the kinetic triplet and thermodynamic analysis. Bioresource Technology, 317, 124007.

Venkata, M. S., (2014a). Reorienting waste remediation towards harnessing bioenergy: A paradigm shift. Industrial Wastewater Treatment, Recycling, and Reuse (p. 235).

Wang, K., Kim, K. H., & Brown, R. C., (2014). Catalytic pyrolysis of individual components of lignocellulosic biomass. Green Chemistry, 16(2), 727–735.

Wicker, R. J., Kumar, G., Khan, E., & Bhatnagar, A., (2021). Emergent green technologies for cost-effective valorization of microalgal biomass to renewable fuel products under a biorefinery scheme. Chemical Engineering Journal, 415, 128932.

Woiciechowski, A. L., Neto, C. J. D., De Souza, V. L. P., De Carvalho, N. D. P., Sydney, A. C. N., Letti, L. A. J., & Soccol, C. R., (2020). Lignocellulosic biomass: Acid and alkaline pretreatments and their effects on biomass recalcitrance–Conventional processing and recent advances. Bioresource Technology, 304, 122848.

Wyman, C. E., (2013). Aqueous Pretreatment of Plant Biomass for Biological and Chemical Conversion to Fuels and Chemicals. John Wiley & Sons.

Yaashikaa, P. R., Kumar, P. S., & Varjani, S., (2022). Valorization of agro-industrial wastes for biorefinery process and circular bioeconomy: A critical review. Bioresource Technology, 343, 126126.

Yadav, G., Dash, S. K., & Sen, R., (2019). A biorefinery for valorization of industrial waste-water and flue gas by microalgae for waste mitigation, carbon-dioxide sequestration, and algal biomass production. Science of the Total Environment, 688, 129–135.

Yaman, S., (2004). Pyrolysis of biomass to produce fuels and chemical feedstocks. Energy Conversion and Management, 45(5), 651–671.

Yentekakis, I. V., & Goula, G., (2017). Biogas management: Advanced utilization for production of renewable energy and added-value chemicals. Frontiers in Environmental Science, 5, 7.

Zakzeski, J., Bruijnincx, P. C., Jongerius, A. L., & Weckhuysen, B. M., (2010). The catalytic valorization of lignin for the production of renewable chemicals. Chemical Reviews, 110(6), 3552–3599.

Zhang, M., Qi, W., Liu, R., Su, R., Wu, S., & He, Z., (2010). Fractionating lignocellulose by formic acid: Characterization of major components. Biomass and Bioenergy, 34(4), 525–532.

Zhang, Y. P., (2008). Reviving the carbohydrate economy via multi-product lignocellulose biorefineries. Journal of Industrial Microbiology and Biotechnology, 35(5), 367–375.

Zheng, Y., Zhao, J., Xu, F., & Li, Y., (2014). Pretreatment of lignocellulosic biomass for enhanced biogas production. Progress in Energy and Combustion Science, 42, 35–53.

Zuin, V. G., & Ramin, L. Z., (2018). Green and sustainable separation of natural products from agro-industrial waste: Challenges, potentialities, and perspectives on emerging approaches. Chemistry and Chemical Technologies in Waste Valorization, 229–282.

Index

D

Q

R